INTEGRATING POPULATION OUTCOMES, BIOLOGICAL MECHANISMS AND RESEARCH METHODS IN THE STUDY OF HUMAN MILK AND LACTATION

ADVANCES IN EXPERIMENTAL MEDICINE AND BIOLOGY

Recent Volumes in this Series

Volume 494
THE NIDOVIRUSES (CORONAVIRUSES AND ARTERIVIRUSES)
Edited by Ehud Lavi, Susan R. Weiss, and Susan T. Hingley

Volume 495
PROGRESS IN BASIC AND CLINICAL IMMUNOLOGY
Edited by Andrzej Mackiewicz, Maciej Kurpisz, and Jan Żeromski

Volume 496
NONINVASIVE ASSESSMENT OF TRABECULAR BONE ARCHITECTURE AND THE COMPETENCE OF BONE
Edited by Sharmila Majumdar, Ph.D., and Brian K. Bay, Ph.D.

Volume 497
INTRACTABLE SEIZURES: Diagnosis, Treatment, and Prevention
Edited by W. McIntyre Burnham, Peter L. Carlen, and Paul A. Hwang

Volume 498
DIABETES AND CARDIOVASCULAR DISEASE: Etiology, Treatment, and Outcomes
Edited by Aubie Angel, Naranjan Dhalla, Grant Pierce, and Pawan Singal

Volume 499
FRONTIERS IN MODELING AND CONTROL OF BREATHING
Edited by Chi-Sang Poon and Homayoun Kazemi

Volume 500
BIOLOGICAL REACTIVE INTERMEDIATES VI: Chemical and Biological Mechanisms of Susceptibility to and Prevention of Environmental Diseases
Edited by Patrick M. Dansette, Robert Snyder, Marcel Delaforge, G. Gordon Gibson, Helmut Greim, David J. Jollow, Terrence J. Monks, and I. Glenn Sipes

Volume 501
BIOACTIVE COMPONENTS OF HUMAN MILK
Edited by David S. Newburg

Volume 502
HYPOXIA: From Genes to the Bedside
Edited by Robert C. Roach, Peter D. Wagner, and Peter H. Hackett

Volume 503
INTEGRATING POPULATION OUTCOMES, BIOLOGICAL MECHANISMS AND RESEARCH METHODS IN THE STUDY OF HUMAN MILK AND LACTATION
Edited by Margarett K. Davis, Charles E. Isaacs, Lars Å. Hanson, and Anne L. Wright

---

A Continuation Order Plan is available for this series. A continuation order will bring delivery of each new volume immediately upon publication. Volumes are billed only upon actual shipment. For further information please contact the publisher.

INTEGRATING POPULATION OUTCOMES, BIOLOGICAL MECHANISMS AND RESEARCH METHODS IN THE STUDY OF HUMAN MILK AND LACTATION

Edited by

Margarett K. Davis

Centers for Disease Control and Prevention
Atlanta, Georgia

Charles E. Isaacs

New York State Institute for Basic Research in Developmental Disabilities
Staten Island, New York

Lars Å. Hanson

University of Göteborg
Göteborg, Sweden

and

Anne L. Wright

University of Arizona
Tucson, Arizona

Springer Science+Business Media, LLC

Library of Congress Cataloging-in-Publication Data

International Society for Research on Human Milk and Lactation. International Conference (10th: 2000: Tucson, Ariz.)
Integrating population outcomes, biological mechanisms and research methods in the study of human milk and lactation/edited by Margarett K. Davis ... [et al.].
p. ; cm. — (Advances in experimental medicine and biology; v. 503)
"From September 15–19, 2000, members of the International Society for Research on Human Milk and Lactation (ISRHML) gathered in Tucson, Arizona for the 10th International Conference of the Society"—Pref.
Includes bibliographical references and index.
ISBN 978-1-4613-5132-0 ISBN 978-1-4615-0559-4 (eBook)
DOI 10.1007/978-1-4615-0559-4
1. Breast milk—Congresses. 2. Breast feeding—Congresses. Lactation—Congresses. I. Davis, Margarett K. II. Title. III. Series.
[DNLM: 1. Lactation—Congresses. 2. Milk, Human—Congresses. 3. Biological Transport—Congresses. 4. Breast Feeding—Congresses. 5. Infant, Premature—Congresses. 6. Micronutrients—deficiency—Congresses. 7. Virus Diseases—transmission—Congresses. WP 825 I587i 2001]
QP246 .I57 2000
613.2′69—dc21

2002022219

Proceedings of the 10th International Society for Research on Human Milk and Lactation Symposium, held September 15–19, 2000, Tucson, Arizona

ISBN 978-1-4613-5132-0

Originally published by Kluwer Academic Publishers in 2002
Softcover reprint of the hardcover 1st edition 2002
http://www.wkap.nl/

10 9 8 7 6 5 4 3 2 1

A C.I.P. record for this book is available from the Library of Congress

PREFACE

From September 15-19, 2000, members of the International Society for Research on Human Milk and Lactation (ISRHML) gathered in Tucson, Arizona for the 10th International Conference of the Society. The title of this book, Integrating Population Outcomes, Biological Mechanisms, and Research Methods in the Study of Human Milk and Lactation, reflects the title of the conference. Five topical sessions and 77 posters were presented to more than 100 registrants following a superb keynote presentation by Professor Per Brandtzaeg on the regulation and biological significance of the secretory immunoglobulin system. The sessions were diverse: 1) The impact of micronutrient deficiencies during lactation on maternal and infant health, 2) The premature infant, 3) Developmental immunology, 4) Breastfeeding in the industrialized world, and 5) Viral transmission in milk. Wherever possible, we organized the sessions to include human population research, research showing the biological underpinnings or mechanisms of the effects on human health, and important methodological issues. In some instances, all these 3 areas were included for a topic and in other instances, only the biological findings or only the human population studies were presented, highlighting the need for research in the complementary area. The presentations generated extensive discussion and left the Society, as usual, with a number of questions and recommendations suggesting directions for proceeding with further research.

Review papers in the micronutrient session reflected the relatively advanced stage of research on population health outcomes in the areas of vitamins A and B-12, zinc, and iron. Scientific work on these micronutrients critical to early childhood survival and development raised discussion and questions regarding policy and program implications of the findings that zinc, iron, and vitamin deficiencies are not uncommon among inadequately supplemented exclusively breastfed children in developing countries. Concern about these deficiencies is especially intense because of the controversy regarding recommendations for the duration of exclusive breastfeeding.

Research methods, especially the difficulties in identifying and controlling potential confounding variables, were highlighted in the presentation on the relationship of breastfeeding and cognitive outcomes in premature infants. Confounding is a particularly difficult issue for all studies attempting to find a relationship between infant feeding and late effects such as chronic diseases. Studies of chronic childhood diseases such as cancer, or inflammatory bowel disease cannot address the problem of potential confounding by using randomization, as these health conditions are too rare to study prospectively. As yet, we have inadequate means to identify a concentrated cohort of high-risk infants to follow.

The sessions on the Premature Infant and Breastfeeding in the Industrialized World presented different types of methodological challenges. Data on breastfeeding from the ethnographic, economic, or evolutionary and neuro-endocrine perspectives was presented by Drs. Sellen, Ball, and Winberg, respectively. These perspectives show that observational techniques that elucidate maternal or infant feeding behavior in various socioeconomic or perinatal health care contexts can provide important data for understanding or intervening in the infant feeding process.

A biological finding can provide stimulation for a human population study, or the human population finding can precede the biological research. Some speakers in the Developmental Immunology session presented us with population level findings regarding effects of infant feeding practices on chronic disease risk; others presented biological findings that might provide some explanation for health risk. These provocative findings included associations between infant feeding and type 1 diabetes, celiac disease and childhood cancer as well as the description of tumor cell apoptosis by a common component of human milk called HAMLET (human alpha-lactalbumin made lethal to tumor cells) and the finding that colostral cells may be found in the lymphoid system of newborn pigs and lambs. In aggregate, these studies provide evidence of possible links between disease risk and modulation of the infant immune system by human milk and cry out for further study to either show health outcome effects of the biological findings or describe biological mechanisms that may explain the health outcome data. The need for collaboration in order to accomplish next steps in this and other areas of human milk research is acute.

The case for the possible contribution of infant feeding practices to the risk for type 1 diabetes among genetically susceptible children in this area of research is unique. In contrast to other chronic conditions associated with infant feeding, this area of research has been extremely well studied and boasts all types of studies including the simplest of ecological comparisons, laboratory studies to explore mechanisms, gene-environment interaction studies and a randomized intervention trial. Nevertheless, further biological explanation is needed to fully understand this association. Research has focused on the possible hazards of early introduction to intact cow milk protein but should also address the possible protective effects of adequate exposure to the immune modulating effects of human milk.

Several presentations integrated laboratory-based and population-level investigations. One example was the presentation on infant feeding and type 1 diabetes. Another example was the linked presentations of the Lactational Amenorrhea Method of contraception at both population and endocrinological levels. This method of contraception when practiced according to very simple guidelines in both developing and developed countries has been demonstrated to be a highly effective intervention resulting in the safe introduction of other family planning methods as exclusively breastfeeding women approach the onset of ovulation around 6 months postpartum.

The presentations and discussions that followed them drew attention to a number of themes and specific questions. The need for interdisciplinary collaboration is striking. In order to answer many research questions or advance knowledge on a topic, research must be planned and conducted by more than one discipline. This need is particularly pronounced in research on the relationship between HIV transmission from

HIV-infected mother to child via breastfeeding. To elucidate important aspects of this documented risk in ways that satisfy scientific curiosity and provide information that can be useful for prevention programs, collaboration among experts from several disciplines should occur. HIV and breastfeeding epidemiologists, HIV and human milk biologists, and geneticists might comprise a desirable collaborative team. Other collaborative configurations might be useful. Such collaborative alliances are critical to the optimal design, conduct, analysis, and interpretation of a study. While collaborations often occur within disciplines or across less widely separated disciplines such as biochemistry and microbiology at the *in vitro* bench level, the type of collaboration needed for questions like the relationship between type 1 diabetes and breastfeeding or between HIV transmission and breastfeeding rarely occurs in an effective manner. Research on the transmission of HIV infection through breastfeeding still cries out for collaboration between HIV researchers and human milk experts. Research on this topic and other topics could have proceeded much more efficiently over the past 10 – 15 years through the collaboration of human milk biologists and breastfeeding epidemiologists with topical clinical specialists. In the investigation of the relationship of infant feeding and diabetes, the intense focus on risk associated with early cow milk exposure to the exclusion of studying the possible protective role of exposure to human milk is another example of the way in which inadequate collaboration between disease-oriented researchers and human milk/breastfeeding researchers compromises the advancement of knowledge.

We are grateful to all the speakers and poster presenters for sharing their work with those who attended the conference. We wish to give our special and sincerest appreciation to Anna Parese for her tireless and attentive editorial work on the contributions to this book. This conference was made possible by generous financial support from UNICEF, New York, NY; The World Health Organization, Geneva, Switzerland; The National Institute of Child Health and Human Development, Bethesda, MD; The National Institute of Allergy and Infectious Diseases, Bethesda, MD; The National Institute of Diabetes and Digestive and Kidney Diseases, Bethesda, MD; Medela, Inc., McHenry, IL; and The Swedish Agency for Research Cooperation with Developing Countries, Department for Research Cooperation, 105 25 Stockholm, Sweden.

This preface was written by Margaret K. Davis in her private capacity. No official support or endorsement by Centers for Disease Control and Prevention or the Department of Health and Human Services is intended or should be inferred.

Margarett K. Davis
Charles E. Isaacs
Lars Å. Hanson
Anne Wright

CONTRIBUTORS

Adkins, Yuriko PhD
University of California
Department of Nutrition
One Shields Avenue
Davis, CA US 95616
Telephone - 530 752 8438
FAX - 530 752 8966
Email - ycadkins@ucdavis.edu

Agostoni, Carlo
S. Paolo Hospital
Clinica Pediatrica U
Via A Di Rudini 8
20112 Milano, Italy
Telephone - 39 02 8135375
FAX - 39 02 89122090

Albernaz, Elaine MD, MsC
Catholic University of Pelotas
and Federal University
Gomes Carneiro 1881, 901
Pelotas, RS Brazil 96010-610
Telephone - 053 2782580
FAX - 053 2712442
Email - zanrebla@ufpel.tche.br

Altaye, Mekibib PhD
Center for Pediatric Research
855 West Brambleton Avenue
Norfolk, VA USA 23510
Telephone - 757-668-6444
FAX - 757-668-6475
Email - altayem@chkd.com

Armeni, Maria Ersilia
Via Picco Tre Signori 18
Roma, Italy 00141
Telephone - 0039 06 8273768
FAX - 0039 06 8272186
Email - siliadoc@libero.it

Ashraf, Riffat Nisar MBBS, DCH, DPH, PhD.
King Edward Medical College
17 Danepus Lane COR1
Lahore, Pakistan
Telephone - 092 42 7233509
FAX - 092 42 7233509
Email - yassirashrafi@yahoo.com

Bachrach, Virginia MD, MPH
4179 Oak Hill Avenue
Palo Alto, CA US 94306-3721
Telephone - 650/493-4422
Email - Virginia.Bachrach@stanford.edu

Baerlocher, Kurt Prof. Dr. Med
Ostschweizer Kinderspital
Swiss Group for Promotion of Breast Feeding
Claudiusstrasse 6
St. Gallen, Switzerland CH 900600
Telephone - 41 71 245 81 90
FAX - 0041 71 243 7699
Email - kurt.baerlocher@gd-kispi.sg.ch

Ball, Thomas M. MD, MPH
University of Arizona
Department of Pediatrics
1501 N. Campbell Avenue
Tucson, AZ USA 85724-5073
Telephone - 520/626-4049
Email - tball@u.arizona.edu

Banta-Wright, Sandra RNC, MN, NNP/CNS
Doernbecher Children's Hospital
Oregon Health Sciences University
3181 SW Sam Jackson Pk Rd (NRC5)
Portland, OR USA 97201-3098
Telephone - 503/494-1077
FAX - 503/494-0428
Email - bantawrs@ohsu.edu

Begay, Tina
Phoenix Indian Medical Center
2 Renaissance Square
40 North Central Avenue
Phoenix, AZ USA 85004-4424

Bernath, Sandor DVM, PhD.
Institute for Veterinary Medicinal Products
Szallas U.8.
Budapest, Hungary 1107
Telephone - 36 1 262 8947 x122
FAX - 36 1 262 2839
Email -bernath@oai.hu

Borman, Laraine Lockhart BA. IBCLC Mothers Milk bank
Presbyterian St. Lukes Medical Center
1719 E. 19th Avenue
Denver, CO USA 80218
Telephone - 303/869-1888
FAX - 303/869-2480
Email - mmilkbank@health1.org

Borschel, RD
Marlene W, PhD
Ross Products Division
Abbott Laboratories
625 Cleveland Avenue
Columbus, OH USA 43215
Telephone - 614/624-7578
FAX - 614/624-3453
Email - marlene.borschel@rossnutrition.com

Brandtzaeg, Per Ph.D.
Liipat Institue of Pathology
Riks Hospitalet
Oslo, Norway N-0027
Telephone - 47 23072743
FAX - 47 23071511
Email - per.brandtzaeg@labmed.uio.no

Bryan, Dani-Louise Bsc.
Flinders University of South Australia
Sturt Rd, Bedford Park
South Australia, Australia
Telephone - 61 8 8204 5668
FAX - 61 8 8204 3945
Email - dani.bryan@flinders.edu.au

Butte, Nancy PhD
Baylor College of Medicine
100 Bates Street
Houston, TX USA 77004
Telephone - 713 665 4398
Email - nbutte@bcm.tmc.edu

Caire, Graciela PhD., student
Centro de Investigacion en Alimentacion y Desarrol
Km 0.6 Carr a La Victoria
POB 1735
Hermosillo
Sonora, Mexico 83000
Telephone - 62 80 0057
FAX - 62 80 0055
Email - amc@cascabel.ciad.mx

Calderon, de la Barca Ana M. Ph.D.
Centro de Investigacion en Alimentacion y Desarrol
Km 0.6 Carr. A La Victoria
Hermosillo
Sonora, Mexico 83000
Telephone - 62 80 00 57
FAX - 62 80 00 55
Email - amc@cascabel.ciad.mx

Canfield, Louise Ph.D.
University of Arizona
1501 N. Campbell Avenue
Tucson, AZ USA 85718
Telephone - 520/621-9368
FAX - 520/626-2110
Email - lmcanfie@u.arizona.edu

Caplan, Michael MD
Evanston Northwestern Healthcare
2650 Ridge Avenue
Evanston, IL USA 60201
Telephone - 847/570-2530
FAX - 847/570-0231
Email - mca113@northwestern.edu

Castetbon, Katia Ph.D.
Universite V. Segaler
Bordeaux 2
A46 Lue Leo Saignat
Inserm U330
Bordeaus, Codex France 33076
Telephone - 00 33 5 57 57 17 65
FAX - 00 33 5 57 57 11 72
Email - katia.castetbon@u-bordeaux2.fr

Chapman, Donna Ph.D.
University of Connecticut
U-17, Department of Nutritional Sciences
Storrs, CT USA 06095
Telephone - 860/486-5073
FAX - 860/486-3674
Email - djc@discovernet.net

Cisneros, Silva Ignacia MPH, MD
Hospital Luis Castelazo
Ayala Instituto Mexicano Del Soho Social
Fortin Squina Coatzacoaloos,
Col San Jeronimo Acul
Mexico D.F., Mexico C.P. 10400

Clevenger, Charles MD, PhD.
University of Pennsylvania Medical Ctr
Department of Pathology & Lab Med
513 SC Labs
422 Curie Blvd
Philadelphia, PA USA 19104
Telephone - 215/898-0734
FAX - 215/573-8944
Email - clevengc@mail.med.upenn.edu

Covington, Chandice RN, Ph.D, CPNP
Wayne State University
5557 Cass Avenue
Detroit, MI 48202 USA
Telephone - 313/577-4092
FAX - 313/577-4188
Email - ad4975@wayne.edu

Da Costa, Teresa Helena Ph.D.
University of Brasilia
Universidade de Brasilia
C.P. 04511
Brasilia, DF Brazil 70919-970
Telephone - 55-61-347-9746
FAX - 55-61-349-6286
Email - hdacosta@unb.br

Davanzo, Riccardo MD
Divisione de Neonatologia e Terapia Intensiva Neo
IRCCS Burlo
Via dell'Istria 65/1
Trieste, Italy 34100
Telephone - 040 309067
Email - davanzor@burlo.trieste.it

Davis, Margarett MD, MPH
Ctrs for Disease Control and Prevention
1761 Inverness Avenue
Atlanta, GA USA 30306
Telephone - 770/488-3213
FAX - 770/488-3112
Email - mkd1@cdc.gov

De La Ossa, Lydia
University of Arizona
Arizona Respiratory Sciences Center
1501 N. Campbell Avenue
Tucson, AZ 85724

Dewey, Kathryn G. Ph.D.
University of California
Department of Nutrition
One Shields Avenue
Davis, CA USA 95616-8669
Telephone - 530/752-0851
FAX - 530/752-3406
Email - kgdewey@ucdavis.edu

Dominguez, Jessica
University of Arizona
Department of Pediatrics
1501 N. Campbell Avenue
Tucson, AZ USA 85724
Telephone - 520/626-6553
FAX - 520/626-5009
Email - jessicad@u.Arizona.edu

Dosch, Hans-Michael MD
The Hospital for Sick Children
555 University Avenue
Toronto, Ontario, Canada M5G1X8
Telephone - 416/813-6200
FAX - 416/813-6255
Email - hmdosch@sickkids.on.ca

Dowling, Donna A. MN, Ph.D.
Case Western Reserve University
FPB School of Nursing
10900 Euclid Avenue
Cleveland, OH USA 44106
Telephone - 216/368-1869
FAX - 216/368-3542
Email - dad10@po.cwru.edu

Dvorak, Bohuslav Ph.D.
University of Arizona
Department of Pediatrics
1501 N. Campbell Avenue
Tucson, AZ USA 85724
Telephone - 520/626-6553
FAX - 520/626-5009
Email - dvorakb@peds.arizona.edu

Edde, Edith Lynn DO
The University of Arizona
1501 N. Campbell
POB 245073
Tucson, AZ USA 85724-5073
Telephone - 520 626-6627
Email - ledde@peds.arizona.edu

Eidelman, Arthur I. MD
Shaare Zedek Medical Center
POB 3235
Jerusalem, Israel 92182
Telephone - 972 2 655 5643
FAX - 972 2 652 0689
Email - eidel@cc.huji.ac.il

Filteau, Suzanne Ph.D.
Centre for International Child Health
Institute of Child Health
30 Guilford Street
London, UK WC1N 1EH
Telephone - 44 020 7905 2352
FAX - 44 020 7404 2062
Email - sfilteau@ich.ucl.ac.uk

Fituch, Camellia MD
Baylor College of Medicine
Department of Neonatology
One Baylor Plaza
Houston, TX USA 77030-3498
Telephone - 713/770-1380
FAX - 713/770-2799
Email - cfituch@neo.bcm.tmc.edu

Gartner, Lawrence M. MD
University of Chicago
28398 Alamar Road
Valley Center, CA USA 92082-6452
Telephone - 760/751-9479
FAX - 760/749-1244
Email - gart@midway.uchicago

Garza, Cutberto MD, Ph.D.
Cornell University
317 Savage Hall
Ithaca, NY USA 14853
Telephone - 617/254-5144
FAX - 617/255-1033
Email - cg30@cornell.edu

Georgeson, Jennifer MSc
Centre for International Child Health
Institute of Child Health
30 Guilford Street
London, UK WC1N 1EH
Telephone - 44 020 7905 2352
FAX - 44 020 7404 2062
Email - jengeorgeson@hotmail.com

Giovannini, Marcello
S. Paolo Hospital
Pediatrics Dept 5
Via A Di Rudini 8
Milano, Italy 20122
Telephone - 39 02 8135375
FAX - 39 02 89122090

Giugliani, Elsa Regina MD, Ph.D.
Federal University of Rio Grande do Sul
Rua Itaborai, 1477
Porto Alegre
Rio Grande do S, Brazil 90670-030
Telephone - 55 51 3360282
FAX - 55 51 3165119
Email - elsag@vortex.ufrgs.br

Glew, Robert H. Ph.D.
University of New Mexico
School of Medicine
915 Camino de Salud
Albuquerque, NM USA 87131
Telephone - 505/272-2362
FAX - 505/272-6587
Email - rglew@salud.unm.edu

Goga, Ameena Ebrahim MBChB, DTMH, DCH
University of Natal Medical School
Paediatrics/Africa Ctr for Population Studies
3 Merton Gardens, Reservoir Hills
Durban, KwaZulu Natal, South Africa 4091
Telephone - (002731) 2622690
FAX - 035 5501674
Email - ameena.goga@mrc.ac.za

Goldman, Armond S. MD
The University of Texas Medical Branch
Pediatric Immunology/Allergy/Rheumatology
301 University Blvd.
Galveston, TX USA 77555-0369
Telephone - 409/772-2658
FAX - 409/772-6622
Email - agoldman@utmb.edu

Gonzalez-Cossio, Teresa Ph.D.
Instituto Nacional de Salud Publica, Mexico
Ave. Universidad
655 Col. Sta. Ma. Ahuacatitian
Cuernavaca, Morelos, Mexico CP62508
Telephone - 527 329 30 09
FAX - 527 311 22 19
Email - tgonzale@insp3.insp.mx

Green, Patricia W. BA, BSRN
University of Wisconsin
202 South Park Meriter Hospital
Madison, WI USA 53715
Telephone - 608/267-6362
Email - pwgreen@facstaff.wisc.edu

Greer, Frank R. MD
University of Wisconsin
Center for Perinatal Care
202 S. Park Street
Madison, WI USA 53715
Telephone - 608/262-6561
FAX - 608/267-6377
Email - frgreer@facstaff.wisc.edu

Grummer-Strawn, Laurence Ph.D.
Centers for Disease Control & Prevention
4770 Buford Hwy, Mailstop K25
Atlanta, GA USA 30341
Telephone - 770/488-6048
FAX - 770/488-5369
Email - lgrummer-strawn@cdc.gov

Gutierrez, Irene
University of Arizona
Arizona Respiratory Sciences Center
1501 N. Campbell Avenue
Tucson, AZ 85724

Hamosh, Margit Ph.D, MSc
Georgetown University Medical Center
9410 Balfour Ct.
Bethesda, MD USA 20814
Telephone - 301 530 6788
FAX - 301 530 6788
Email – phamosh@excite.com

Hamosh, Paul MD
Georgetown University Medical Center
9410 Balfour Ct
Bethesda, MD USA 20814
Telephone - 301 530 6788
FAX - 301 530 6788
Email - phamosh@excite.com

Haney, Peter MD, Ph.D.
Baylor College of Medicine
Children's Nutrition Research Center
1100 Bates St/ Room 10018
Houston, TX USA 77030
Telephone - 713/798-7067
FAX - 713/798-7057
Email - phaney@neo.bcm.tmc.edu

Hansen, James MD, Ph.D.
Mead Johnson Nutritionals
2400 W. Lloyd Expressway
Evansville, IN USA 47721
Telephone - 812/429-7417
FAX - 812/429-7483
Email - james.hansen@bms.com

Hanson, Lars MD, Ph.D.
University of Göteborg
Guldhedsgatan 10
Göteborg, Sweden S-413-46
Telephone -46-31604916
FAX - 8-93801148
Email - lars.a.hanson@immuno.gu.se

Hartmann, Peter BRurSc, Ph.D.
The University of Western Australia
Nedlands, WA, Australia 6907
Telephone - 61 8 93803327
FAX - 61 8 93801148
Email – hartmanp@cyllene.uwa.edu.au

Hegger, Tami
Medela, Inc.
1101 Corporate Drive
McHenry, IL USA 60050
Telephone - 800/435-8316 x 535
FAX - 714/721-1909
Email - tami.hegger@medelaine.com

Heinig, Jane Ph.D.
UC Davis
Department of Nutrition
One Shields Avenue
Davis, CA USA 95616
Telephone - 530/752-8681
FAX - 530/752-7582
Email - mjheinig@ucdavis.edu

Hill, Pamela D. Ph.D.
University of Illinois at Chicago
555 6th Street, Ste 500
Moline, IL USA 51265
Telephone - 309/757-9467
FAX - 309/757-9473
Email - phill@uic.edu

Holubec, Hana
University of Arizona
Dept. of Microbiology and Immunology
1501 N. Campbell Avenue
Tucson, AZ USA 85724
Telephone - 520/626-7541
FAX - 520/626-5009
Email - holubec@u.arizona.edu

Hopkinson, Judy Ph.D.
USDA/ARS Baylor College of Medicine
Children's Nutrition Research Ctr.
1100 Bates
Houston, TX USA 77030
Telephone - 713 798 7008
FAX - 713 798 7098
Email - judyh@bcm.tmc.edu

Hörnell, Agneta
International Maternal & Child Health
University Hospital
Entrance 11
Uppsala, Sweden SE-751 85
Telephone - 46 18 611 59 96
FAX - 46 (0) 18 50 80 13
Email - agnehorn@spray.se

Howie, Peter W. MD, FRCOG, FRSE
University of Dundee
Department of Ob/Gyn
Level 5 Ninewells, Hospital & Medical School
Dundee, UK DD1 9SY
Telephone - 01382 632147
FAX - 01382 633847
Email - b.c.charnley@dundee.ac.uk

Hurst, Nancy RN, MSN
Texas Children's Hospital
Baylor College of Medicine
7039 Centre Oaks Drive
Houston, TX USA 77069
Telephone - 713/770-3612
FAX - 713/770-3633
Email - nmhurst@texaschildrenshospital.org

Iizuka, Tadashi MD
Kihoku Hospital, Wakayama Medical College
219 Myoji, Katsuragi-town, Ito-gun
Wakayama, Japan 649-7113
Telephone - 81 736 22 0066
FAX - 81 736 22 2579
Email - taiizuka@wakayama-med.ac.jp

Isaacs, Charles E. Ph.D.
NYS Institute for Basic Research
1050 Forest Hill Road
Staten Island, NY USA 10314
Telephone - 718/494-5227
FAX - 718/698-3803
Email - chisi@cunyvm.cuny.edu

Ito, Marina Kiyomi Ph.D.
University of Brasilia
Universidade de Brasilia, C.P. 04354
Brasilia, DF Brazil 70919-970
Telephone - 55-61-347-9100
Email - marina@unb.br

Jakobsen, Marianne Skytte MD, Ph.D.
Bandim Health Project
Apartado 1004, 861 Bissau Codex
Guinea Bissau, West Africa
Telephone - 245 20 25 34
FAX - 245 20 16 72
E-mail - marianne@sol.gtelecom.gw

Jarvenpää, Anna-Liisa MD, Ph.D.
HUCH, Hospital for Children and Adolescents
Department of Pediatrics, Neonatology
Sofianlehdonkatu 5 A
PostBox 610 HUS
Helsinki, Finland
Telephone - 358 9 47161160
FAX - 358 8 47165440
Email - anna-liisa.jarvenpaa@hus.fi

Jarvinen, Kirsi-Marjut MD
Helsinki University Central Hospital Skin
and Allergy Hospital
833 Lexington Avenue, Apt #3B
New York, NY USA 10021
Telephone - 212/426-1902
FAX - 212/426-1902
Email - kirsi_jarvinen@hotmail.com

Jensen, Robert G. Ph.D.
University of Connecticut
186 Chaffeeville Road
Storrs-Mansfield, CT USA 06268-2673
Telephone - 860/423-5361
FAX - 860/423-5361
Email - rjensen@uconnvm.uconn.edu

John, Grace MD, MPH
Univ. of Washington and Nairobi
BOX 359909, IARTP, UW
Harborview Med. Ctr, 325 9th Avenue
Seattle, WA USA 98195
Telephone - 206/731-2822
FAX - 206/731-2427
Email - gjohn@u.washington.edu

Kelleher, Shannon Ph.D., Student
University of California Davis
1 Sheilds Avenue
Davis, CA USA 95616
Telephone - 530/752-8438
FAX - 530/752-8966
Email - slkelleher@ucdavis.edu

Kennedy, Kathy Irene DrPH, MA
University of Colorado Hlth Sci Ctr
2201 So. Fillmore Street
Denver, CO USA 80210
Telephone - 303/758-5494
FAX - 303/758-5660
Email - kkennedy@du.edu

Kirsten, Gert F. FCP(SA), MD
Tygerberg Hosp & Univ. of Stellenbosch
Department of Paediatrics
39 Chavonne Str, Welgemoed
Bellville 7530
Cape, South Africa
Telephone - 27 21 938640
FAX - 27 21 938 9138
Email - gfk@gerga.sun.ac.za

Klacanska, Sarka Southern Ph.D.
University of California, San Diego
SWF Science Center
8604 La Jolla Shores Drive
La Jolla, CA USA 92037
Telephone - 858 546 5678
FAX - 858 546 7003
Email -sarka@caliban.ucsd.edu

Kling, Pamela J. MD
University of Arizona
Department of Pediatrics
POB 245073
Tucson, AZ USA 85724-5073
Telephone - 520/626-6627
FAX - 520/626-5009
Email - pkling@peds.arizona.edu

Krebs, Nancy F. MD, MS
University of Colorado School of Medicine
4200 E. 9th Avenue/Box C225
Denver, CO USA 80262
Telephone - 303/315-7037
FAX - 303/315-3273
Email - nancy.krebs@uchsc

Kurtz, Debra BA, BS, MBA
Medela, Inc.
1101 Corporate Drive
McHenry, IL USA 60050
Telephone - 800/435-8316 x440
FAX - 815/363-9941
Email - debra.kurtz@medelaine.com

Labbok, Miriam MD, MPH
USAID, Adjunct Hopkins and Tulane
4707 Connecticut Ave #301
Washington, DC USA 20008
Telephone - 202 363 3872
FAX - 202 216 4702
Email - mlabbok@usaid.gov

Larsson, Michael MBA
Medela, Inc.
1101 Corporate Drive
McHenry, IL USA 60050
Telephone - 800/435-8316
FAX - 815/363-9941
Email - miclar@swissonline.ch

Lau, Chantal Ph.D.
Baylor College of Medicine
Department of Pediatrics
One Baylor Plaza
Houston, TX USA 77030
Telephone - 713/798-6710
FAX - 713/798-7187
Email - clau@bcm.tmc.edu

Lawrence, Ruth A. MD
University of Rochester School of Medicine
601 Elmwood Avenue
Rochester, NY USA 14642
Telephone - 716/275-4354
FAX - 716/461-3614
Email - ruth_lawrence@URMC.rochester.edu

Lee, Yi-Kyoung MS
Iowa State University
1127 HNSB
Dept of Food Science and Human Nutrition
Ames, IA USA 50014
Telephone - 515/294-9377
FAX - 515/294-6193
Email - ehowl@iastate.edu

Leufkens, Paul PharmD
Pharming B.V.
Archimedesweg 4
Leiden, The Netherlands 2533 CN
Telephone - 31 71 5267675
FAX - 31 71 5267433
Email - p.leufkens@pharming.com

Li, Ruowei (Rosie), MD, Ph.D.
Centers for Disease Control & Prevention
4770 Buford Hwy, NE
Atlanta, GA USA 30338
Telephone - 770/488-6033
FAX - 770/488-5369
Email - ril6@cdc.gov

Lien, Eric L. Ph.D.
Wyeth Nutritionals International
POB 42528
Philadelphia, PA USA 19101
Telephone - 610/341-2383
FAX - 610/989-4856
Email - liene@war.wyeth.com

Lindell, Marilyn
University of Arizona
Arizona Respiratory Sciences Center
1501 N. Campbell Avenue
Tucson, AZ 85724

Lohman, Carla
University of Arizona
Arizona Respiratory Sciences Center
1501 N. Campbell Avenue
Tucson, AZ 85724

Lonnerdal, Bo Ph.D.
University of California
Department of Nutrition
One Shields Avenue
Davis, CA USA 95616
Telephone - 530 752 8347
FAX - 530 752 3564
Email - bllonnerdal@ucdavis.edu

Lope, Leena MD
Jorvi Hospital
Apollonkatu 3 D 33
Helsinki, Finland FIN-00100
Telephone - 358 9 861 81 5460
FAX - 358 9 861 5945
Email -leena.lope@helsinki.fi

Lopez-Alarcon. Mardya MD, Ph.D.
Cornell University
Institute Mexicano Sequo Social
118 Snyder Hill Rd.
Ithaca, NY USA 14850
Telephone - 607/277-0112
Email - mgl5@cornell.edu

Lovelady, Cheryl Ph.D.
University of North Carolina
Department of Nutrition & FSS
POB 26170
Greensboro, NC USA 27402-6170
Telephone - 336/334-5313
FAX - 336/334-4129
Email - cheryl_lovelady@uncg.edu

Makokha, Ernest MSc
Kenyatta University
Kemri Infectious Disease Research
Mbagathi Road
Nairobi, Africa
Telephone -254 2 713679
FAX - 254 2 719269
Email - epmakokha@yahoo.com

Mallie, Jody
University of Arizona
Arizona Respiratory Sciences Center
1501 N. Campbell Avenue
Tucson, AZ 85724

Marquis, Grace S. Ph.D.
Iowa State University
1127 Human Nutritional Sciences Bldg.
Ames, IA USA 50014
Telephone - 515 294 9231
FAX - 515 294 6193
Email - gmarquis@iastate.edu

Martin-Calama, Jesus MD
Hospital General de Teruel
Ronda del Parque 14, 8F
Teruel, Spain 44002
Telephone - 978 610825
FAX - 978 610825
Email - martinc@nexo.es

Mayorga, Evelyn Maria
Ctr Studies of Sensory Impairment
Aging & Metabol
17 ave 16-89 (interior), Zona 11
Guatemala City, Guatemala 01011
Telephone - +502 4733942
FAX - +502 4733942
Email -cessiam@guate.net

McGuire, Shelley Ph.D.
Washington State University
POB 646376
Pullman, WA USA 99164-6376
Telephone - 509/335-3896
FAX -509/335-4815
Email - smcguire@wsu.edu

McNeilly, Alan S. BSc Ph.D DSc FRSE
Medical Research Council
Human Reproductive Sciences Unit
37 Chalmers Street
Edinburgh, Scotland UK EH3 9ET
Telephone - 0044 131 229 2575
FAX - 0044 131 228 5571
Email - a.mcneilly@ed-rbu.mrc.ac.uk

McWilliam, Debra L.
University of Arizona
Department of Pediatrics
1501 N. Campbell Avenue
Tucson, AZ USA 85724
Telephone - 520/626-6450
FAX - 520/626-5009
Email - dmcwilli@peds.arizona.edu

Meier, Paula DNSc.
Rush Children's Hospital
1753 W. Congress Parkway
Room 625 Jones
Chicago, IL USA 60612
Telephone - 312/942-4932
FAX - 312/942-3355
Email -pmeier@rush.edu

Mikiel-Kostyra, Krystyna
Institute of Mother and Child
01-211 Warsaw
Ul. Kasprzaka 17A
Poland
Telephone - 48 22 632-36-74
FAX - 48 22 632-94-54
Email - breastfeed@imid.med.pl

Miracle, Donna Jo MSN
Rush University
818 Longford Way
Noblesville, IN USA 46060
Telephone - 317/877-2407
Email - dmiracle@rushu.rush.edu

Monaco, Marcia H. Ph.D.
University of Illinois
338 Bevier Hall
905 S. Goodwin Avenue
Urbana, IL USA 61801
Telephone - 217 244 2873
FAX - 217 333 9368
Email - monaco@uiuc.edu

Morley, Ruth MB Bchir FRCPCH
Univ. of Melbourne, Pediatrics
Flemington Road
Parkville Melbourne
Victoria, Australia VIC 3052
Telephone - +61 3 9345 6552
FAX - +61 3 9345 6667
Email -morleyr@cryptic.rch.unimelb.edu.au

Morrill, Jimi Francis BS, MS
UC Davis
864 Glen Meadow Drive
Sparks, NV USA 89434
Telephone - 775-762-4025
FAX - 775-356-9163
Email - jfhninc@nvbell.net

Murphy, Sue
Phoenix Indian Medical Center
2 Renaissance Square
40 North Central Avenue
Phoenix, AZ 8504-4424

Naylor, Audrey J. MD, Dr.P.H.
Wellstart International
4062 First Avenue
San Diego, CA USA 92103
Telephone - 619/574-8174
FAX - 619/574-8159
Email - naylor@wellstart.org

Neville, Margaret C. Ph.D.
University of Colorado Health Sciences Ctr.
Box C240
Denver, CO USA 80262
Telephone - 303/315-8230
FAX - 303/315-8110
Email - peggy.neville@UCHSC.edu

Newburg, David S. Ph.D.
Shriver Center, 200 Trapelo Road
Waltham, MA USA 02452
Telephone - 781 642 0025
FAX - 781 893 4018
Email - dnewburg@shriver.org

Ntourntoufi, Agathi MD
14-16 Lycourgou Str.
Athens, Greece 10552
Telephone - 301 9985145
FAX - 301 9935561
Email - everest@ath.forthnet.gr

O'Connor, Deborah L. Ph.D.
The Hospital for Sick Children
University of Toronto, Room 4103B
555 University Avenue
Toronto, Ontario, Canada M5G 1X8
Telephone - 416 813 7844
FAX - 416 813 7849
Email - deborah_l.o'connor@sickkids.on.ca

O'Hara, MaryAnn MD, MPH, MSt
University of Washington, Robert Wood Johnson
Clinical Scholars Program
1628 E. McGraw St.
Seattle, WA USA 98112
Telephone - 206/616-8724
Email - maryanno@u.washington.edu

Oberle, Doris MSc
Ludwig Maximilians University
Institute for Social Pediatrics & Adolescent Med
Heiglhofstr.63
Munchen, Germany 81337
Telephone - 39 89 7009 120
FAX - 49 7009 315
Email - Doris.Oberle@lrz.uni-muenchen.de

Onyango, Adelheid W. Ph.D.
World Health Organization
Department of Nutrition
20 Avenue Appia
Geneva, 27 Switzerland CH-1211
Telephone - 41 22 791 34 95
FAX - 41 22 791 41 56
Email - onyangoa@who.ch

Pardo, Dona Ph.D, RN
University of Arizona, CME
1501 N. Campbell Avenue
Tucson, AZ 85724

Payne, Claire M. Ph.D.
University of Arizona
Dept. of Microbiology and Immunology
1501 N. Campbell Avenue
Tucson, AZ USA 85724
Telephone - 520/626-2870
FAX - 520/626-2100
Email - claire-payne@ns.arizona.edu

Pérez-Escamilla, Rafael Ph.D.
The University of Connecticut
3624 Horsebarn Rd. Extension
NUSC, U-4017, Storrs, CT USA 06269
Telephone - 860 486 5073
FAX - 860 486 3674
Email -rperez@canr.uconn.edu

Persson, Lars Ake MD, Ph.D.
ICDDRB
Box 128, Dhaka 1000
Telephone - 880 2 988 5155
FAX - 880 2 882 6050
Email - persson@icddrb.org

Philipps, Anthony F. MD
University of California, Davis
2516 Stockton Blvd., #318
Sacramento, CA USA 95817
Telephone - 916-734-5178
FAX - 916-456-2236
Email - a.philipps@ucdmc.ucdavis.edu

Picciano, Mary Frances Ph.D.
The Pennsylvania State University
Department of Nutrition
126 Henderson Building South
University Park, PA USA 16802-6504
Telephone - 814/863-2919
FAX - 814/863-6103
Email - mfp4@psu.edu

Powers, Nancy G. MD
Wesley Medical Center
550 N. Hillside
Wichita, KS USA 67214
Telephone - 316 651 8580
Email - nancy.powers@HCAhealthcare.com

Rasmussen, Kathleen M. ScD
Cornell University
Div. of Nutritional Sciences
111 Savage Hall
Ithaca, NY USA 14853-6301
Telephone - 607/255-2290
FAX - 607/255-2290
Email - kmr5@cornell.edu

Ruiz-Palacios, Guillermo M. MD
Instituto Nacional de Ciencias
Medicas y Nutricion
Vasco de Quiroga 15
Mexico, DF Mexico 14000
Telephone - 525 6559675
FAX - 525 51300010
Email - gmrps@servidor.unam.mx

Schanler, Richard MD
Baylor College of Medicine
1100 Bates Street, Ste 8072
Houston, TX USA 77030
Telephone - 713/798-7176
FAX - 713/798-7187
Email - schanler@bcm.tmc.edu

Sellen, Daniel William MA, Ph.D.
Emory University
Department of Anthropology
1557 Pierce Dr., NE
Atlanta, GA USA 30322
Telephone - 404/727-4777
FAX - 404/727-2860
Email - dsellen@emory.edu

Shinohara, Hisae Ph.D.
University of Arizona
Department of Pediatrics
1501 N. Campbell Avenue
Tucson, AZ USA 85724
Telephone - 520/626-6553
FAX - 520/626-5009
Email - dvorakb@peds.arizona.edu

Silfverdal, Sven-Arne MD, MPH
Orebro Medical Centre Hospital
Department of Pediatrics
RSO Orebro
Sweden, SE-701 85
Telephone - 46-19-6021000
FAX - 46-6023122
Email - sven-arne.silfverdal@orebroll.se

Simondon, Kirsten MD, Ph.D.
Institut de Rechesche pour le Developpement
(IRD) BP 5045
Montpellier, France 34032
Telephone - 33 4 67 41 6190
FAX - 33 4 67 547800
Email - kirsten.simondon@mpl.ird.fr

Sisk, Paula M. B.S.
Forsyth Medical Center
3333 Silas Creek Parkway
Winston-Salem, NC USA 127103
Telephone - 336/718-3277
FAX - 336/718-9765
Email - pmsisk@novanthealth.org

Smit, Ella N. DRS
University Hospital Groningen
POB 30.001
Groningen, The Netherlands 9700 RB
Telephone - 31-50 3611060
FAX - 31-50 3611062
Email - pip@bih.net.ba

Springer, Skadi MD
University of Leipzig
Children's Hospital
Oststrabe 21-25
Leipzig, Germany D-04317
Telephone - 00 49 341 9726075
FAX - 00 49 341 9726039
Email - skspr@medizin.uni-leipzig.de

Stoltzfus, Rebecca J. Ph.D.
The Johns Hopkins University
Center for Human Nutrition
615 N. Wolfe Street
Baltimore, MD USA 21209
Telephone - 410/955-2786
FAX - 410/955-0196
Email - rstoltzf@jhsph.edu

Svensson, Malin MSc
Inst. Laboratory Medicine
Dept. MIG
Solvegatan 23
Lund, Sweden SE-223 62
Telephone - 46 46 173933
FAX - 46 46 137468
Email - malin.svensson@mig.lu.se

Torres, Alexandre Guedes MSc
Universidade Federal do Rio de Janeiro
Rua Timoteo da Costa 1001/3L-2/402
Rio de Janeiro RJ, Brazil 22.450-130
Telephone - 55 21 249 6880
FAX - 55 21 224 6285
Email - agtorres@domain.com.br

Trugo, Nadia M.F. Ph.D.
Universidade Federal do Rio de Janeiro
Instituto de Quimica
CT bloco A, Cidade Universitaria
Rio de Janeiro RJ, Brazil 21949-900
Telephone - 0055 21 5627352
FAX - 0055 21 2956876
Email -trugo@iq.ufrj.br

Tuboly, Sandor Ph.D., DSC
University of Veterinary Science
Hungaria krt. 23-25
Budapest, Hungary 1581
Telephone - 36 1 251 99 00
FAX - 36 1 251 92 60
Email - tubolys@novell.vmri.hu

Urquieta, Barbara Chemistrician
Av. Cuauhtemoc #330, Col Doctores
Mexico City, Mexico 06720
Telephone - 52 5627 6944
Email - saumar2@prodigy.net.mx

Van de Perre, Philippe MD, Ph.D.
Centre Muraz/OCCGE
2054 rue Mamdou Konate
Bobo-Dioulasso, Burkina Faso 01 BP 153
Telephone - 226-971341
FAX - 226-970457
Email - direction.muraz@fasonet.bf

Vanderjagt, Dorothy Ph.D.
University of New Mexico, School of Medicine
915 Camino de Salud NE
Albuquerque, NM USA 87131
Telephone - 505/272-5799
FAX - 505/272-6587
Email - dvanderjagt@salud.unm.edu

Vasan, Ushanalini MBS, DCH
Rush Presbyterian St. Lukes Medical Ctr.
1653 W. Congress Parkway
Chicago, IL USA 60612
Telephone - 312 942 2826
FAX - 312 942 4370
Email -uvasan@rush.edu

Vaucher, Yvonne E. MD, MPH
University of California, San Diego
607 Fern Glen, La Jolla, CA USA 92037
Telephone - 619/543-3759
FAX - 619/543-3812
Email - yvaucher@uscd.edu

Villalpando, Salvador MD, Ph.D.
Unidad de Investigacion en Nutricion
Hospital de Pediatria CMN, Av Cuautemoc 330
Mexico City, Mexico 06720
Telephone - 525/627-6944
Email - svnutri@data.net.mx

Wagner, Carol L. MD
Medical University of South Carolina
171 Ashley Avenue
Charleston, SC USA 29425
Telephone - 843/792-2401
FAX - 843/792-8801
Email - wagnercl@musc.edu

Wight, Nancy E. MD
Children's Hospital San Diego, 1230 Trieste Drive
San Diego, CA USA 92107
Telephone – 619/222-0442
FAX - 619/222-0443
Email - wightsd@aol.com

Willeitner, Andrea MD
University of Munich
Children's Hospital
Elisabeth Str., 23
Munich, Germany 80796
Telephone - 49 89 2723702
FAX - 49 89 2723702
Email - awilleitner@lrz.uni-muenchen.de

Williams, Catherine S.
University of Arizona
Department of Pediatrics
1501 N. Campbell Avenue
Tucson, AZ USA 85724
Telephone - 520/626-6450
FAX - 520/626-5009
Email - cwilliam@peds.arizona.edu

Williams-Arnold, Lois D. MPH
National Commission on Donor Milk Banking
8 Jan Sebastian Way, #11
Sandwich, MA 02563 USA
Telephone - 508 888 9366
FAX - 508 888 8050
Email - milkbank@capecod.net

Winberg, Jan Prof Emeritus, Ph.D.
Karolinska Hospital
Department of Pediatrics
Stockholm, Sweden 17176
Telephone - 46 8 517 729 85
FAX - 46 8 517 740 34

Wright, Anne Ph.D.
University of Arizona
Respiratory Sciences Center
1501 N. Campbell Avenue
Tucson, AZ USA 85724
Telephone - 520/626-6686
FAX - 520/626-6970
Email - awright@resp-sci.arizona.edu

Zaman, Shakila MBBS, MLPS, Ph.D.
King Edward Medical College
85-K Ste 77, Defence Housing Authority
Lahore, Pakistan
Telephone - 092-42-5726851-2
FAX - 092-42-7233509
Email - kemc@kemc.lho.sdnpk.org

Ziegler, Edhard E. MD
University of Iowa
Department of Pediatrics
200 Hawkins Drive
Iowa City, IA USA 52242
Telephone - 319/356-2836
FAX - 319/356-8669
Email - ekhard-ziegler@uiowa.edu

CONTENTS

SECTION I

THE IMPACT OF MICRONUTRIENT DEFICIENCIES DURING LACTATION ON MATERNAL AND INFANT HEALTH

SECTION II

The Premature Infant

SECTION III

Developmental Immunology

SECTION IV

Breastfeeding in the Industrialized World

SECTION V

VIRAL TRANSMISSION IN HUMAN MILK AND COLOSTRUM: MECHANISMS, POTENTIAL INTERVENTIONS, AND IMPLICATIONS FOR BREASTFEEDING PROMOTION

SECTION VI

BREASTFEEDING AND REPRODUCTIVE HEALTH

SECTION VII

ABSTRACTS

THE SECRETORY IMMUNOGLOBULIN SYSTEM: REGULATION AND BIOLOGICAL SIGNIFICANCE

Focusing on human mammary glands

Per Brandtzaeg*

1. INTRODUCTION

The surface of mucosal epithelia amounts to some 400 m^2 in an adult individual. This extensive and generally quite vulnerable barrier is protected by numerous innate mechanisms that cooperate intimately with specific mucosal immunity. The main humoral mediators of this local immune system are secretory IgA (SIgA) and IgM (SIgM); the former class of antibodies constitutes the largest noninflammatory defense system of the body.[1, 2] Although the secretory antibody system is mainly directed against colonization of pathogens and penetration of "dangerous" antigens, it is also involved in immune exclusion of innocuous soluble proteins present in food (Figure 1). However, the latter type of antigen, as well as components of the indigenous microflora, generally induce poorly understood suppressive mechanisms collectively called oral tolerance when induced via the gut.[3, 4] This complex phenomenon of mucosally induced immunological downregulation apparently explains why most individuals show no adverse immune reactions to persistent contact with food proteins and the normal microbial flora.

Successful interactions between local innate and specific immunity is a prerequisite for health because the various mucosae are favored as portals of entry by most infectious agents, allergens, and carcinogens. The neonatal period is particularly critical in this respect, because the newborn is immediately exposed to a large number of micro-organisms, foreign proteins, and chemicals. This review deals with mechanisms involved in the regulation and function of the adaptive first line defense system. Its contribution to immune exclusion is virtually lacking during a variable period after birth; therefore, breast-feeding is important, not only as a natural immunological "substitution therapy", but most likely also because immune-modulating factors in breast milk influence the development of the suckling`s immune system.[4, 5]

*Per Brandtzaeg, Laboratory for Immunohistochemistry and Immunopathology (LIIPAT), Institute of Pathology, University of Oslo, Rikshospitalet, N-0027, Norway

Integrating Population Outcomes, Biological Mechanisms and Research Methods in the Study of Human Milk and Lactation
Edited by Davis *et al.*, Kluwer Academic/Plenum Publishers, 2002

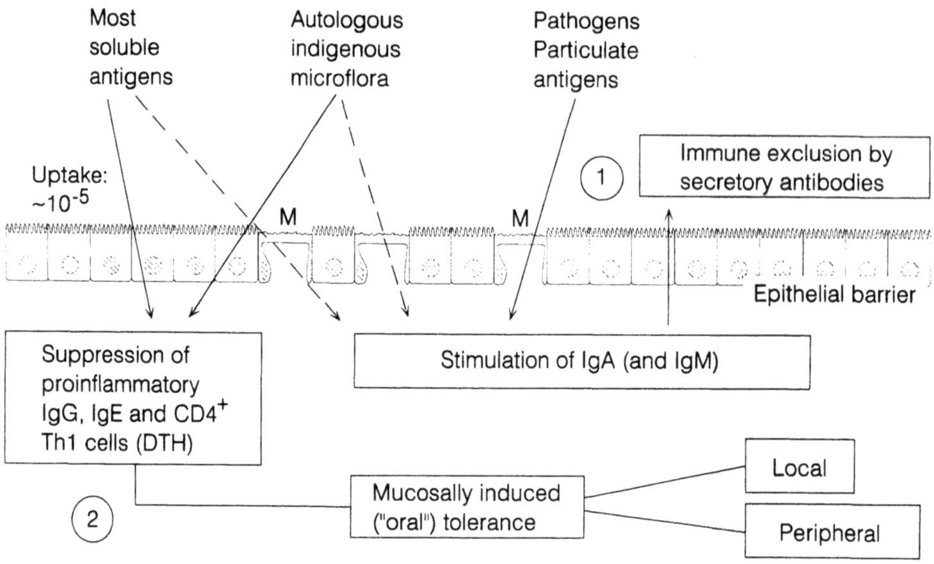

Figure 1. Schematic depiction of two major adaptive immune strategies for mucosal defense: (1) Immune exclusion limits epithelial colonization of microbes and inhibits penetration of harmful foreign material; it is principally mediated by secretory antibodies of the IgA (and IgM) class in co-operation with various non-specific innate protective factors (not shown). Secretory immunity is preferentially stimulated by pathogens and other particulate antigens taken up through thin M cells (M) located in the dome epithelium covering inductive mucosa-associated lymphoid tissue. (2) Penetrating harmless soluble environmental antigens, including dietary proteins (magnitude of uptake indicated) and the normal autologous microbial flora, are less stimulatory for secretory immunity (self-limiting responses; broken arrows) and induce suppression of proinflammatory humoral immune responses (IgG and IgE antibodies) as well as delayed-type hypersensitivity (DTH) mediated by activated helper T cells (CD4+) of the interferon-γ-producing Th1 subset. This complex and poorly defined phenomenon is called mucosal or "oral" tolerance; it may exert down-regulatory effects both locally and in the periphery.

2. Induction and Dissemination of Mucosal Immunity

After the first period of maternally derived IgG-mediated immunity, the survival of the infant depends on its own adaptive immune responses. In this respect, mucosal immunity is crucial,[6] and the relative resistance of SIgA against many microbial and endogenous proteases makes it well suited for surface protection.[7] SIgM may also exert protection, particularly in early infancy and in selective IgA deficiency, but antibodies of this isotype are more easily degraded in the gut lumen than SIgA.[6]

Primary B-cell responses that give rise to secretory antibodies are elicited mainly in organized lymphoepithelial structures where antigens are sampled from the mucosal surface.[6,9,10] Such gut-associated lymphoid tissue (GALT) includes aggregated (Peyer's patches) and scattered B-cell follicles. In humans, the Peyer's patches are mainly found in the distal ileum, whereas most of the solitary follicles occur in the appendix and distal large bowel. All these components of GALT appear functionally similar; they contain a characteristic follicle-associated epithelium with "membrane" (M) cells capable of transporting live and dead antigens into the underlying lymphoid tissue.[6, 9, 10]

Although GALT constitutes the major part of mucosa-associated lymphoid tissue (MALT), induction of mucosal immune responses can also take place in the palatine tonsils and other lymphoepithelial structures of Waldeyer's pharyngeal ring, including nasal-associated lymphoid tissue such as the adenoids.[6, 11] Accumulating evidence suggests that a certain regionalization exists in the mucosal immune system, especially a dichotomy between the gut and the upper aerodigestive tract with regard to homing properties and terminal differentiation of B cells.[6, 10] This disparity may be explained by microenvironmental differences in the antigenic repertoire as well as in the lymphoid and vascular adhesion molecules involved in local leukocyte extravasation. It appears that primed immune cells preferentially home to effector sites corresponding to the inductive sites where they initially responded to antigens.

Lactating mammary glands are also part of the integrated mucosal immune system, and milk antibodies reflect antigenic stimulation of MALT in the gut as well as in the airways. This fact has been documented by showing that SIgA from breast milk exhibits specificity for an array of common intestinal and respiratory pathogens.[12] The secretory antibodies are thus highly targeted against infectious agents in the mother's environment, which are those likely to be encountered by the infant during its first weeks of life. Therefore, breast-feeding represents an ingenious immunological integration of mother and child (Figure 2). Although the protection provided by this defense mechanism is most readily demonstrable in populations living in poor sanitary conditions,[13, 14] a beneficial clinical effect is also apparent in the industrialized world,[15] even in relation to relatively common diseases such as otitis media and acute lower respiratory tract infections.[16, 17] Antibodies to various dietary antigens, such as cow's milk proteins[18] and gluten,[19] are also present in breast-milk. The possible role of such antibodies for tolerance induction against food is discussed below.

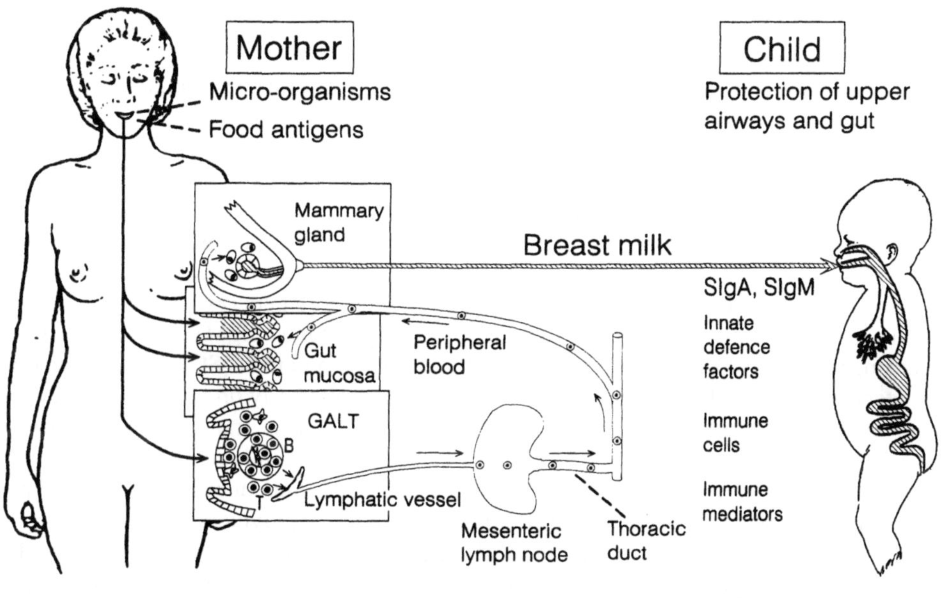

Figure 2. Integration of mucosal immunity between mother and the newborn, with emphasis on migration of primed B (and probably T) cells from Peyer's patch via lymph and peripheral blood to the lactating mammary gland. Such distribution (arrows) beyond the gut of precursors for IgA plasma cells is crucial for glandular production and subsequent occurrence in breast milk of secretory antibodies (SIgA and SIgM) specific for enteric antigens (microorganisms and food proteins). By this mechanism, the breast-fed infant will receive relevant secretory antibodies directed against the microbiota colonizing its mucosae (initially reflecting the microflora of the mother) and hence be better protected both in the gut and in the upper airways (hatched areas) in the same way as the mother's gut mucosa is protected by similar antibodies (hatched areas).

3. Critical Role of Secretory Immunity in Infancy

Most full-term babies growing up under privileged conditions exhibit satisfactory resistance to mucosal infections, provided that the innate defense mechanisms are normal. Adequate systemic antibody protection of their mucosae is provided by circulating maternal IgG, which is distributed extravascularly at a ratio of 50%-60%. Placental transfer of IgG in the fetus is unique for primates, and postnatal uptake of breast milk-derived macromolecules such as SIgA seems to be of minor importance in supporting early immunity in humans,[8, 20] except perhaps in pre-term babies.[21, 22]

When much of the maternal IgG has been catabolized around 2 months, the protective value of breast-feeding is highlighted in relation to severe infections, particularly in the developing countries. At least 90% of micro-organisms infecting humans, use the mucosae as portals of entry; such pathogens are a major killer of children below the age of 5 years, being responsible for more than 14 million deaths of children annually. Diarrheal disease alone claims a toll of 5 million children per year, or about 500 deaths every hour. These sad figures document the need for mucosal vaccines to enhance surface defense against infectious agents, in addition to advocating breast-feeding. Convincing epidemiological documentation suggests that the risk of dying from diarrhea is reduced 14-24 times in nursed children.[13, 14] Indeed, exclusively breast-fed infants are better protected against a variety of infections,[15-17, 23] asthma and atopy,[24, 25] and also celiac disease.[26] Moreover, recent experiments in neonatal rabbits strongly suggest that SIgA is a crucial protective component of breast milk.[27] The role of secretory antibodies for mucosal homeostasis is furthermore supported by the fact that knock-out mice lacking SIgA and SIgM show increased mucosal leakiness.[28]

4. Efficiency and Regulation of Epithelial Receptor- Mediated Secretory Antibody Transport

The remarkable magnitude of GALT as an inductive site for B cells is documented by the fact that at least 80% of all Ig-producing blasts and plasma cells (collectively called immunocytes) in an adult are located in the intestinal lamina propria.[6] Some 90% of these terminally differentiated mucosal B cells normally produce dimers or larger polymers of IgA, collectively called pIgA.[6] Such polymers (as well as pentameric IgM) are efficiently transported externally as SIgA (and SIgM) antibodies by a transmembrane ~100-kD glycoprotein called secretory component (SC) or the polymeric Ig receptor (pIgR), which is constitutively expressed basolaterally on intestinal crypt cells and other serous types of glandular epithelial cells.[6, 29] This transport mechanism is shared by pIgA and pentameric IgM because they contain a common 15-kD polypeptide called J (joining) chain produced preferentially by mucosal immunocytes.[6, 30] The J chain constitutes an essential part of the pIgR binding site in the Ig polymers.[31]

After transcytosis to the luminal surface, SIgA and SIgM are released by cleavage of the pIgR, and only the C-terminal smaller receptor domain remains apically for degradation in the epithelial cell (Figure 3); the 80-kD extracellular part is incorporated into the secretory antibodies as bound SC, thereby providing protection against proteolytic degradation, particularly to SIgA in which SC becomes covalently linked. In adult humans, more pIgA (~40 mg/kg body weight) is translocated to the intestinal secretions by this receptor-mediated mechanism every day than the total daily

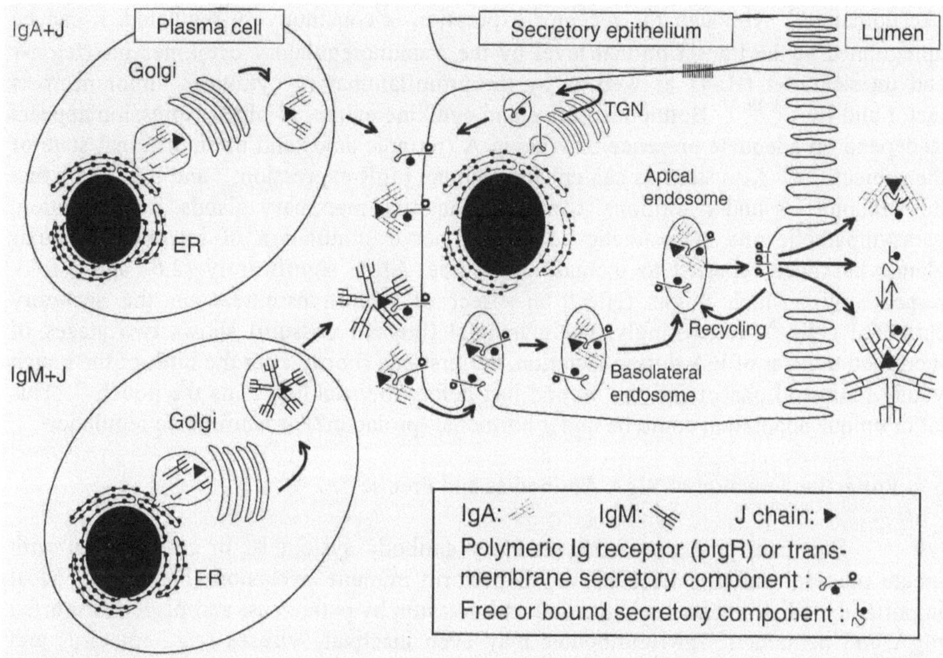

Figure 3. Model for local generation of human SIgA (top right) and SIgM (bottom right). J-chain-containing dimeric IgA (IgA+J) and pentameric IgM (IgM+J) are produced by local plasma cells (left). Polymeric Ig receptor (pIgR), or transmembrane secretory component, is synthesized by secretory epithelial cells in the rough endoplasmic reticulum and matures in the Golgi complex by terminal glycosylation (•-). In the trans-Golgi network (TGN), pIgR is sorted for delivery to the basolateral plasma membrane. The receptor becomes phosphorylated (o-) on a serine residue in the cytoplasmic tail. After endocytosis, ligand-complexed and unoccupied pIgR is delivered to basolateral endosomes and sorted for transcytosis to apical endosomes. However, some recycling from basolateral endosomes to the basolateral surface may take place for unoccupied pIgR (not shown), and receptor recycling also occurs at the apical cell surface as indicated, although most pIgR is cleaved and SIgA, SIgM, and free SC are released to the lumen. During epithelial translocation, covalent stabilization of SIgA regularly occurs (disulfide bond between bound SC and one IgA subunit indicated), whereas free SC in secretions apparently stabilizes the noncovalently bound SC in SIgM (dynamic equilibrium indicated).

IgG production (~30 mg/kg).[32] Excess of unoccupied pIgR is released to the lumen by proteolytic cleavage in the same manner as SIgA and SIgM to form so-called free SC (Figure 3). This 80-kD fragment (identical to bound SC) occurs in most secretions and, by equilibrium with the bound component, it exerts a stabilizing effect on the quaternary structure of SIgM in which SC remains non-covalently linked.[6, 29] In various ways, free SC may also contribute to innate mucosal defense (see below).

The pIgR belongs to the Ig super family and binds its two ligands (Figure 3) non-covalently in a somewhat different manner via the first of its five extracellular Ig-like domains.[29] Although the receptor expression is constitutively regulated, it can be upregulated at the transcriptional level by the immunoregulatory cytokines interferon-γ and interleukin-4 (IL-4), as well as by the proinflammatory cytokines tumor necrosis factor and IL-1.[6, 29, 33] Both constitutive and cytokine-enhanced pIgR expression appears to depend on adequate presence of vitamin A (retinoic acid) and the nutritional state of the subject.[34, 35] Also steroids can enhance human pIgR expression,[36] and it is likely that this receptor is under hormonal control in lactating mammary glands.[37] In addition, parasympathetic and sympathetic autonomic nerve stimulation of rat submandibular glands has been reported to increase the output SIgA significantly (2.6- and 6-fold, respectively), which might reflect an effect of neurotransmitters on the secretory epithelial cells.[38] Interestingly, the marsupial (brustail possum) shows two stages of increased transfer of IgA during lactation, the first one shortly after the birth of the pouch young (colostral phase) and the second just before the suckling exits the pouch.[39] This latter unique adaptation could be under hormonal (prolactin?) or neurogenic regulation.

5. Protective Function of SIgA Antibodies and Free sc

The main purpose of the secretory antibody system is, in co-operation with innate mucosal defense mechanisms, to perform immune exclusion (Figure 1). Most importantly, SIgA inhibits colonization and invasion by pathogens, and pIgR-transported pIgA and pentameric IgM antibodies may even inactivate viruses (e.g., rotavirus and influenza virus) inside secretory epithelial cells and carry the pathogens and their products back to the lumen, thus avoiding cytolytic damage to the epithelium.[29] Both the agglutinating and virus-neutralizing antibody effect of pIgA is superior compared with monomeric antibodies,[1] and SIgA antibodies may block microbial invasion quite efficiently. This has been particularly well documented in relation to human immunodeficiency virus (HIV),[40] and specific SIgA antibodies isolated from human colostrum has been shown to be more efficient in this respect than comparable IgG antibodies.[41]

Induction of SIgA responses has likewise been shown to interfere significantly with mucosal uptake of soluble macromolecules in experimental animals.[1] Collectively, therefore, the function of locally produced pIgA, including antibodies in breast milk, would be to inhibit mucosal colonization of microorganisms as well as penetration of antigens; this effect is most likely enhanced by the relatively high levels of polyreactive SIgA antibodies.[42] In the gut, interaction of SIgA with the intestinal superantigen protein Fv (Fv fragment binding protein) may, moreover, build an immune fortress by forming large complexes of intact or degraded antibodies with different specificities,[43] thereby reinforcing immune exclusion. It has also been claimed that SIgA can enhance the sticking of certain bacteria to mucus, interfere with growth factors (e.g., iron) and enzymes necessary for pathogenic bacteria and parasites,[1, 6] and exert positive influences on the inductive phase of mucosal immunity by promoting antigen uptake in GALT via

IgA receptors on the M cells.[44] The latter possibility adds to the importance of breast-feeding in providing a supply of relevant SIgA antibodies to the infant's gut. Interestingly, free SC may on its own be able to block epithelial adhesion of *Escherichia coli*,[45] and a pneumococcal surface protein (SpsA) has been shown to interact with both free and bound SC.[46] Such observations suggest that SC phylogenetically has originated from the innate defense system before being exploited by the adaptive secretory immune system to function as pIgR.

6. Other Immune Components of Human Breast Milk

Numerous constituents of breast milk, in addition to antibodies and free SC, are though to protect the suckling. These include innate defense factors such as lysozyme, lactoferrin, peroxidase, complex oligosaccharides (receptor analogues), fatty acids (lipids) and mucins.[12, 15, 16] Moreover, a variety of leukocytes occur in colostrum (~4x10^6/ml) and later milk (~10^5/ml). Macrophages (55-60%) and neutrophilic granulocytes (30-40%) dominate over lymphocytes (5-10%), the latter being mainly (75-80%) T lymphocytes.[12, 15] Oral administration of macrophages in newborn mice showed survival of these cells for several hours in the gut and even some mucosal uptake.[47] The macrophages contain engulfed SIgA, which they may release on contact with bacteria in the gut,[48] and they may also secrete an array of important immunoregulatory factors. Thus, it has been reported that unfractionated supernatants of breast milk cell cultures preferentially stimulate IgA production by peripheral blood lymphocytes.[48] An explanation for this effect may be the various cytokines that are secreted by stimulated milk macrophages.[15] The same soluble cytokines are found in breast milk,[12, 15] and the presence of transforming growth factor-β (TGF-β), IL-6, and IL-10 is of particular interest for the development and differentiation of IgA-producing cells.[6] Direct evidence to this end has been provided for colostral IL-6.[49] Even if these cytokines are unable to survive the passage through the gastrointestinal tract, they may be locally released from milk macrophages stimulated in the neonatal gut and thereby enhance the development of mucosal immunity.

7. Postnatal Development of Secretory Immunity

7.1. Induction of the Local B-Cell System

Peyer's patches and other GALT structures are well developed at birth, discrete T- and B-cell areas being apparent as early as at 19 weeks' gestation.[50] However, secondary follicles with germinal centers signifying B-cell activation, do not occur until some weeks after birth; this reflects their dependency on exogenous environmental stimulation. The germinal-center B cells of murine Peyer's patches express small amounts of surface IgA along with less IgM or IgG.[51] Such isotype skewing reflects B-cell switching in the course of clonal differentiation to precursors for IgA-producing immunocytes, whose preferential induction is the hallmark of GALT.[6, 20]

The fact that the postnatal immune activation of GALT is retarded, parallels the temporary immaturity of systemic immunocompetence observed in the newborn period.[4, 52, 53] Thus, very few B cells with IgA-producing capacity (presumably GALT-derived) are present in peripheral blood of newborns (<8/10^6 lymphocytes), but after 1 month this number is remarkably increased (~600/10^6 lymphocytes), reflecting the progressive environmental stimulation of GALT.[54] An initial early elevation of positive cells can be seen in pre-term infants, especially in those with intrauterine infections, although IgM production dominates in these cases. In agreement with these observations in peripheral blood, only occasional IgM- and IgG-producing intestinal immunocytes are present at birth, and local IgA immunocytes are either absent or extremely rare even until after 10 days of age.[8] The numbers of mucosal IgM- and IgA-producing cells increase rapidly after 2-4 weeks, the latter class becoming predominant at 1-2 months and usually peaking at about 12 months.

An early SIgM antibody response is probably of protective value, but it is known that specific immunity to certain bacterial capsular polysaccharides is poor or lacking before 2 years of age. This creates a window of susceptibility at the time of disappearance of protective maternal IgG antibodies, especially when combined with weaning (which naturally means deprivation of breast milk SIgA). The basis for the impaired immune response to polysaccharides is unclear, but reduced levels of complement receptor 2 (CR2, CD21) expression on B cells and follicular dendritic cells, together with low complement activity in newborns, may result in lack of CR2/B-cell receptor synergy and thereby suboptimal B-cell activation.[55] Compelling evidence shows that interaction of the complement split product C3d with CR2 is an extremely important link between innate immunity and specific B-cell responses.[56]

7.2. Role of Antigen Exposure in the Development of Mucosal Immunity

The degree of antigenic and mitogenic exposure appears to be decisive for the postnatal development of the secretory immune system. Antigenic constituents of food clearly exert a stimulatory effect as suggested by fewer lamina propria IgA-producing cells in mice fed on hydrolysed milk proteins[57] as well as in parenterally fed babies.[58] However, the indigenous microbial flora is most important as shown by the fact that the intestinal IgA system of germ-free or specific pathogen-free mice is normalized after about 4 weeks of conventionalization.[59, 60] *Bacteroides* and *E. coli* strains seem to be particularly stimulatory for the development of intestinal IgA immunocytes.[61, 62] The large dietary and bacterial antigenic load in the gut lumen therefore explains that the greatest density of IgA immunocytes is seen in the intestinal lamina propria, amounting to some 10^{10} cells per meter of adult gut.[6] In the human lactating mammary glands the density is much less, one gland having an IgA-producing capacity similar to 1 meter of intestine.[63] Thus, the daily output of IgA/ kilogram wet weight parenchymal tissue is similar for salivary and lactating mammary glands, and it is not clear how the terminal plasma-cell differentiation is driven in these effector organs at considerable distances from mucosal surfaces.[6] Large capacity for storage of pIgA/SIgA in the mammary gland epithelium and duct system, rather than high immunocyte density, thus explains the striking output of SIgA during feeding.[63]

In keeping with these observations, decreased amounts of both dietary and microbial antigens resulted in a 50% reduction of the colonic numbers of IgA and IgM immunocytes after 2-11 months of defunctioning colostomies in children.[64] Prolonged studies of defunctioned ileal segments in lambs did even more strikingly reveal a scarcity of lamina propria immunocytes; this was caused by decreased local proliferation and differentiation of B-cell blasts and perhaps reduced homing from GALT.[65] Accordingly, the postnatal development of the mucosal IgA system usually is much faster in developing countries than in the industrialized part of the world,[8] a difference that apparently holds true even in undernourished children.[66]

Also interestingly, it has been reported that undernourished children respond to bacterial overgrowth in the gut with enhanced synthesis as well as upregulated external transport of IgA.[67] It is of further great clinical importance that detrimental effects of severe malnutrition exerted on the SIgA system can be reversed with nutritional rehabilitation.[68] In a recent study based on whole gut lavage obtained from healthy adult volunteers in Dhaka, Bangladesh, the intestinal concentration of IgA was found to be almost 50% higher than that of comparable samples collected in Edinburgh, UK, and the intestinal IgA antibody titre against lipopolysaccharide (LPS) core types of *E. coli* was almost seven times higher in the former group of subjects, in contrast to the lower levels of ovalbumin antibodies.[69]

In view of the above information, the possibility exists that suboptimal stimulatory reinforcement of the SIgA-dependent mucosal barrier function might contribute to the increased frequency of certain diseases in industrialized countries, particularly allergies and other inflammatory disorders. The potential beneficial effect of probiotic preparations has therefore been evaluated in several experimental and clinical studies. Especially viable preparations containing species common to the normal intestinal microflora, such as lactobacillus and bifidobacteria, have been reported to provide bystander enhancement of IgA responses, both in humans and experimental animals,[70-74] apparently in a T cell-dependent manner.[75] Interestingly, early colonization of infants with a non-enteropathogenic strain of *E. coli* has recently been reported to have a long-term beneficial effect in reducing both infections and allergies.[76]

7.3. Individual Variations in the Development of Mucosal Immunity

The postnatal mucosal B-cell development shows large individual variations, even within the same population.[8] To some extent this might be genetically determined and it could exert an important impact on children's health because of reduced antigen exclusion. Thus, on the basis of IgA measurements in serum, it has been suggested that infants and children at hereditary risk of atopy have a retarded postnatal development of their IgA system.[77, 78] This notion was later supported by enumeration of jejunal immunocytes; a significantly reduced IgA response to luminal antigens, without any IgM compensation, was indicated in the mucosa of atopic children.[79] Another study showed an inverse relationship between the serum IgE concentration and the number of IgA-producing cells in jejunal mucosa of food-allergic children.[80] It has also been reported that infants born to atopic parents have a significantly higher prevalence of salivary IgA deficiency than age-matched control infants.[81] Interestingly, Kilian *et al.*[82] more recently found that the throats of 18-month-old infants with presumably IgE-mediated clinical problems, contained significantly higher proportions of IgA1 protease-producing bacteria than age-matched healthy controls.

Altogether, a reduced SIgA-dependent epithelial barrier function combined with a hereditary elevated IgE responsiveness could often underlie the pathogenesis of mucosal hypersensitivity, at least in atopic children. This notion accords with the increased frequency of infections, atopic allergies, and gluten-dependent enteropathy (celiac disease) seen in subjects with permanent selective IgA deficiency, although compensatory overproduction of SIgM to some extent may counteract the adverse consequences of their absent mucosal IgA responses.[8, 83]

7.4. Effect of Breast-Feeding on Active Mucosal Immunity Development

In addition to the remarkable reinforcement of mucosal defense provided by maternal SIgA (and SIgM) antibodies as a natural immunological "substitution therapy", it is important to emphasize the positive nutritional effect of breast-feeding on immune development.[3, 14] Also, as mentioned above, breast milk contains a number of immune cells, cytokines, and growth factors that may exert a significant biological effect in the suckling infant's gut, apparently enhancing in an indirect way even the subsequent health of the individual.[5, 15, 16]

Numerous studies of the effect of breast-feeding on secretory immunity have been performed with salivary IgA measurements as a read-out system. Discrepant observations have been made and the influence of contaminating the sample with milk SIgA, shielding of the suckling`s mucosal immune system by maternal SIgA antibodies, and altered growth and composition of the infant's gut flora have been discussed as possible uncontrollable variables.[8] However, evidence does suggest that breast-feeding promotes the postnatal development of secretory immunity,[5, 15] apparently even in the urinary tract[16]; and there are reports on enhanced secretory as well as systemic immune responses to oral and parenteral vaccines in breast-fed babies.[84, 85]

Nevertheless, several prospective studies have reported that the early physiological increase of salivary IgA (and IgM) is more prominent in formula-fed than in solely breast-fed infants,[8, 86, 87] although this difference apparently disappears after weaning.[88] It likewise appears that breast-feeding, in comparison with formula-feeding, reduces the salivary IgA antibody titers to cow's milk proteins;[89] this decrease was seen after a nursing period of only 3 weeks and appeared also in infants receiving mixed feeding.[87, 89] Altogether, therefore although breast-feeding initially appears to reduce induction of salivary IgA, it will later on in infancy (up to 8 months) boost this response.[90, 91] Experiments in mice have demonstrated that SIgA antibodies from breast milk affect the normal gut flora in the suckling by retarding its contact with the developing GALT.[92] When the host's mucosal immune response subsequently is successfully elicited, GALT will be further shielded by the SIgA antibodies produced in the gut; local immunostimulation is thereby attenuated despite the continued presence of microorganisms.[92] This could partly explain the hyporesponsiveness that normally exists towards members of the indigenous gut bacteria, resembling some sort of oral tolerance phenomenon both in rodents and in humans.[93] So-called “knock-in” mice with IgH insertion of defined VDJ segment specificity (“quasimonoclonal”) are now available for more precise studies of antibody diversification in response to environmental antigens.[94]

7.5. Effect of Breast-Feeding on Oral Tolerance Development

Oral tolerance unquestionably involves more than one immune mechanism, and available data do indeed suggest an overwhelming complexity. Identifiable variables in experimental animals are genetics, age, feeding dose and timing, antigenic structure, epithelial barrier integrity, and the degree of concurrent local immune activation (as reflected by co-stimulatory molecules expressed by antigen-presenting cells and microenvironmental cytokine profiles).[3, 4]

Through avoidance of too early immune activation (e.g., upregulation of the co-stimulatory molecules B7),[4, 9, 95] the shielding effect exerted by SIgA from breast milk on the suckling`s GALT (see above)[92] may likewise contribute to the establishment of oral tolerance not only against the indigenous microflora, but also against dietary antigens such as gluten. Antibodies to gluten peptides are present in breast milk,[19] and breast-feeding has in fact been shown to protect significantly against the development of celiac disease in children unrelated to the time of solid food introduction.[26] On the basis of such studies, it can be tentatively concluded that mixed feeding, rather than abrupt weaning, appears to promote tolerance to food proteins. This notion is supported by a report suggesting that cow's milk allergy is more likely to develop in infants whose mothers have relatively low levels SIgA antibodies to cow's milk proteins in their breast milk.[18] The presence of TGF-β in breast milk might further contribute to its tolerogenic properties[96] because this cytokine exerts a pronounced immunosuppressive effect on GALT[97] and enhances the epithelial barrier function.[98] Although still a quite controversial issue, the balance of epidemiological studies supports the view that breast-feeding also protects against atopic allergy[24] and asthma.[25]

8. Conclusions

Several more or less well-defined variables influence the development of active mucosal immunity as well as oral tolerance, and some of these components are reciprocally modulated by the immune system (Figure 4). Epithelial permeability for luminal antigens is likely an important primary or secondary event in the pathogenesis of several mucosal diseases, including adverse immune reactions to foods. This variable is determined by the individual's age (e.g., preterm versus term infant), interactions between mast cells, nerves and neuropeptides, concurrent infection, and the shielding effect of

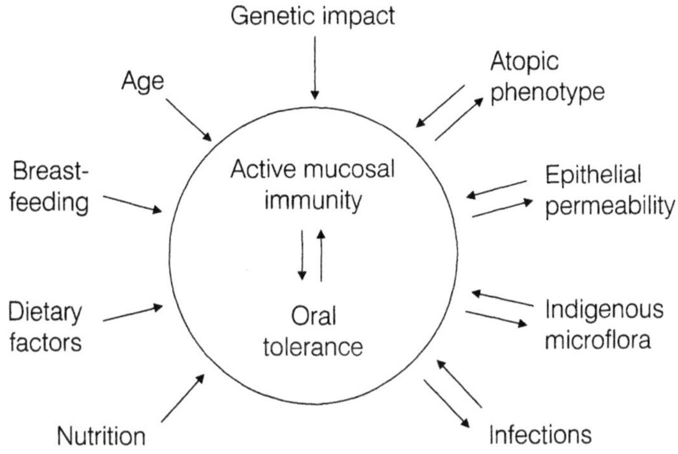

Figure 4. Summary of variables influencing the development of active mucosal immunity and oral tolerance. Several of the components are reciprocally modulated as indicated by bidirectional arrows.

SIgA provided by breast milk or produced in the infant's gut. The consequences will depend on how fast "closure" of the epithelial barrier can be attained or re-established, which is influenced both by the age of the infant and by its successful mounting of adaptive SIgA responses as well as generation of oral tolerance towards innocuous antigens from the diet and from the normal indigenous microbiota. SIgA is the best defined effector component of the mucosal immune system, and much knowledge has recently been obtained at the molecular level about the constitutive and induced transcriptional regulation of pIgR-mediated epithelial antibody transport. Large capacity for storage of pIgA in the mammary gland epithelium and duct system apparently explains the remarkable output of SIgA during feeding. Altogether, the secretory immune system is of considerable clinical interest because SIgA and SIgM form the first line of specific immunological defense against infectious agents and other harmful substances. Human studies and characterization of a recently generated pIgR knock-out mouse, support the notion that secretory antibodies are important in maintaining the barrier function of mucosal epithelia.

Acknowledgements

Studies in the author's laboratory are supported by the Norwegian Cancer Society, the Research Council of Norway, and Anders Jahre's foundation. Erik K. Hagen is thanked for excellent assistance with the preparation of this manuscript.

References

1. P. Brandtzaeg, K. Baklien, K Bjerke, T.O. Rognum, H. Scott, and K. Valnes, Nature and properties of the human gastrointestinal immune system, in: *Immunology of the Gastrointestinal Tract,* Vol. I, edited by K. Miller and S. Nicklin (CRC Press, Boca Raton, Florida, 1987), pp. 1-85.
2. M.W. Russell, J. Reinholdt, and M. Kilian, Anti-inflammatory activity of human IgA antibodies and their Fabα fragments: inhibition of IgG-mediated complement activation, Eur. J. Immunol. 19:2243-2249 (1989).
3. P. Brandtzaeg, History of oral tolerance and mucosal immunity, Ann. N.Y. Acad. Sci. 778:1-27 (1996).
4. P. Brandtzaeg, Development and basic mechanisms of human gut immunity, Nutr. Rev. 56:S5-18 (1998).
5. P. Brandtzaeg, Development of the mucosal immune system in humans, in: *Recent Developments in Infant Nutrition,* edited by J. G. Bindels, A. C. Goedhart, and H. K. A. Visser (Kluwer Academic Publishers, London, 1996), pp. 349-376.
6. P. Brandtzaeg, I.N. Farstad, F.-E. Johansen, H.C. Morton, I.N. Norderhaug, and T. Yamanaka, The B-cell system of human mucosae and exocrine glands, Immunol. Rev. 171:45-87 (1999).
7. M. Kilian, J. Reinholdt, H. Lomholt, K. Poulsen, and E.V. Frandsen, Biological significance of IgA1 proteases in bacterial colonization and pathogenesis: critical evaluation of experimental evidence, APMIS 104:321-338 (1996).
8. P. Brandtzaeg, D.E. Nilssen, T.O. Rognum, and P.S. Thrane, Ontogeny of the mucosal immune system and IgA deficiency, Gastroenterol. Clin. North. Am. 20:397-439 (1991).
9. P. Brandtzaeg, E.S. Baekkevold, I.N. Farstad, F.L. Jahnsen, F-E. Johansen, E.M. Nilsen and T. Yamanaka, Immunol. Today 20:141-151 (1999).
10. P. Brandtzaeg, I.N. Farstad, and G. Haraldsen, Regional specialization in the mucosal immune system: primed cells do not always home along the same track, Immunol. Today 20:267-277 (1999).
11. P. Brandtzaeg, Regionalized immune function of tonsils and adenoids, Immunol. Today 20:383-384 (1999).
12. A.S. Goldman, The immune system of human milk: antimicrobial, antiinflammatory and immunomodulating properties, Pediatr. Infect. Dis. J. 12:664-671 (1993).
13. L.Å. Hanson, R.Ashraf, B. Carlsson, F. Jalil, J. Karlberg, B.S. Lindblad, S.R. Khan, and S. Zaman, Child health and the population increase, in: *Peace, Health and Development,* edited by L. Å. Hanson and L. Köhler (University of Göteborg and the Nordic School of Public Health, NHV Report 4, 1993), pp. 31-38.
14. Anonymous, A warm chain for breastfeeding, Lancet 344:1239-1241 (1994).
15. A. E. Wold and L.Å. Hanson, Defence factors in human milk, Curr. Opin. Gastroenterol. 10:652-658 (1994).
16. J. Newman, How breast milk protects newborns, Sci. Am. 273:76-79 (1995).

17. A. Pisacane, L. Graziano, G. Zona, G. Granata, H. Dolezalova, M. Cafiero, A. Coppola, B. Scarpellino, M. Ummarino, and G. Mazzarella, Breast feeding and acute lower respiratory infection, Acta. Paediatr. 83:714-718 (1994).
18. E. Savilahti, V.M. Tainio, L. Salmenpera, P. Arjomaa, M. Kallio, J. Perheentupa, and M.A. Siimes., Low colostral IgA associated with cow's milk allergy, Acta. Paediatr. Scand. 80:1207-1213 (1991).
19. P. Juto and S. Holm., Gliadin-specific and cow's milk protein-specific IgA in human milk, J. Pediatr. Gastroenterol. Nutr. 15:159-162 (1992).
20. S.S. Ogra, D. Weintraub, and P.L. Ogra, Immunologic aspects of human colostrum and milk. III. Fate and absorption of cellular and soluble components in the gastrointestinal tract of the newborn, J. Immunol. 119:245-248 (1977).
21. L.T. Weaver, N. Wadd, C.E. Taylor, J. Greenwell, and G.L. Toms, The ontogeny of serum IgA in the newborn, Pediatr. Allergy Immunol. 2:72-75 (1991).
22. I.R. Sanderson and W.A. Walker, Uptake and transport of macromolecules by the intestine: possible role in clinical disorders (an update), Gastroenterology 104:622-639 (1993).
23. A.L. Wright, M. Bauer, A. Naylor, E. Sutcliffe, and L. Clark, Increasing breastfeeding rates to reduce infant illness at the community level, Pediatrics 101:837-844 (1998).
24. U.M. Saarinen and M. Kajosaari, Breast feeding as prophylaxis against atopic disease: prospective follow-up study until 17 years old, Lancet. 346:1065-1069 (1995).
25. W.H. Oddy, Association between breast feeding and asthma in 6 year old children: findings of a prospective birth cohort study, BMJ 319:815-819 (1999).
26. P. Brandtzaeg, Development of the intestinal immune system and its relation to coeliac disease, in: *Coeliac Disease. Proceedings of the Seventh International Symposium on Coeliac Disease*, edited by M. Mäki, P. Collin, and J. K. Visakorpi (Coeliac Disease Study Group, Tampere, 1997), pp. 221-244.
27. E.C. Dickinson, J.C. Gorga, M. Garrett, R. Tuncer, P. Boyle, S.C. Watkins, S.M. Alber, M. Parizhskaya, M. Trucco, M.I. Rowe, and H.R. Ford, Immunoglobulin A supplementation abrogates bacterial translocation and preserves the architecture of the intestinal epithelium, Surgery 124:284-290 (1998).
28. F.-E. Johansen, M. Pekna, I.N. Norderhaug, B. Haneberg, M.A. Hietala, P. Krajci, C. Betsholtz, and P. Brandtzaeg, Absence of epithelial immunoglobulin A transport, with increased mucosal leakiness, in polymeric immunoglobulin receptor/secretory component-deficient mice. J. Exp. Med. 190:915-922 (1999).
29. I.N. Norderhaug, F.-E. Johansen, H. Schjerven, and P. Brandtzaeg, Regulation of the formation and external transport of secretory immunoglobulins, Crit. Rev. Immunol. 19:481-508 (1999).
30. P. Brandtzaeg, Presence of J chain in human immunocytes containing various immunoglobulin classes, Nature 252:418-420 (1974).
31. P. Brandtzaeg and H. Prydz, Direct evidence for an integrated function of J chain and secretory component in epithelial transport of immunoglobulins, Nature 311:71-73 (1984).
32. M.E. Conley and D.L. Delacroix, Intravascular and mucosal immunoglobulin A: two separate but related systems of immune defense? Ann. Intern. Med. 106:892-899 (1987).
33. H. Schjerven, P. Brandtzaeg, and F.-E. Johansen, Mechamism of IL-4-mediated up-regulation of the polymeric Ig receptor: role of STAT6 in cell type-specific delayed transcriptional response. J. Immunol. 165:3898-3906 (2000).
34. J. Sarkar, N.N. Gangopadhyay, Z. Moldveanu, J. Mestecky, and C.B. Stephenson, Vitamin A is required for regulation of polymeric immunoglobulin receptor (pIgR) expression by interleukin-4 and inerferon-gamma in a human intestinal epithelial cell line, J. Nutr. 128:1063-1069 (1998).
35. C.L. Ha and B. Woodward, Depression in the quantity of intestinal secretory IgA and in the expression of the polymeric immunoglobulin receptor in caloric deficiency of the weanling mouse, Lab. Invest. 78:1255-1266 (1998).
36. A. Haelens, G. Verrijdt, E. Schoenmakers, P. Alen, B. Peeters, W. Rombauts and F. Claessens, The first exon of the human sc gene contains an androgen responsive unit and an interferon regulatory factor element, Mol. Cell. Endocrinol. 153:91-102 (1999).
37. R. Rosato, H. Jammes, L. Belair, C. Puissant, J.P. Kraehenbuhl, and J. Djiane, Polymeric-Ig receptor gene expression in rabbit mammary gland during pregnancy and lactation: evolution and hormonal regulation, Mol. Cell. Endocrinol. 110:81-87 (1995).
38. G.H. Carpenter, J.R. Garrett, R.. Hartley, and G.B. Proctor, The influence of nerves on the secretion of immunoglobulin A into submandibular saliva in rats, J. Physiol. (Lond.) 512:567-573 (1998).
39. F.M. Adamski and J. Demmer, Two stages of increased IgA transfer during lactation in the marsupial, trichosurus vulpecula (Brushtail possum), J Immunol. 162:6009-6015 (1999).
40. S. Mazzoli, D. Trabattoni, S. Lo Caputo, S. Piconi, C. Blé, F. Meacci, S. Ruzzante, A. Salvi, F. Semplici, R. Longhi, M.L. Fusi, N. Tofani, M. Biasin, M.L. Villa, F. Mazzotta, and M. Clerici, HIV-specific mucosal and cellular immunity in HIV-seronegative partners of HIV-seropositive individuals, *Nature Med.* **3**, 1250-1257 (1997).

41. H. Hocini and M. Bomsel, Infectious human immunodeficiency virus can rapidly penetrate a tight human epithelial barrier by transcytosis in a process impaired by mucosal immunoglobulins, J. Infect. Dis. 179:(Suppl. 3), S448-453 (1999).
42. C.P. Quan, A. Berneman, R. Pires, S. Avrameas and J.P. Bouvet, Natural polyreactive secretory immunoglobulin A autoantibodies as a possible barrier to infection in humans, Infect. Immun. 65:3997-4004 (1997).
43. J.-P. Bouvet, R. Pirès, and C.P. Quan, Protein Fv (Fv fragment binding protein): a mucosal human superantigen reacting with normal immunoglobulins, in: *Human B Cell Superantigens*, edited by M. Zouali (R. G. Landes, Austin, 1996), pp. 179-187.
44. A. Frey and M.R. Neutra, Targeting of mucosal vaccines to Peyer's patch M cells, Behring Inst. Mitt. ***No.*** 98:376-389 (1997).
45. L.G. Giugliano, S.T. G. Ribeiro, M.H. Vainstein, and C.J. Ulhoa, Free secretory component and lactoferrin of human milk inhibit the adhesion of enterotoxigenic *Esherichia coli,* J. Med. Microbiol. 42:3-9 (1995).
46. S. Hammerschmidt, S.R. Talay, P. Brandtzaeg, and G.S. Chhatwal, SpsA, a novel pneumococcal surface protein with specific binding to secretory immunoglobulin A and secretory component, Mol. Microbiol. 25:1113-1124 (1997).
47. A. Huges, J.H. Brock, D.M. Parrott, and F. Cockburn, The interaction of infant formula with macrophages: effect on phagocytic activity, relationship to expression of class II MHC antigen and survival of orally administered macrophages in the neonatal gut, Immunology 64:213-218 (1988).
48. H.B. Slade and S.A. Schwartz, Mucosal immunity: the immunology of breast milk, J. Allergy Clin. Immunol. 80:346-356 (1987).
49. S. Saito, M. Maruyama, Y. Kato, I. Moriyama, and M. Ichijo, Detection of IL-6 in human milk and its involvement in IgA production, J. Reprod. Immunol. 20:267-276 (1991).
50. J. Spencer and T.T. MacDonald, Ontogeny of human mucosal immunity, in: *Ontogeny of the Immune System of the Gut,* edited by T. T. MacDonald (CRC Press, Boca Raton, Florida, 1990), pp. 23-50.
51. E.C. Butcher, R.V.Rouse, R.L. Coffman, C.N. Nottenburg, R.R.Hardy and I.L. Weissman, Surface phenotype of Peyer's patch germinal center cells: implications for the role of germinal centers in B cell differentiation, J. Immunol. 129:2698-2707 (1982).
52. P.G Holt, Postnatal maturation of immune competence during infancy and childhood, Pediatr. *Allergy* Immunol. 6:59-70 (1995).
53. T.T. MacDonald and J. Spencer, Development of gastrointestinal immune function and its relationship to intestinal disease, Curr. Opin. Gastroenterol. 9:946-952 (1993).
54. B.J. Stoll, F.K. Lee, E. Hale, D. Schwartz, R. Holmes, R. Ashby, C. Czerkinsky, and A.J. Nahmias, Immunoglobulin secretion by the normal and the infected newborn infant, J. Pediatr. 122:780-786 (1993).
55. A. W. Griffioen, S.W. Franklin, B.J. Zegers, and G.T. Rijkers, Expression and functional characteristics of the complement receptor type 2 on adult and neonatal B lymphocytes, Clin. Immunol. Immunopahtol. 69:1-8 (1993).
56. P.W. Dempsey, M.E. Allison, S. Akkaraju, C.C. Goodnow, and D.T. Fearon, C3d of complement as a molecular adjuvant: bridging innate and acquired immunity, Science 19:348-350 (1996).
57. E. Sagie, J. Tarabulus, D.M. Maeir, and S. Freier, Diet and development of intestinal IgA in the mouse, Isr. J. Med. Sci. 10:532-534 (1974).
58. W.F. Knox, Restricted feeding and human intestinal plasma cell development, Arch. Dis. Child. 61, 744-749 (1986).
59. P.A. Crabbé, D.R. Nash, H. Bazin, H. Eyssen, and J.F. Heremans, Immunohistochemical observations on lymphoid tissues from conventional and germ-free mice, Lab. Invest. 22:448-457 (1970).
60. D.J. Horsfall, J.M. Cooper, and D. Rowley, Changes in the immunoglobulin levels of the mouse gut and serum during conventionalisation and following administration of *Salmonella typhimurium*, Aust. J. Exp. Biol. Med. Sci. 56:727-735 (1978).
61. R. Lodinova, V. Jouja and V. Wagner, Serum immunoglobulins and coproantibody formation in infants after artificial intestinal colonization with *Escherichia coli* 083 and oral lysozyme administration, Pedriatr. Res. 7:659-669 (1973).
62. M.C. Moreau, R. Ducluzeau, D. Guy-Grand, and M.C. Muller, Increase in the population of duodenal immunoglobulin A plasmocytes in axenic mice assosiated with different living or dead bacterial strains of intestinal origin, Infect. Immun. 121:532-539 (1978).
63. P. Brandtzaeg, The secretory immune system of lactating human mammary glands compared with other exocrine organs, Ann. N. Y. Acad. Sci. 30:353-382 (1983).
64. S.S. Wijesinha and H.W. Steer, Studies of the immunoglobulin-producing cells of the human intestine: the defunctioned bowel, Gut 23:211-214 (1982).
65. J.D. Reynolds and B. Morris, The influence of gut function on lymphoid cell populations in the intestinal mucosa of lambs, Immunology 49:501-509 (1983).

66. A.T. Nagao, M.I. Pilagallo, and A.B. Pereira, Quantitation of salivary, urinary and faecal sIgA in children living in different conditions of antigenic exposure, J. Trop. Pediatr. 39:278-283 (1993).
67. D.W. Beatty, B. Napier, C.C. Sinclair-Smith, K. McCabe and E. J. Hughes, Secretory IgA synthesis in Kwashiorkor, J. Clin. Lab. Immunol. 12:31-36 (1983).
68. R.R. Watson, D.N. McMurray, P. Martin, and M.A. Reyes, Effect of age, malnutrition and renutrition on free secretory component and IgA in secretions, Am. J. Clin. Nutr. 42:281-288 (1985).
69. S.S. Hoque, S. Ghosh, and I.R. Poxton, Differences in intestinal humoral immunity between healthy volunteers from UK and Bangladesh, Eur. J. Gastroenterol. Hepatol. 12:1185-1193 (2000).
70. M. Kaila, E. Isolauri, E. Soppi, E. Virtanen, S. Laine, and H. Arvilommi, Enhancement of the circulating antibody secreting cell response in human diarrhea by a human Lactobacillus strain, Pediatr. Res. 32:141-144 (1992).
71. M. Kaila, E. Isolauri, M. Saxelin, H. Arvilommi, and T. Vesikari, Viable versus inactivated lactobacillus strain GG in acute rotavirus diarrhoea, Arch. Dis. Child. 72:51-53 (1995).
72. E. Isolauri, J. Joensuu, H. Suomalainen, M. Luomala, and T. Vesikari, Improved immunogenicity of oral D x RRV reassortant rotavirus vaccine by *Lactobacillus casei GG*, Vaccine 13:310-312 (1995).
73. H. Yasui, J. Kiyoshima, and H. Ushijima, Passive protection against rotavirus-induced diarrhea of mouse pups born to and nursed by dams fed *Bifidobacterium breve* YIT4064, J. Infect. Dis. 172:403-409 (1995).
74. M. Malin, H. Suomalainen, M. Saxelin, and E. Isolauri, Promotion of IgA immune response in patients with Crohn's disease by oral bacteriotherapy with Lactobacillus GG, Ann. Nutr. Metab. 40:137-145 (1996).
75. L. Prokesová, P. Ladmanová, D. Cechova, R. Stepánková, H. Kozáková, Å. Mlcková, R. Kuklik, and M. Mára, Stimulatory effects of *Bacillus firmus* on IgA production in human and mice, Abstract 11.5, Immunol. Lett. 69:55-56 (1999).
76. R. Lodinova-Zádniková and B. Cukrowská, Influence of oral colonization of the intestine with a non-enteropathogenic *E. coli* strain after birth on the frequency of infectious and allergic diseases after 10 and 20 years, Abstract 12.6, Immunol. Lett. 69: 64 (1999).
77. B. Taylor, A.P. Norman, H.A. Orgel, C.R. Stokes, M.W. Turner, and J.F. Soothill, Transient IgA deficiency and pathogenesis of infantile atopy, Lancet 2:111-113 (1973).
78. J.F. Soothill, Some intrinsic and extrinsic factors predisposing to allergy, Proc. R. Soc. Med. 69:439-442 (1976).
79. K.S. Sloper, C.G. Brook, D. Kingston, J.R. Pearson, and M. Shiner, Eczema and atopy in early childhood: low IgA plasma cell counts in the jejunal mucosa, Arch. Dis. Child. 56:939-942 (1981).
80. M. Perkkiö, Immunohistochemical study of intestinal biopsies from children with atopic eczema due to food allergy, Allergy 35:573-580 (1980).
81. P.P. van Asperen, M. Gleeson, A.S. Kemp, A.W. Cripps, S.B. Geraghty, C.M. Mellis, and R.L. Clancy, The relationship between atopy and salivary IgA deficiency in infancy, Clin. Exp. Immunol. 62:753-757 (1985).
82. M. Kilian, S. Husby, A. Host, and S. Halken, Increased proportions of bacteria capable of cleaving IgA1 in the pharynx of infants with atopic disease, Pediatr. Res. 38:182-186 (1995).
83. P. Brandtzaeg and D.E. Nilssen, Mucosal aspects of primary B-cell deficiency and gastrointestinal infections, Curr. Opin. Gastroenterol. 11:532-540 (1995).
84. M. Hahn-Zoric, F. Fulconis, I. Minoli, G. Moro, B. Carlsson, M. Bottiger, N. Raiha, and L.Å. Hanson, Antibody responses to parenteral and oral vaccines are impaired by conventional and low protein formulas as compared to breast-feeding, Acta Paediatr. Scand. 79:1137-1142 (1990).
85. H.F. Pabst and D.W. Spady, Effect of breast-feeding on antibody response to conjugate vaccine, Lancet 336:269-270 (1990).
86. S. Stephens, Development of secretory immunity in breast fed and bottle fed infants, Arch. Dis. Child. 61:263-269 (1986).
87. M. Gleeson, A.W. Cripps, R.L. Clancy, M.J. Hensley, A.J. Dobson, and D.W. Firman, Breast feeding conditions a differential developmental pattern of mucosal immunity, Clin. Exp. Immunol. 66:216-222 (1986).
88. A.R Tappuni and S.J. Challacombe, A comparison of salivary immunoglobulin A (IgA) and IgA subclass concentrations in predentate and dentate children and adults, Oral Microbiol. Immunol. 9:142-145 (1994).
89. H. Renz, C. Brehler, S. Petzoldt, H. Prinz, and C.H. Rieger, Breast feeding modifies production of SIgA cow's milk-antibodies in infants, Acta Paediatr. Scand. 80:149-154 (1991).
90. M.A. Avanzini, A. Plebani, V. Monafo, G. Pasinetti, M. Teani, A. Colombo, L. Mellander, B. Carlsson, L. Å. Hanson, and A.G. Ugazio, A comparison of secretory antibodies in breast-fed and formula-fed infants over the first six months of life, Acta Paediatr. 81:296-301 (1992).

91. S.P. Fitzsimmons, M.K. Evans, C.L. Pearce, M.J. Sheridan, R. Wientzen and M.F. Cole, Immunoglobulin A subclasses in infants' saliva and in saliva and milk from their mothers, J. Pediatr. 124:566-573 (1994).
92. J.J. Cebra, H.Q. Jiang, J. Strerzl, and H. Tlaskalová-Hogenová, The role of mucosal microbiota in the development and maintenance of the mucosal immune system, in: *Mucosal Immunology*, 2nd Ed., edited by P. L Ogra, J. Mestecky, M. E. Lamm, W. Strober, J. Bienenstock, and J. R. McGhee (Academic Press, London, 1999), pp. 267-280.
93. T.T. MacDonald, Breakdown of tolerance to the intestinal bacterial flora in inflammatory bowel disease, Clin. Exp. Immunol. 102:445-447 (1995).
94. M. Cascalho, J. Wong, and M. Wabl, VH gene replacement in hyperselected B cells of the quasimonoclonal mouse, J. Immunol. 159:5795-5801 (1997).
95. Y. Chen, K. Song, and S.L. Eck, An intra-Peyer's patch gene transfer model for studying mucosal tolerance: distinct roles of B7 and IL-12 in mucosal T cell tolerance, J. Immunol. 165:3145-3153 (2000).
96. M. Kalliomäki, A. Ouwehand, H. Arvilommi, P. Kero, and E. Isolauri, Transforming growth factor-β in breast milk: a potential regulator of atopic disease at an early age, J. Allergy Clin. Immunol. 104:1251-1257 (1999).
97. S. Ishizaka, M. Kimoto, T. Tsujii, and S. Saito, Antibody production system modulated by oral administration of human milk and TGF-β, Cell. Immunol. 159:77-84 (1994).
98. S.M. Planchon, C.A. Martins, R.L. Guerrant, and J.K. Roche, Regulation of intestinal epithelial barrier function by TGF-β1. Evidence for its role in abrogating the effect of a T cell cytokine, J. Immunol. 153:5730-5739 (1994).

THE MILKY WAY: FROM MAMMARY GLAND TO MILK TO NEWBORN

Macy-Gyorgy Award Presentation (1999)

Margit Hamosh*

1. PHYSIOLOGY OF LACTATION

2. Lipoprotein Lipase

My interest in this topic started during studies on the role of lipoprotein lipase in pregnancy and lactation. These studies, carried out in the late 60's, showed that prior to parturition and throughout lactation lipoprotein lipase (LPL) activity (the key enzyme that regulates the uptake of chylomicron and VLDL triglyceride-fatty acids by various organs) is markedly increased in rat mammary gland. Concomitantly, there was a decrease in adipose tissue LPL activity, which resulted in an effective channeling of dietary fat from storage in adipose tissue to the lactating mammary gland for the synthesis of milk fat. These studies also showed that suckling is essential for the maintenance of high LPL activity in the mammary gland and its suppression in adipose tissue.[1] Changes in LPL activity in both tissues were effected rapidly starting at 4 hours of cessation of suckling. Initial studies attributed these changes in LPL activity to hormonal changes associated with lactation such as prolactin secretion.[2] This was the first study to show that regulation of LPL activity in individual organs can affect their ability to utilize circulating triglyceride-fatty acids for their immediate needs.

Lipoprotein lipase activity in lactating and suckling gray seals is probably responsible for the transfer of large quantities of triglycerides from blubber to milk (35 Kg fat during 16 days of lactation in a species that lactates exclusively under fasting conditions) and the rapid deposition of blubber (1.9 and 2.7 Kg/day) during early and later lactation in suckling pups.[3]

*Margit Hamosh, Professor of Pediatrics, Georgetown University Medical Center, Department of Pediatrics, 3800 Reservoir Road, NW, Washington, DC 20007

3. Synthesis of medium chain fatty acids

Synthesis of medium chain fatty acids is specific to the mammary gland because only this tissue contains thioesterase II, which terminates chain elongation at 6-14 carbon atoms. In all other tissues, lipogenesis results in long-chain fatty acids associated with the enzyme thioesterase I, which terminates chain elongation at or above 16 carbons.

Comparison of prepartum mammary secretion collected at 6-10 weeks before full term delivery, with colostrum and milk secreted by women who delivered 10 to 14 weeks prematurely or after a full term pregnancy showed marked differences in medium-chain fatty acid content between the prepartum secretions and the post-partum colostrum and milk. Thus, medium-chain fatty acid concentration was significantly lower in prepartum as compared to postpartum secretions.[4] There was, however, no difference between the concentration of these fatty acids in colostrum or milk of mothers of preterm or full term infants.[5, 6] One can, therefore, conclude that parturition, irrespective of length of pregnancy, seems to be the trigger for the increase in de novo fatty acid synthesis within the mammary gland. It is also possible that milk removal, whether by the newborn's sucking or by milk expression, after full term or premature delivery, respectively, is an important factor in the increase of lipogenesis within the human mammary gland. In animal studies milk removal has been shown to be essential for the increase in lipogenic enzymes immediately after parturition.

In contrast to the higher level of medium-chain fatty acids in colostrum than in serum,[4, 7] long-chain polyunsaturated fatty acids (LCPUFA) such as arachidonic (C20:4n6) and docosohexaenoic (C22:6n3) were present at lower levels in colostrum than in serum (milk:serum ratio 0.18 for both fatty acids).[4, 7]

The low fat concentration in prepartum secretion as compared to colostrum and milk can be attributed to low prepartum de novo synthesis of medium-chain fatty acids as well as to the absence of lipoprotein lipase, which would limit the uptake of serum triglyceride fatty acids by he mammary gland.

4. Lipid Composition of Milk Secreted by Mothers of Preterm or Full Term Infants

The increase in survival rate of very premature infants (born at 26-30 weeks gestation) as well as an increased awareness of the beneficial effect of human milk on the newborn led to a rise in feeding of mother's expressed milk to these infants. The absence of mammary secretory activity before term delivery as well as the absence of premature delivery in many species, led to numerous studies of the composition of "preterm" human milk. Our studies have focused on lipid composition and on the activity levels of specific milk enzymes in "preterm" milk compared to milk secreted after full term delivery.

Milk fat composition was shown to differ during the first month of lactation between preterm and full term milk. Thus, medium-chain and long-chain polyunsaturated fatty acid level was higher in the milk of mothers who delivered at 26-37 weeks gestation.[6] Furthermore, the difference in composition caused by the differences in size of milk fat globules was greater in preterm than in full term milk. Milk fat is contained within specific structures, the milk fat globules, which have a core of non-polar lipid, the triglycerides, and an outer membrane composed of polar lipids such as phospholipids and cholesterol. The milk fat globules grow in size during early lactation, resulting in a decrease in the polar lipid content.[6] Since the milk fat globules in preterm milk are smaller during early lactation than in full term milk, the phospholipid and cholesterol content of preterm milk is higher than that of full term milk at this time.[8, 9] Indeed, prepartum mammary secretions of women who deliver at term had the highest phospholipid and cholesterol and lowest triglyceride content.[6] The percentage of individual phospholipids and of free cholesterol in these lipid classes was, however, similar between preterm and full term milk.[8, 9]

We have also investigated the effect of maternal diseases that affect lipid metabolism on the amount and composition of milk fat. Pathologic changes in maternal lipid metabolism are also reflected in the secreted milk. Increased fatty acid chain elongation and changes in the desaturation (especially Δ6 desaturase) as well as lower linoleic acid content were evident in fatty acid profiles of milk from mothers with cystic fibrosis[10]. These changes are consistent with essential fatty acid deficiency. Changes consistent with decreased fatty acid esterification, impairment of fatty acid synthesis within the mammary gland and increased chain elongation were seen in milk of diabetic mothers[11]. Compensatory increases in medium chain fatty acids have been described in hypobetalipoproteinemia and type I hyperlipoproteinemia(reviewed in reference [12]).

5. Enzymes in Milk

Milk contains a large number of enzymes, some with specific functions in the newborn.[13] Among the latter are bile salt dependent lipase (BSDL) and α-amylase. These enzymes compensate in the newborn for low neonatal pancreatic exocrine function, which results in very low levels of pancreatic colipase dependent lipase (the major intestinal fat digesting enzyme) and absence of α-amylase. BSDL is identical to the pancreatic cholesterol ester lipase (CEL). These identical enzymes are expressed in two secretory organs, the pancreas and the mammary gland. Because of the even greater need for compensatory digestive enzymes in the premature infant, we have quantified the activity and assessed the characteristics of BSDL and α-amylase in the milk of mothers of preterm (26-37) weeks gestation, and full term infants, during the first three months of lactation. These studies have shown identical quantitative and qualitative characteristics of these digestive enzymes irrespective of length of pregnancy.[14-16] Lipoprotein lipase, an enzyme that functions in the mammary gland is also present in milk but, because of its characteristics, probably does not function in the infant. The major difference between the milk compensatory digestive enzymes and LPL is their resistance to the low pH of the stomach (2.0-5.0) and absence of a requirement for specific cofactors such as apoprotein C for LPL. The studies of LPL, BSDL and α-amylase in milk as a function of length of gestation and duration of lactation led to further inquiry into certain aspects of their secretion, concentration and function.

A comparison between activity levels of LPL and BSDL in prepartum mammary secretion and in mature milk collected during weaning showed that whereas LPL activity was completely absent before parturition, BSDL was present in prepartum secretions collected as early at 10 weeks before full-term delivery. Furthermore, during weaning, LPL activity was strongly related to the volume of milk secreted, decreasing 30-40 fold with milk volume changes from 600-1200 ml/day to 1-99 ml per day. These differences suggest that BSDL is a constitutive mammary gland enzyme, whereas LPL expression is conditioned by parturition and lactation.[17]

Activity levels of these lipases differ greatly among species. Thus, early studies showed that LPL activity is several fold higher in guinea pig than in rat milk.[18] BSDL was initially thought to be found exclusively in primate milk, but subsequent studies[19] showed the enzyme to be present in the milk of carnivore species, the highest levels being in ferret milk.[20] Studies during the last decade have shown BSDL to be present in numerous other species.[21] Our recent studies have shown that expression of both pancreatic colipase dependent lipase and cholesterol ester lipase is very low in newborn ferret pancreas, whereas expression of BSDL is high in ferret lactating mammary gland, indicating that the latter compensates for both pancreatic lipases in fat digestion during the neonatal period.[22] Further studies in the ferret, carried out in collaboration with Dr. Dominique Lombardo and his team in Marseille led to the recent molecular cloning of mammary gland BSDL and to the identification of several functional domains.[21] Ferret mammary gland BSDL has high homology with the enzyme cloned from pancreas or mammary gland of other species at residues 1-484, whereas the C-terminal has much less homology to other species.[21] Mammary gland BSDL has been cloned so far only from mouse, human and ferret, in contrast to pancreatic CEL which has been cloned from a greater number of species.[21] While relatively little work has been done on milk α-amylase, except for its initial description in milk and subsequent studies that showed activity in the newborn (reviewed in ref.[9]), the role of BSDL in fat digestion has been studied more extensively.[13] Thus, for hydrolysis of the triglyceride contained within the core of milk fat globules, partial predigestion of the latter by the infant's gastric lipase is essential for the subsequent action of BSDL.[13, 23] This partial predigestion is essential even in species with very high milk BSDL activity, such as the ferret,[23,24] probably because of the inability of BSDL to penetrate through the milk fat globule membrane. Access to the core triglyceride by gastric lipase is due to the hydrophobic nature of this enzyme.[25] The more extensive fat digestion in the stomach[26] of mother's milk than formula by very premature infants is probably due to the easier access of gastric lipase to the milk fat globule core than to the fat particles of formula fat. In the newborn, and especially in premature infants, the joint action of gastric lipase and BSDL acting in the stomach and in the intestine, respectively, results in normal absorption of fat even in the absence or at minimal activity of the major pancreatic lipase.[26-31]

Studies in the dog, a species similar to the human in that milk contains BSDL and milk fat is composed mainly of long chain fatty acids as well as the fact that gastric lipase is the predominant lipolytic enzyme acting in the stomach, have permitted to compare in vivo studies with the above mentioned in vitro studies [9,19]. Gastric lipolysis of milk fat was extensive (14-42% during 30 minutes) and this initial fat hydrolysis led to a six fold higher digestive action by BSDL, 30% fat hydrolysis compared to only 5% hydrolysis without prior intragastric lipolysis.[32]

6. Protective Function against Infection by Milk Fat Globule Components

The hydrolysis of milk fat globule core triglyceride produces free fatty acids and monoglycerides which are potent detergents with considerable lytic activity against enveloped viruses, bacteria and protozoa.[29, 30] The milk fat globule membrane surface varies among species, in human and mare milk filaments extend as far as 1 mm from the surface whereas cow, sheep and goat milk fat globules are devoid of such filaments.[31] These filaments have been shown to contain the mucin MUC-1 which binds fimbriated E.

coli, a major cause of meningitis in newborns, [30, 31] and lactadherin that protects against rotavirus diarrhea in infants.[30, 31] We have collaborated with Dr. Jerry Peterson and his group to investigate the concentration of these glycoproteins in milk secreted by mothers of premature infants during the first three months of lactation and to assess whether they may remain active in the newborn. These studies have shown that the concentration of these glycoproteins (MUC-1, mol. Wt. 400 kD and lactadherin mol. Wt. 46 kD) is highest during the first two weeks of lactation when milk fat globule size is smaller and, therefore, total membrane content of milk is higher than in later lactation.[5, 8] The study also showed that these glycoproteins are stable in the stomach, i.e., are not affected by low pH and proteolytic enzymes[33] and could provide protection against microorganisms throughout the entire gastrointestinal tract. These studies on milk fat globule function indicate that both components, the core and the membrane play important roles in the protection against neonatal infection: more efficient hydrolysis of core tirglyceride[26] would provide greater detergent-like free fatty acid lytic action against microorganisms, whereas bacterial and viral ligand action of the membrane glycoproteins prevents infection even in the very premature infants.[30, 33]

7. Long Chain Polyunsaturated Fatty Acids

The controversy over the need for supplementation of infants with LCPUFA, because of their potential contribution to brain and retinal development, has resulted in a large number of studies that have compared these functions in preterm and full term breast fed and LCPUFA supplemented formula fed infants (reviewed in refs [34, 35]). Most studies showed only slight and transient effects of LCPUFA supplementation in both groups of infants. Our observation that LCPUFA decrease markedly in human milk after 2-3 months of lactation,[6, 36] suggests that by this age the infants might be self sufficient in the elongation and desaturation of essential fatty acids and might not need such supplementation. The decrease in milk LCPUFA in well-nourished women relatively early in lactation [6, 36] raised the question whether concentration of milk LCPUFA might be regulated by the mammary gland. While most studies that have measured LCPUFA accretion in the human are based on postmortem data or only on blood data (serum and erythrocytes) of living infants, comprehensive studies to assess maternal LCPUFA stores and infant tissue accretion rates can not be carried out in humans for ethical reasons.

We have therefore, decided to assess maternal LCPUFA status during late pregnancy and throughout lactation, to quantify milk lipids and fatty acid profiles with special emphasis on n3 and n6 LCPUFA and to assess late fetal and postnatal accretion of LCPUFA prior to weaning in an animal species. We have chosen the ferret because maternal diet is adequate in LCPUFA content and because neonatal digestive function is similar to that in the newborn infant.[20, 23, 24, 25, 27, 37] Maternal and newborn ferrets were studied from late pregnancy (wk 5 of a 6 wk gestation period) throughout exclusive milk feeding (up to 4 wks postpartum). Data from pregnant or lactating females were compared to those from virgin animals who served as controls. Maternal and fetal/newborn tissues were analyzed for total lipid and fatty acid content. I will discuss briefly here only the findings on docosahexaenoic acid (DHA, C22:6 n3 essential for normal functional development of brain and retina) and arachidonic acid (AA, C20:4 n6 the precursor of eicosanoids with an important role in neonatal growth).[34, 35]

Our study found that maternal adipose tissue n3 and n6 LCPUFA as well as DHA and AA remained constant during pregnancy and the entire lactation period and were identical to those of virgin controls.[38]

Although the dietary intake and adipose tissue levels of DHA and AA remained constant during pregnancy and lactation, as did the concentration of these fatty acids, milk DHA and AA levels decreased throughout lactation. The data suggest that milk LCPUFA levels are intrinsically regulated within the lactating mammary gland, the concentration of DHA and AA in milk secreted during early lactation being 2 and 3 fold higher than dietary content, and 3 and 4 fold higher than adipose tissue content. The decrease of DHA and AA in late milk is very similar in the ferret and in the human,[6, 36] although absolute values are higher in the ferret.[38] LCPUFA accretion in the offspring showed marked organ differences. No adipose tissues were detectable in the fetus at the start of the last week of gestation, however, one week after birth the newborn had well-developed adipose depots at all sites examined. As in the adult, there was no difference in LCPUFA profiles among various adipose tissues examined. DHA and AA were highest in adipose tissue in the first week after birth and were identical to milk DHA levels, indicating rapid deposition of milk DHA into adipose depots during early postnatal development. While not exactly identical, adipose tissue AA closely paralleled milk AA concentrations. The rapid postnatal development of adipose tissue and the rapid accretion of DHA and AA in fat depots suggests that these might serve as reservoir for the LCPUFA accretion in tissues during later periods when milk supply of DHA and AA decreases.

Fatty acid profiles in brain, liver and kidney showed a very different developmental pattern from that in adipose tissue. Brain DHA levels increased 6 fold from the late fetus to the first week after birth and at four weeks after birth had only reached 50% of the DHA level of adult brain. The percentage of DHA was much higher in brain than in adipose tissue. AA levels were higher than DHA levels in fetal brain. AA increased three fold from the late fetus to 1 week after birth and reached adult levels four weeks after birth. Accretion of DHA and AA in the liver closely paralleled that of the brain. DHA levels increased more than two fold from the late fetal period to the early neonatal period, but amounted to only 50% of adult levels 4 weeks after birth, whereas AA levels were much higher in fetal liver than DHA levels and by one week after birth had reached adult levels. Entirely different developmental patterns exist in the kidney, an organ that depends upon eicosanoids for its functional maturation. In this organ DHA actually decreased two fold from the late fetus to the 4 week old newborn whereas AA levels, already higher than in all other organs in the fetus, continued to rise during the postnatal period.

In general, these studies show that milk LCPUFA levels are independent of dietary intake or adipose tissue reserves and seem to be regulated at the mammary gland level throughout lactation. In the newborn greater n6 than n3 LCPUFA accretion might be due to earlier development of endogenous synthesis of the former (as shown recently in newborn infants). The differences in accretion rates among tissues indicates that this process is regulated by specific organ demands for DHA or AA.[38]

8. Effect of Storage on Human Milk

Last, but not least, I would like to summarize studies on the stability of milk nutrients and digestive enzymes during storage. These studies were conducted in response to numerous questions about the adequacy of feeding stored milk to the infant. In the US, the only developed country without government-paid maternity leave, these questions are especially pertinent since mothers return to work shortly after delivery and continue to feed milk pumped at work to their infants. The topic is also of interest to mothers of premature or sick infants who are separated during hospitalization of the newborns.

Briefly, these studies have shown that milk components are stable during storage at -70°C, whereas prolonged storage (more than one month) at -20°C results in progressive breakdown of milk triglyceride and production of free fatty acids.[39] This might actually be beneficial because of the lytic effect of free fatty acids against microorganisms. Studies designed to mimic storage conditions as they might occur in the work environment, i.e., in a cooler at 15°C, at room temperature, in moderate (25°C) or hot (38°C) environments, have shown that it is safe to store milk at 15°C for 24 hours and at 25°C for 4 hours but that milk deteriorates rapidly at 38°C.[40]

Proteolysis was minimal during storage, whereas lipolysis was rapid and extensive. This might be the reason that bacterial growth, restricted mainly to non-pathogens, was minimal at 15°C and low at 25°C, due to the protective function of many human milk proteins and the lytic effect against microorganisms of products of lipolysis.[40] It is interesting that LPL, an extremely labile enzyme, was relatively stable in milk, retaining 85% and 55% of initial activity after 24 hours of storage at 15°C and 25°C, respectively.[41] The compensatory digestive milk enzymes, BSDL and α-amylase, retain full activity after 24 hours of storage at 15, 25 or 38°C.[41]

Differences between BSDL and amylase were, however, evident during Holder pasteurization which completely inactivated BSDL, while reducing amylase activity only 15%.[42] Holder pasteurization did not affect milk fat composition indicating that LCPUFA are not oxidized during the process, probably because of the activity of antioxidative components of human milk.

In conclusion, our studies and those of many others in this field have led to a better understanding of the role of human milk in infant development, a role that extends beyond the nutritional value of its components.
Much more remains to be done!

Acknowledgements

None of these studies would have been possible without the collaboration with my husband Dr. Paul Hamosh. We thank the many colleagues and associates who have contributed to these studies, and express our special gratitude to Drs. Nitin R. Mehta, Joel Bitman and Teresa R. Henderson for their substantial input.

References

1. M. Hamosh, T.R. Clary, S.S. Chernick, and R.O. Scow, Lipoprotein lipase activity of adipose and mammary tissue and plasma triglyceride in pregnant and lactating rats, Biochim. Biophys. Acta 210:473-482 (1971).
2. O. Zinder, M. Hamosh, T.R. Clary-Fleck, and R.O. Scow, Effect of prolactin on lipoprotein lipase in mammary gland and adipose tissue of rats, Am. J. Physiol. 226:744-748 (1974).
3. S.J. Iverson, M. Hamosh, and W.D. Bowen, Lipoprotein lipase activity and its relationship to high milk fat transfer during lactation in gray seals, J. Comp. Phys. B 165:384-395 (1995).

4. M.L. Spear, J. Bitman, M. Hamosh, D.L. Wood, D. Gavula, and P. Hamosh, Human mammary gland function at the onset of lactation: medium chain fatty acid synthesis, Lipids 27:908-911 (1992).
5. J. Bitman, L.M. Freed, M.C. Neville, D.L. Wood, P. Hamosh, and M. Hamosh, Lipid composition of prepartum human mammary secretion and postpartum milk, J. Pediatr. Gastroenterol. Nutr. 5:608-615 (1986).
6. J. Bitman, D.L. Wood, M. Hamosh, P. Hamosh, and N.R. Mehta, Comparison of the lipid composition of breast milk from mothers of term and preterm infants, Am. J. Clin. Nutr. 38:300-312 (1983).
7. M.L. Spear, M. Hamosh, J. Bitman, M.L. Spear, and D.L. Wood, Milk and blood fatty acid composition during two lactations in the same woman, Am. J. Clin. Nutr. 56:65-70 (1992).
8. J. Bitman, D.L. Wood, N.R. Mehta, P. Hamosh, and M. Hamosh, Comparison of the phospholipid composition of breast milk from mothers of term and preterm infants during lactation, Am. J. Clin. Nutr. 40:1103-1119 (1984).
9. J. Bitman, D.L. Wood, N.R. Mehta, P. Hamosh, and M. Hamosh, Comparison of the cholesterylester composition of breast milk from preterm and term mothers, J. Pediatr. Gastroenterol. Nutr. 5:780-786 (1986).
10. J. Bitman, M. Hamosh, D.L. Wood, L.M. Freed and P. Hamosh, Lipid composition of milk from mothers with cystic fibrosis, Pediatrics 80:927-932 (1987).
11. J. Bitman, M. Hamosh, P. Hamosh, V. Lutes, M.C. Neville, J. Seacat, and D.L. Wood, Milk composition and volume during the onset of lactation in a diabetic mother, Am. J. Clin. Nutr. 50:1364-1369 (1989).
12. M. Hamosh and J. Bitman, Human milk in disease: lipid composition, Lipids 27:848-857 (1992).
13. M. Hamosh, Enzymes in human milk: characteristics and physiologic functions, in: Handbook of Milk Composition, edited by R.G. Jensen (Academic Press, San Diego, 1995), pp 388-427.
14. N.R. Mehta, J.B. Jones, and M. Hamosh, Lipases in human milk: ontogeny and physiologic significance, J. Pediatr. Gastroenterol. Nutr. 1:317-326 (1982).
15. J.B. Jones, N.R. Mehta, and M. Hamosh, α-Amylase in preterm human milk, J. Pediatr. Gastroenterol. Nutr. 1:43-48 (1982).
16. L.M. Freed, S.E. Berkow, P. Hamosh, C.M. York, N.R. Mehta, and M. Hamosh, Lipases in human milk: effect of gestational age and length of lactation on enzyme activity, J. Am. Coll. Nutr. 8:143-150 (1989).
17. M. Hamosh, Enzymes in human milk, in: Human Milk in Infant Nutrition and Health, edited by R.R. Howell, E.H. Morris, and L.K. Pickering (Charles C. Thomas, Springfield, IL 1986), pp 66-97.
18. M. Hamosh, and R.O. Scow, Lipoprotein lipase activity in guinea pig and rat milk, Biochim. Biophys. Acta 321:282-289 (1971).
19. L.M. Freed, C.M. York, M. Hamosh, J.A. Sturman, and P. Hamosh, Bile salt-stimulated lipase in non-primate milk: longitudinal variation and lipase characteristics in cat and dog milk, Biochim. Biophys. Acta 878:209-215 (1992).
20. L.A. Ellis, and M. Hamosh, Bile salt stimulated lipase: comparative studies in ferret milk and lactating mammary gland, Lipids 27:917-922 (1992).
21. V. Sbarra, N. Bruneau, E. Mas, M. Hamosh, D. Lombardo, and P. Hamosh, Molecular cloning of the bile salt dependent lipase of ferret lactating mammary gland, Biochim. Biophys. Acta 1939:80-89 (1998).
22. V. Sbarra, E. Mas, T.R. Henderson, M. Hamosh, D. Lombardo, and P. Hamosh, Digestive lipases of the newborn ferret: compensatory role of milk bile salt dependent lipase, Pediatr. Res. 40:263-268 (1996).
23. M. Hamosh, and P. Hamosh, Selectivity of lipases: developmental physiology aspects, in: Engineering of/with lipases, edited by F.X. Malcata, NATO ASI Series E Applied Sciences (Kluwer Academic Publ. Dardrecht, Vol 317, 1996), pp 31-39.
24. M. Hamosh, S.J. Iverson, C.L. Kirk, and P. Hamosh, Milk lipids and neonatal fat digestion: relationship between fatty acid composition, endogenous and exogenous digestive enzymes and digestion of milk fat, in: Fatty Acids and Lipids: Biological Aspects, edited by C. Galli, A.P. Simopoulos, and E. Tremoli, World Rev. Nutr. Diet 75:86-91 (1994).
25. M. Hamosh, Lingual and Gastric Lipases: Their Role in Fat Digestion, C.R.C. Press, Boca Raton, FL (1990).
26. M. Armand, M. Hamosh, N.R. Mehta, P.A. Angelus, J.R. Philpott, T.R. Henderson, N.K. Dwyer, D. Lairon, and P. Hamosh, Effect of human milk or formula on gastric function and fat digestion in the premature infant, Pediatr. Res. 40:429-437 (1996).
27. M. Hamosh, Digestion in the premature infant: the effects of human milk, Sem. Perinatol. 18:485-494 (1994).
28. M. Hamosh, Lipid metabolism in pediatric nutrition, Pediatric Clin. N. Am. 42:839-859 (1995).
29. M. Hamosh, Free fatty acids and monoglycerides: Antiinfective agents produced during the digestion of milk fat by the newborn, in: Immunology of Milk and the Neonate, edited by J. Mestecky, C. Blair, and P.L. Ogra (Plenum Press, NY 1991), pp 151-158.

30. M. Hamosh, J.A. Peterson, T.R. Henderson, C.D. Scallen, P. Kirwan, R.L. Ceriani, M. Armand, N.R. Mehta, and P. Hamosh, Protective functions of human milk: the milk fat globule, Sem. Perinatol. 23:242-249 (1999).
31. J.A. Peterson, S. Patton, and M. Hamosh, Glycoproteins of the human milk fat globule in the protection of the breast-fed infant against infections, in: Human Milk and Infant Development, Special issue edited by M. Hamosh, Biol. Neonate 74:143-162 (1998).
32. S.J. Iverson, C.L. Kirk, M. Hamosh and J. Newsome, Milk lipid digestion in the neonatal dog: the combined action of gastric lipase and bile salt stimulated lipase, Biochim. Biophys. Acta 1083:109-119 (1991).
33. J.A. Peterson, M. Hamosh, C. Scallan, R.L. Ceriani, T.R. Henderson, N.R. Mehta, M. Armand and P. Hamosh, Fat globule glycoproteins in human milk and in gastric aspirates of mother's milk-fed premature infants, Pediatr. Res. 44:1-9 (1998).
34. M. Hamosh, Long chain polyunsaturated fatty acids – who needs them, Biochem. Soc. Trans. 26:96-103 (1998).
35. M. Hamosh, and N. Salem Jr., Long chain polyunsaturated fatty acids, in: Human Milk and Infant Development Special issue, edited by M. Hamosh, Biol. Neonate 74:106-120 (1998).
36. M. Hamosh, T.R. Henderson, and L. Hayman, Long chain unsaturated fatty acids in human milk during prolonged lactation, FASEB J. 10:A516 (1996).
37. M. Hamosh, Gastric and lingual lipases, in: Physiology of the Gastrointestinal Tract, edited by L.R. Johnson, 3rd edition (Raven Press, NY 1994), pp 1239-1253.
38. M. Hamosh, T.R. Henderson, M.A. Kemper, N.M. Orr, A. Gil, and P. Hamosh, Long chain polyunsaturated fatty acids during early development: contributions of milk LC-PUFA to accretion rates varies among organs, in: Bioactive Components of Human Milk, edited by D.S. Newburg (Plenum Press, NY 2001), (in press).
39. J. Bitman, D.L. Wood, N.R. Mehta, P. Hamosh, and M. Hamosh, Lipolysis of triglyceride of human milk during storage at low temperatures: a note of caution, J. Pediatr. Gastroenterol. Nutr. 2:521-524 (1983).
40. M. Hamosh, L.A. Ellis, D.R. Pollock, T.R. Henderson, and P. Hamosh, Breastfeeding and the working mother: effect of time and temperature on short term storage on proteolysis, lipolysis and bacterial growth, Pediatrics 97:492-498 (1996).
41. M. Hamosh, T.R. Henderson, L.A. Ellis, J-I Mao, and P. Hamosh, Digestive enzymes in human milk: stability at suboptimal storage temperatures, J. Pediatr. Gastroenterol. Nutr. 24:38-43 (1997).
42. T.R. Henderson, T.N. Fay, and M. Hamosh, Effect of pasteurization on long chain polyunsaturated fatty acid levels and enzyme activity in human milk, J. Pediatr. 132:876-878 (1998).

MACY-GYÖRGY AWARD ADDRESS – YEAR 2000
A HALF-CENTURY INQUIRY INTO THE IMMUNOBIOLOGY OF HUMAN MILK

Armond S. Goldman*

It is an honor to have been chosen by the International Society for Research in Human Milk and Lactation to be the recipient of the Macy-György Award for Research in Human Milk and Lactation for the year 2000. The Award is of special significance because Icey Macy and Paul György were pioneers in research concerning human milk and lactation and because the Society is the leading organization that deals with research in this aspect of human biology.

In accepting this award, I wish to acknowledge colleagues from the University of Texas Medical Branch, Baylor College of Medicine, Cornell University, and the University of Göteborg who collaborated with me over the past four decades in my quest to understand the origins, structure, components, and biological effects of the immune system in human milk. In a sense, this award is as much theirs as is mine.

1. How My Interests Developed

I have been asked to summarize my work in human milk over the past 40 years. The inception of my interests in the immune system in human milk preceded my initial research. Indeed, the scientific questions that would consume my adult life began in an amusing, but poignant way during my second year of medical school. My wife, Barbara, and I were expecting the birth of our first child. I had innocently obtained a free supply of a well-known infant formula for the child. When the case of milk arrived at our apartment a month beforehand, Barbara asked what was the purpose of the milk. When I told her, she looked quizzical and perhaps disappointed. I hurriedly asked her whether it was the wrong brand. After all, Barbara was an excellent professional nurse and I, a lowly, ignorant medical student. She replied in a good-natured, but slightly plaintive way – 'You might say so,' and went on to tell me that she would breastfeed the baby because breastfed babies were more resistant to infections than those who were not breastfed.

*Armond S. Goldman, Professor Emeritus, Department of Pediatrics, Division of Immunology/Allergy/Rheumatology, The University of Texas Medical Branch, Galveston, Texas 77555-0369

This was a surprise to me since breastfeeding and human lactation were not presented in the curriculum.

Our first borne arrived in perfect health and breastfeeding was initiated. The child and mother bonded, the infant thrived, Barbara was joyful and healthy, and I was mystified by the process. Barbara's prediction was correct. The child was free of the common respiratory and gastrointestinal infections that were prevalent in other young infants in the community. Of course, at that time the vast majority of them were artificially fed. That experience was repeated with our other children. I wondered what was the basis of this protection by breastfeeding as well as other mysteries concerning human lactation.

After graduating from medical school and completing an internship, I began a pediatric residency at the medical school in Galveston. Dr. Arild E. Hansen was the Chair of Pediatrics at that time. During the first few months of my training, I was fortunate to participate in his research that established that linoleic acid was an essential human nutrient.[1] Indeed, human infants deprived of that polyunsaturated fatty acid developed an extensive dermatitis that was reversed by the introduction of that nutrient into the diet. Hansen emphasized that linoleic acid deficiency was rare in breastfed human infants because of the occurrence of the fatty acid in human milk. Those facts strengthened my idea that human milk might be ideally suited for the human infant.

A few years after I joined the faculty of that same medical school, the new Chair of Pediatrics, C.W. (Bill) Daeschner, was concerned about my growing interest in the then nebulous field of clinical immunology. Indeed, only a small number of academic pediatricians and internists had entered into that new field. My lack of education in that scientific discipline heightened my insecurity. Daeschner called to my attention the possibility of joining a study by the Borden Company to ascertain whether the diagnosis of cow's milk allergy could be verified by oral challenge with purified cow's milk proteins (casein, β-lactoglobulin, α-lactalbumin, and serum albumin) and to determine which of the proteins were most allergenic. I was surprised when I was asked to head the clinical part of the multi-physician study. The results of this study demonstrated that cow's milk allergy was a disorder with a spectrum of clinical manifestations.[2] Although the design of the study did not pose the question as to whether these infants had been breastfed, I prospectively determined that none of them had been. Was that fortuitous or did it suggest further evidence that human milk was protective not only against infections, but also perchance against allergic reactions due to food allergens? Of course it seemed equally plausible that breastfeeding protected by avoiding microbial pathogens and foreign antigens in artificial foods. Nevertheless, the possibility that the protection was due to agents in milk was intriguing.

I was not the first to consider possibility that human milk might have protective effects. The eminent pioneering immunologist, Paul Ehrlich, who won a Nobel Prize in 1905 for his research into basic immunology, also discovered in elegant cross-fostering experiments with mice that protection against ricin toxin was transferred by breastfeeding.[3] The 1919 Nobel Prize winner, Jules Bordet, not only ascertained the roles of complement in bacterial killing,[4] but he also discovered lytic activity in human milk[5] that mimicked that of lysozyme that Arthur Fleming had found a year before in other external secretions.[6] In addition, epidemiological studies in the first part of the 20th century suggested that breastfeeding protected against diarrheal diseases and atopic dermatitis.[7-9] But these and other key observations by excellent scientists concerning human milk and breastfeeding were forgotten by a new generation of physicians and nutritionists.[10]

In the late 1950s – early 1960s when I began to think more seriously about the question of an immune system in human milk, comparatively little was known about the human immune system and even less about the possible immune properties of human milk. The origins of antibodies from plasma cells, the chemistry of antibodies and complement, the nature of opsonins, the separation of humoral from cellular immunity, the potential role of the thymus in immunity, and the basic biologic functions for the major types of leukocytes were being unraveled. Paul György was publishing his studies on the bifidus growth factor[11] and the anti-staphylococcal activity in human milk.[12] A few years later a young Swedish pediatrician, Lars Å. Hanson, demonstrated by immunoelectrophoresis that proteins in human milk were distinct from human serum.[13] One of those complex precipitin arcs proved to be secretory IgA, the dominant immunoglobulin in external secretions including human milk. By that time, lactoferrin as well as lysozyme were also known to be in human milk.

2. Initial Research. Leukocytes in Human Milk – 1960s and 1970s

In the 1960s, I and an excellent, young medical student, C. Wayne Smith, began to examine whether certain known proteins in human milk acted synergistically to kill bacterial enteropathogens. As we centrifuged milk to remove the most buoyant lipids, we became curious about the least buoyant material that was also discarded. To our surprise, phase microscopy revealed that there were moving bodies in the preparation. We had no idea that the first microscopist, Van Leeuwenhoek, had made similar observations 1.5 centuries beforehand.[14] Stained preparations revealed many cells in human milk, most of which were vacuolated neutrophils, monocytes, and macrophages.[15] Fewer of them were lymphocytes.[15] The neutrophils, monocytes, and macrophages were phagocytic.[15] The lymphocytes underwent blastogenesis in response to a phytomitogen, phytohemagglutinin.[15] The macrophages and lymphocytes were quite motile.[15, 16] Electron microscopy suggested that the macrophages were activated.[17]

3. Formulation of the Concept of an Immune System in Human Milk –1970s

Thus we were the first to discover living, functioning leukocytes in human milk and perhaps in any mammalian milk. That finding and other discoveries that have been previously mentioned crystallized the concept of an immune system in human milk. Where there were leukocytes and soluble immune factors, there should be an immune system. In addition, the finding of many macrophages and the types of soluble immune factors strongly suggested that the immune system in human milk was unique. It was, however, easier to conceive of the idea than to publish the manuscript. After a rapid rejection from the editors of the Journal of Pediatrics and a subsequent brief reactive depression, I asked the editor, Waldo Nelson, not to reconsider the paper but to tell me the reasons for the turndown. He reread the paper and pointed out why American physicians were reluctant to believe that human milk was important. He then suggested how the paper could be improved by pointing out why there were doubts about the protection afforded by human milk and how research might further clarify whether there was an immune system in human milk. The changes were few, but important. The revision was accepted and the 1973 publication helped to establish the paradigm of an immune system in human milk.[18]

4. Biological Factors that Influence the Concentrations of Immune Factors in Human milk – the 1980s

Although I continued investigating the leukocytes in human milk,[19-21] it was impossible to extend those studies into other aspects of the immune system in human milk because of a lack of funding from federal granting agencies. It was particularly depressing because my first research proposal on human milk to the National Institutes of Health (probably the best one I ever wrote) was rejected without comment. But toward the end of that decade I was approached by members of the Children's Nutritional Institute at Baylor College of Medicine to help to prepare a proposal concerning the effects of banking procedures upon key components of human milk. The proposal was in response to a request from the National Institutes of Child Health and Human Development. The investigators at Baylor would be responsible for the general organization of the study, the collection of milk, experimental banking procedures, and the nutrient analyses and I would undertake with a colleague, Randall M. Goldblum, to examine the concentrations of selected immunological factors in the specimens of human milk.

We submitted our part of the grant proposal even though I was not very interested in human milk banking, and I did not think that the project would be funded. However, I decided to participate because I wished to continue to investigate human milk and because of the quality of the scientists at Baylor who wished to do the study. To my surprise, the grant was funded and the study commenced.

Cutberto Garza from Baylor spearheaded the investigation. He became an important colleague for many years to come and a luminary in the field of human milk and lactation. Besides new information concerning human milk banking,[22-24] we began to understand that the duration of pregnancy[25] and lactation[26, 27] and gradual weaning[28] significantly affected the concentrations of leukocytes, secretory IgA, lactoferrin, and lysozyme in human milk. The first quantitative assessment of the amounts of antimicrobial factors in human milk ingested by the recipient infant was made.[29] Nancy F. Butte from Baylor, who later became one of the leaders of this research society, was the key investigator in those quantitative investigations.

Once the basic patterns of antimicrobial agents in human milk were determined, we investigated their fate in recipient premature infants. The current President of this research society, Richard J. Schanler from Baylor, organized and implemented the balance studies that permitted those investigations. Far more secretory IgA, lactoferrin, and lysozyme in human milk were found in the stools of low-birth-weight infants fed a human milk formula than those fed a cow's milk formula.[30] Those finding were anticipated because those human milk proteins were known to be comparatively resistant to digestive enzymes found in the gastrointestinal tract. The results from a companion study that we carried out were, however, unanticipated. Much more lactoferrin was found in the urine of those same infants who were fed human milk than those fed the bovine milk preparation.[31] Subsequently, we found that fragments as well as whole molecules of lactoferrin were in the stools and urine of those human milk fed infants.[32] The origin of the urinary lactoferrin from the ingested human milk was later verified in a study where human milk lactoferrin was labeled with a stable-isotope.[33] Thus it appeared that at least in low-birth weight infants that certain bioactive factors from human milk or fragments of them created during partial digestion in the recipient's gastrointestinal tract not only acted at mucosal sites but were also absorbed into the systemic circulation and then excreted into the urinary tract.

5. Expansion of the Concept of the Immune System in Human Milk. Antiinflammatory Aspects – the 1980s

More basic features of the immune system in human milk emerged from those and other investigations. The immune system in human milk proved to be a dynamic system made up of biochemically heterogeneous agents, many of which were common to and adapted to survive at mucosal sites.[34] A reciprocal relationship between the production of these immune factors by the mammary gland and the developing infant also began to emerge.[34] Based upon in vitro studies, it also appeared that many of the host defense agents in human milk were multifunctional and interactive.

While I was continuing the investigations with my colleagues at Baylor, I also began to consider why breastfed infants who were exposed to microbial pathogens failed to develop any significant clinical evidence of inflammation at the site of the infection. I was particularly intrigued with the studies of Leonardo Mata and his colleagues in Central America.[35-37] Despite the ingestion of bacterial enteropathogens from the nipples and areolae of the breast, the infants remained asymptomatic. Was this due simply to the antimicrobial effects of human milk or was inflammation generated by the infection also inhibited by human milk?

Since computerized searches were not yet available, an inquiry into the question required several weeks of library study. My readings revealed that many investigators who were primarily interested in inflammation had found that human milk contain a spectrum of antiinflammatory agents, but few inflammatory mediators or their precursors. Although the information was known, it was not considered in the context of the biology of human milk. I presented the concept of an antiinflammatory wing of the immune system in human milk to Lars Å. Hanson from Goteborg, Sweden during his visit to our institution. He was intrigued with and helped to develop the idea. Once again the concept was easy to formulate, but difficult to publish. One journal rejected it quickly and without comment. A more receptive one accepted it without comment.[37] As with the antimicrobial factors in human milk, the antiinflammatory agents were biochemically heterogenous, multi-functional, and adapted to the hostile environment of the gastrointestinal tract. The concept was expanded upon in a subsequent publication.[38] Both works served as a catalyst for the discovery of other antiinflammatory agents in human milk.

6. Activated Leukocytes in Human Milk-the 1980s and 1990s

The knowledge concerning the antiinflammatory aspects of human milk also forced me to reconsider our initial discovery of living leukocytes in human milk, particularly the neutrophils, since they are the hallmark of acute inflammation. This seemed to be at cross-purposes with our concept of antiinflammatory effects of human milk. Were the inflammatory activities of those cells in human milk reduced? That question was addressed in an *in vitro* study that revealed that the adherence, polarity and motility of human milk neutrophils in two-dimensional systems was reduced as compared to blood neutrophils.[39] In addition, the motility of human milk neutrophils in an adherence-independent system was decreased.[40]

We then questioned whether the decrease in certain inflammatory properties of human milk neutrophils was found to be due to activation, rather than inhibition. Activation of neutrophils in human milk was demonstrated by flow cytometry.[41] Fresh milk neutrophils as well as blood neutrophils activated *in vitro* displayed an increased expression of CD18 and a decreased expression of CD62L (L-selectin). Furthermore, human milk was found to activate blood neutrophils.[41] Thus by activation, certain inflammatory properties of neutrophils, such as the ability to adhere to vascular endothelium because of a paucity of CD62L, was diminished.

In one of the aforementioned experiments,[40] the motility of human milk macrophages was significantly increased as compared to blood monocytes. The phenotype of those cells indicated that they were activated.[41] The increased motility of human milk macrophages was found to be due to a known macrophage activator, tumor necrosis factor-α.[42, 43]

We had previously found that human milk was able to abrogate the serum inhibition of uropod formation (a morphologic indicator of motility) by human blood T cells.[21] This activation was probably responsible in part for our earlier observations that human milk lymphocytes were quite motile, as compared to blood lymphocytes.[16] It therefore came as no surprise that the phenotypic features of T cells in human milk indicated that they were activated, memory cells.[44] Thus, human milk was distinguished by leukocytes that were not only living, but activated. These finding added to the realization of the uniqueness of the immune system in human milk and suggested that immunomodulation was another major feature of the immune system in human milk.

7. Cytokines in Human Milk – the 1990s

Because of the findings of activated leukocytes and leukocyte activators in human milk, I became intrigued with the notion that immunomodulating agents might comprise a third wing of the immune system in human milk. The investigators that helped to establish that concept included Frank C. Schmalstieg, Beth Rudloff, Kimberly H. Palkowitz, Roberto Garofalo, and Sadhana Chheda. We and others found evidence by immunochemical and biological assays that TNF-α,[42, 43] IL-6,[45, 46] IL-8,[46] and transforming growth factor-β[46] were prominent in early human milk secretions. Indeed, some of them were produced by human mammary epithelial cells.[46] Moreover, it was apparent that some of these cytokines were bound to other components in human milk.[43, 44] Some of those binding elements were later found by others to be soluble receptors.[47]

The precise activities of these bound forms of cytokines in human milk remain unclear, but it is possible that their functions, particularly the inflammatory ones, may be reduced. It also seemed possible that the binding might protect the agents against digestive enzymes in the alimentary tract of the recipient infant. These discoveries were a prelude to the uncovering of many other cytokines in human milk with either chemotactic, growth promoting, or differentiating activities.[64, 48-50] Indeed the premise of a third wing of the immune system in human milk comprised of immunomodulating agents proved to be correct.

8. Evolution of Mammary Gland's Immune Functions and Early Postnatal Development of the Immune System. 1990s - 2000

The research that I conducted with my colleagues over four decades gradually led me to consider how the immune system in human milk evolved and how that evolution was intertwined with the early postnatal development of the immune system of the recipient infant. I was very intrigued by the reciprocal relationship between the production of many of the immune factors by the mammary gland and the production of those same factors at mucosal sites of the infant. The demonstration of that relationship by our discoveries of high concentrations of IL-10, an antiinflammatory cytokine, in human milk[50] and a developmental delay in its production by monocytes and T cells in the newborn infant[51] was important to my thinking.

An extensive review of the research literature revealed that this reciprocal relationship was found in many prototherian, metatherian, and eutherian species that had been investigated.[52] Certain immune agents in human milk were also found to be represented in sebum, the secretions of the same epidermal glands from which the mammary gland had evolved many millions of years ago. The evolutionary links were also evident in that certain antimicrobial agents found in milk from prototherians were also prominent in milks from metatherian, and eutherian species. Furthermore, the immune profile of human milk most closely resembled that found in closely related primates such as the chimpanzee.

In a subsequent publication, I extensively reviewed the evolution of the immune system of the gastrointestinal tract[53] and analyzed the relationships between the evolutionary aspects of the defense system provided by the human mammary gland and the postnatal development of mucosal defenses in greater detail. Six overlapping evolutionary strategies regarding the relationships between the immune functions of the mammary gland and the infant's gastrointestinal tract were identified. 1) Certain immunological agents in human milk directly compensate for developmental delays in those same agents in the recipient infant. 2) Other agents in human milk do not directly compensate for developmental delays in the production of those same agents, but nevertheless protect the recipient. 3) Agents in human milk enhance functions that are poorly expressed in the recipient. 4) Agents in human milk change the physiological state of the intestines from one adapted to intrauterine life to one suited to extrauterine life. 5) Some agents in human milk prevent inflammation in the recipient's gastrointestinal tract. 6) Survival of human milk defense agents in the gastrointestinal tract is enhanced because of delayed production of pancreatic proteases and gastric acid by newborn infants, antiproteases and inhibitors of gastric acid production in human milk, inherent resistance of some human milk agents to proteolysis, and protective binding of other factors in human milk. 7) Growth factors in human milk aid in establishing a commensal enteric microflora, which protects against certain bacterial enteropathogens. The statement of these points in an evolutionary context raised the issue of how these processes arose by natural selection. This evolutionary, conceptual framework should allow additional questions to be posed concerning all bioactive agents in human milk and their effects upon the recipient infant.

9. Coda

My past 40 years of research into the mysteries of the immune system in human milk is only a prelude to more exciting findings that are yet to come in this field of human biology. Undoubtedly, there is a great deal more to be discovered concerning the formation, extent, physical features, and precise effects of the immune system in human milk. This will require the participation of multidisciplinary groups and the innovation of new technologies. There are many exciting questions concerning the immune system in human milk. A small sampling is as follows.

1. What are the molecular structures of the bioactive factors in human milk?
2. Are those agents modified during their stay in the infant's alimentary tract by partial digestion?
3. What is the extent of the cytokine network in human milk?
4. What bioactive factors in human milk are bound to soluble receptors or to other components in human milk?
5. What are the purposes of those complexes?
6. Are tolerogens, anti-apoptotic factors. or anti-neoplastic agents present in human milk?
7. How is the production of bioactive factors in human milk influenced by the duration of lactation?
8. Is the production of those factors by the mammary gland preset or influenced in a feedback fashion by the recipient infant?
9. How do bioactive factors in human milk affect the recipient infant? There are many specific aspects to this question including the interactions of human milk defense agents with receptors or other moieties on the epithelium of mucosal cells, the effect of human milk cytokines upon the cytokine production by mucosal cells, and the effects upon the genesis and maintenance of mature T cells.
10. What agents in human milk are responsible for the long-term protection found in children who had been breastfed?

Research that deals with the mammary gland and the alimentary tract and respiratory system of human infants is daunting in that invasive procedures to investigate those sites are not ethically permitted. However, I suggest that innovative ways will be devised to test these questions in humans or in suitable animal model systems that have been engineered to mimic the human situation. Of course pitfalls as well as successes are to be expected in this arena of human research, but this duality is perhaps better expressed poetically than in prose.

Discovery

The world has been filled with mystery
Long before our recorded history
Was writ upon tablets made of stone.
It was so before time had flown.

In our brief moment, some would say
There's naught to be discovered, but I say nay
To those so complacent -so self-possessed.
It is only of self-fulfilling ignorance that they are blessed.

The world's a field of gems, but can those gems be mined?
It takes interest, hard work, and some luck to find
One true gem in the span of a single life-time,
Or if you are a cat in nine.

You may be asked, 'What's to be gained from such a search?'
When failures mount, you'll find that you are in a lurch.'
Friends and family will sadly shake their heads,
Wring their hands and ask, 'Why not choose a civilized way to earn your bread?'

They ask, 'Why ponder questions others will not ask?
Why do you call yourself to task?
Come now. Are not your endeavors not just an elaborate mask
To hide from the outside world with vicissitudes so vast?'

And I reply, Perhaps it's the thrill of the hunt - first and last
That winds back to our Paleolithic past.
Perhaps we're selected in an evolutionary way
To grapple with strange ideas - perchance we think it's play.

But I would add. Beware - ideas may clash – your hopes may be dashed-
Hypotheses may falter or crash.
Rejection, well it's all too commonplace.
If you're lucky, only honest disbelief will slow your pace.

And if success comes, despite obstacles, many rather tall,
Give thanks to the fates that made the call.
But also remember an ancient Grecian fable,
If that 's what it was - hold on to it for all you are able.

It was the tale of Narcissus who peered into a glistening pool so transparently clear,
It reflected a visage so exquisitely dear.
To Narcissus it was more desirable than love,
More precious than the star-studded heavens above.

And as he reached down to embrace the image tantalizing,
He was swallowed up, while fantasizing,
Into the depths from whence the image shown.
Narcissus was forever more entombed.

So beware of self-engulfing pride.
In that way many of promise have died,
Swallowed up and digested in an auto-phagolysosome.
My word, what a ghastly but instructive way to end this poem.

But poetry aside, I wish to again thank the International Society for Research in Human Milk and Lactation for honoring me with the Macy-György Award. It will be an honor that I will treasure for years to come.

References

1. A.E. Hanson, H.F. Wiese, M. Lawlis, D.J.D. Adams, A.S. Goldman, and M.A. Baughan, Fat in diet of infants in relation to caloric consumption, growth, and serum levels for specific fatty acids. A.M.A. J. Dis. Child. 90:621 (1955).
2. A.S. Goldman, D.W. Anderson, W.A. Sellars, S. Saperstein, W.T. Kniker, and S.R. Halpern, Milk allergy. I. Oral Challenge with milk andisolated milk proteins in allergic children. Pediatrics 32:425-443 (1963).
3. P. Ehrlich, Über Immunitat durch Verebung und Saugung. Z. Hug. InfektKr. 12:183-208 (1892).
4. J. Bordet, Les sérums hemolytiques, leurs antitoxines et les théories des sérums cytolytiques. Ann. Inst. Pasteur. 14:257-297 (1900).
5. J. Bordet and M. Mordet, M. Le pouvoir bactériolytique du colostrum et du lait. Comptes Rendus des Séaces de L'Acedemie des Sciences 179:1109-1113 (1924).
6. A Fleming, On a remarkable bacteriolytic element found in tissues and secretions. Proc. Roy Soc., London B93:306-317 (1922).
7. R.M. Woodbury, The relation between breast and artificial feeding and infant mortality. Am. J. Hyg. 2:668 (1922).
8. C.G. Grulee, H.N. Sanford, and P.H. Heron, Breast and artificially-fed infants. Influence on morbidity and mortality of twenty thousant infants. JAMA 103:735-739 (1934).
9. C.G. Grulee, H.N. Sanford, and H. Schwartz, Breast and artificially-fed infants. A study of the age incidence in the morbidity and mortality in twenty thousand cases. JAMA 104:1986-1988 (1935).
10. A.S. Goldman, The immunological system in human milk: the past - a pathway to the future, in: *Advances in Nutritional Research*, In Press.
11. P. György, A hitherto unrecognized biochemical difference between human milk and cow's milk. Pediatrics 11:98-107 (1953).
12. P. György, S. Dhanamitta, and E. Steers, Protective effects of human milk in experimental Staphylococcus infection. Science 137:338-340 (1962).
13. L.Å Hanson, Comparative studies of the immune globulins of human milk and blood serum. Int. Arch Allergy 18:241-267 (1961).
14. A. Van Leeuwenhoek, Epistola 106. Arcana naturae detecta delphis batavorum. Apud Henricum a Krooneveld (1965).
15. C.W. Smith, and A.S. Goldman, The cells of human colostrum. I. _In vitro_ studies of morphology and functions. Pediatr. Res. 2:103-109 (1968).
16. C.W. Smith, and A.S. Goldman, Interactions of lymphocytes and macrophages from human colostrum: Characteristics of the interacting lymphocyte. Res. J. Reticuloendothelial Soc. 8:91-104 (1970).
17. C.W. Smith, A.S. Goldman, and R.D. Yates, Interactions of lymphocytes and macrophages from human colostrum: Electron microscopic studies of the interacting lymphocyte. Exp. Cell Res. 69:409-415 (1971).
18. A.S. Goldman, and C.W. Smith, Host resistance factors in human milk. J. Pediatr. 82:1082-1090 (1973).
19. G.J. Murillo, and A.S. Goldman, The cells of human colostrum. II. Synthesis of IgA and B1c. Pediatr. Res. 4:71-75 (1970).
20. C.W. Smith, and A.S. Goldman, Macrophages from human colostrum: Multinucleated giant cell formation by phytohemagglutinin and concanavalin A. Exp. Cell Res. 66:310-317 (1971).
21. W.D. Dickey, H.B. Rudloff, A.S. Goldman, and F.C. Schmalstieg, Human uropod bearing lymphocytes: Isolation of a factor from human milk that abrogates the uropod inhibitory protein from human serum. Biochem. Biophys. 100:138-145 (1981).
22. R.M. Goldblum, C. Garza, C.A. Johnson, B.L. Nichols, and A.S. Goldman, Human milk banking. I. Effects of container upon immunologic factors in mature milk. Nutrition Research 1:449-459 (1981).
23. R.M. Goldblum, C. Garza, C.A. Johnson, B.L. Nichols, and A.S. Goldman, Human Milk banking. II. Relative stability of immunologic factors in stored colostrum. Acta Paediatr. Scand. 71:143-144 (1982).

24. R.M. Goldblum, C.W. Dill, T.B. Albrecht, E. Alford, C. Garza, and A.S. Goldman, Rapid high temperature treatment of human milk. J. Pediatr. 104:380-385 (1984).
25. A.S. Goldman, C. Garza, B. Nichols, C.A. Johnson, E. Smith, and R.M. Goldblum, The effects of prematurity upon the immunologic system in human milk. J. Pediatr. 101:901-905 (1982).
26. A.S. Goldman, C. Garza, C.A. Johnson, B.L. Nichols, and R.M. Goldblum, Immunologic factors in human milk during the first year of lactation. J. Pediatr. 100:563-567 (1982).
27. A.S. Goldman, R.M. Goldblum, and C. Garza, Immunologic components in human milk during the second year of lactation. Acta Paediatr. Scand. 72:461-462 (1983).
28. A.S. Goldman, C. Garza, C.A. Johnson, B.L. Nichols, and R.M. Goldblum, Immunologic components in human milk during weaning. Acta Paediatr. Scand. 72:133-134 (1983).
29. N.F. Butte, R.M. Goldblum, L.M. Fehl, K. Loftin, E.O. Smith, C. Garza, and A.S. Goldman, Daily ingestion of immunologic components in human milk during the first four months of life. Acta Paediatr. Scand. 73:296-301 (1984).
30. R.J. Schanler, R.M. Goldblum, C. Garza, and A.S. Goldman, Enhanced fecal excretion of secreted immune factors in very low birth weight infants fed fortified human milk. Pediatr. Res. 20:711-715 (1986).
31. A.S. Goldman, C. Garza, R.J. Schanler, and R.M. Goldblum, Molecular forms of lactoferrin in stool and urine from infants fed human milk. Pediatr. Res. 27:252-255 (1990).
32. T.W. Hutchens, et al:, Origin of intact lactoferrin and its DNA-binding fragments found in the urine of human milk-fed preterm infants. Evaluation of stable isotopic enrichment. Pediatr Res 29:243 (1991).
33. A.S. Goldman, and R.M. Goldblum, Immunologic system in human milk: characteristics and effects, in: *Textbook of Gastroenterology and Nutrition In Infancy,* edited by E. Lebenthal, - 2nd. Edition.(Raven Press, New York 1989) pp. 135-142.
34. R.G. Wyatt, and L.J. Mata, Bacteria in colostrum and milk of Guatemalan Indian women. J. Trop. Pediatr. 15:159-162 (1969).
35. L.J. Mata, J.J. Urrutia, and J.E. Gordon, Diarrhoeal disease in a cohort of Guatemalan village children observed from birth to age two years. Trop. Geogr. Med. 19:247-257 (1967).
36. L.J. Mata, et al., Shigella infection in breast-fed Guatemalan Indian neonates. Am. J. Dis. Child. 117:142 (1969).
37. A.S. Goldman, L.W. Thorpe, R.M. Goldblum, and L.A. Hanson, Antiinflammatory properties of human milk. Acta Paediatr. Scand. 75:689-695 (1986).
38. A.S. Goldman, R.M. Goldblum, and L.A. Hanson, Anti-inflammatory systems in human milk, in: *The Antioxidant Nutrition and Immune Functions Symposium of the Agricultural and Food Chemistry,* (Plenum Press, New York and London; 1989) pp. 69-76.
39. L.W. Thorpe, H.E. Rudloff, L.C. Powell, and A.S. Goldman, Decreased response of human milk leukocytes to chemoattractant peptides. Pediatr. Res. 20:373-377 (1986).
40. F. Ozkaragoz, H.E. Rudloff, S. Rajaraman, A.A. Mushtaha, F.C. Schmalstieg, and A.S. Goldman, The motility of human milk macrophages in collagen gels. Pediatr. Res. 23:449-452 (1988).
41. S.E. Keeney, F.C. Schmalstieg, K.H. Palkowetz, H.E. Rudloff, L.E. Binh-Minh, and A.S.Goldman, Activated neutrophils and neutrophil activators in human milk. Increased expression of CD11b and decreased expression of L-selectin. J. Leukocyte Biol. 54 (2):97-104 (1993).
42. A.A. Mushtaha, F.C. Schmaltieg, T.K. Hughes, S. Rajaraman, H.E. Rudloff, and A.S. Goldman, Chemokinetic agents for monocytes in human milk: Possible role of tumor necrosis factor-α. Pediatr. Res. 25:629-633 (1989).
43. H.E. Rudloff, F.C. Schmalstieg, Jr., A.A. Mushtaha, K.H. Palkowetz, S.K. Liu, and A.S. Goldman, Tumor necrosis factor-α in human milk. Pediatr. Res. 31:29-33 (1992).
44. D.P. Wirt, L.T. Adkins, K.H. Palkowetz, F.C. Schmalstieg, and A.S. Goldman, Activated-memory T lymphocytes in human milk. Cytometry 13:282-290 (1992).
45. H.E. Rudloff, F.C. Schmalstieg, Jr., K.H. Palkowetz, and A.S. Goldman, Interleukin-6 in human milk. J. Reprod. Immunol. 23:13-20 (1993).
46. K.H. Palkowetz, C.L. Royer, R. Garofalo, H.E. Rudloff, F.C. Schmalstieg, Jr., and A.S. Goldman, Production of interleukin-6 and interleukin-8 by human mammary gland epithelial cells. J. Reprod. Immunol. 26:57-64 (1994).
47. E.S. Buescher and I. Malinowska, Soluble receptors and cytokine antagonists in human milk. Pediatr. Res. 40:839-844 (1996).
48. A.S. Goldman, S. Chheda, K.H. Palkowetz, H.E. Rudloff, R. Garofalo, and F.C. Schmalstieg, Cytokines in human milk: Their properties and potential effects upon the mammary gland and the recipient infant, in: *Journal of Mammary Gland Biology and Neoplasia*, Third Edition, edited by M.C. Neville, D. Medina, (Plenum Publishing Corporation, New York, 1996) 1:251-258.

49. A.S. Goldman, R. Garofalo, and S. Chheda, Spectrum of immunomodulating agents in human milk. Int. J. Pediatr. Hematol/Oncol. 4: 491-497 (1997).
50. R. Garofalo, S. Chheda, F. Mei, K.H. Palkowetz, H.E. Rudloff, F.C. Schmalstieg, and A.S. Goldman, Interleukin-10 in human milk. Pediatr. Res. 37:444-449 (1995).
51. S. Chheda, K.H. Palkowetz, R. Garofalo, D.K. Rassin, and A.S. Goldman, Decreased interleukin-10 production by neonatal monocytes and T cells: Relationship to decreased production and expression of tumour necrosis factor-α and its receptors. Pediatr. Res. 40:475-483 (1996).
52. A.S. Goldman, S. Chheda, and R. Garofalo, Evolution of immunological functions of the mammary gland and the postnatal development of immunity. Pediatr. Res. 43:155-162 (1998).
53. A.S. Goldman, Modulation of the gastrointestinal tract of infants by human milk. Interfaces and interactions. An evolutionary perspective. J. Nutrition 130(2S Suppl):426S-431S (2000).

VITAMIN A AND THE NURSING MOTHER-INFANT DYAD

Evidence for intervention

Rebecca J. Stoltzfus, Jean H. Humphrey*

1. INTRODUCTION

The role of vitamin A in improving child health and survival in vitamin A-deficient regions of the world was firmly established in the early 1990's, after a series of randomized trials showed that vitamin A supplementation decreased mortality in children 6 months to 5 years of age by around 24%.[1] However the importance of vitamin A to health outcomes in women of reproductive age and infants 0-6 months of age remains a topic of debate and area of active research. The purpose of this paper is to highlight the important questions from a public health perspective, and to discuss the latest ideas and research results related to health outcomes in lactating women and their young infants. In the process, we will make the argument that while our understandings remain imperfect, the evidence at hand is sufficient to justify more aggressive interventions to improve the vitamin A status of pregnant and lactating mothers and young infants.

2. Relation Between Maternal Status and Milk Secretion

The mother's vitamin A status is a strong determinant of the vitamin A content of the milk. This relationship is demonstrated dramatically in recent data from mothers in Central Java, Indonesia (Table 1). In this population, vitamin A status is marginal. Serum and milk vitamin A concentrations are lower than observed in western populations, but means are well above the cutoff of 30 μg/dL used to categorize abnormal values in both serum and milk concentrations of women. To diagnose low vitamin A stores in the mothers we used the Modified Relative Dose Response (MRDR) test, in which a dose of vitamin A2 is given orally, and 4-5 hours later both vitamin A2 and vitamin A are measured in serum. The greater the concentration of vitamin A2 appearing

*Center for Human Nutrition, Department of International Health, The Johns Hopkins School of Public Health, 615 N. Wolfe St., Baltimore MD 21205.

in the serum, the more deficient the individual. Typically the ratio of serum vitamin A2:serum retinol is used to evaluate the test. To avoid the inherent correlation between this ratio and the serum retinol concentration itself (the denominator), we grouped the women by the absolute vitamin A2 concentration in the serum, with values >2.0 μg/dL indicating the deficient group. The mean milk vitamin A concentration of the deficient group was less than half the concentration of the normal group, even though the mean serum retinol concentrations were not dramatically different between the groups. The mammary gland is remarkably sensitive to vitamin A status—even when serum retinol concentrations are relatively adequate.

Table 1. Relation between mother's vitamin stores and milk vitamin A*

	Mother's VA2 dose response		
	≤ 2.0 μg/dL (n=107)	> 2.0 μg/dL (n=16)	P value
Serum retinol (μg/dL)	51.2 ± 19.1	43.6 ± 16.2	0.101
< 30 μg/dL (%)	14.0	25.0	0.271
Milk vitamin A (μg/dL)	39.4 ± 29.3	16.9 ± 13.1	0.000
< 30 μg/dL (%)	45.8	93.8	0.000

*Data from Central Java, Indonesia, lactating mothers at 6 weeks post-partum. Milk samples were obtained by complete expression of one breast not used to nurse the infant for ≥1 hour prior. The vitamin A2 dose response was the serum concentration of vitamin A2 in a MRDR test[2], Ismadi SD, Stoltzfus RJ & Dibley MJ, unpublished data.

3. Effects of Interventions on Milk and Infant Status

It follows that infant vitamin A status can be improved through maternal supplementation, direct supplementation to the infant, or a combination of both. All of these approaches have been evaluated, and the results of several recent trials are summarized in Table 2. The effects of maternal supplementation on milk vitamin A are significant, with larger effects from larger doses. In the Nepal trial, women were provided a total of 1,300,000 IU supplemental vitamin A over 12 months--300,000 IU in the post-partum period alone--and this increased milk secretion of vitamin A nearly two-fold. The degree to which infant vitamin A status was improved in utero in the Nepal trial is unknown, but the majority of this effect likely occurred during lactation, because the transfer of vitamin A from mother to fetus during gestation is very small, even in well-nourished mothers.[3]

The relatively small effects of maternal supplementation on infant vitamin A status are not surprising when one calculates the actual amount of a high-dose supplement that is absorbed and available for secretion. Humphrey & Rice, using data from the Indonesia and Bangladesh supplementation trials, have estimated the amount of supplemental vitamin A that would be needed to bring the vitamin A intake of breastfed babies into the adequate range.[4] The calculated values are close to the empirical results. Given a 200,000 IU dose to the mother plus three 25,000 IU doses to the infant with immunizations (as done in the multi-center trial in India, Ghana and Peru), 50% of infants would theoretically remain vitamin A deficient as defined by the MRDR. Indeed, 43% of infants in the supplementation arm of the multi-center trial were deficient by MRDR. A dose to the mother of 400,000 IU and 4 doses of 50,000 IU to the infant would be needed to bring >90% of infants into adequate status by 6 months of age. These doses have been evaluated and were found to be safe.[5; 6]

Table 2. Effects of vitamin A supplements on breast milk vitamin A (3 mo post-partum) and infant status in recent randomized trials

	Bangladesh[a]	Indonesia[b]	India, Ghana, Peru[c]	Nepal[d]
Form of supplementation	200,000 IU to mother	300,000 IU to mother	200,000 IU to mother + 3 x 25,000 IU to baby	25,000 IU weekly to mother, during pregnancy and post-partum
Milk vitamin A (μmol/L)				
Vitamin A group	1.20 ± 1.00[e] (69)	2.45 ± 1.23 (57)		1.6 (~ 75)
Placebo group	0.83 ± 0.43 (72)	1.82 ± 1.28 (60)	(not reported)	0.87 (~ 75)
	$p < 0.01$	$p = 0.005$		"significant"
Infant deficiency (%)				
Vitamin A group	87% (60)	10% (67)	43% (332)	62% (224)
Placebo group	93% (65)	23% (64)	53% (440)	83% (203)
	N.S.	$p < 0.03$	N.S.	$p < 0.0001$

[a]Milk sampled by full breast expression. Infant deficiency defined by positive MRDR test at 6 months of age. Reference 7. [b]Milk sampled by full breast expression. Infant deficiency defined by positive RDR test at 6 months of age. Reference 8 [c]Infant deficiency defined by positive MRDR test at 6 months of age. Reference 9. [d]Casual (drip) milk samples. Infant deficiency defined by serum retinol < 0.70 μmol/L at 3 months of age. References 10-11[7]. [e]mean ± SD, (n).

Current recommendations by the World Health Organization for supplementation of mothers and infants (Table 3) are lower than the doses used in most research trials, and are now the subject of further discussion. We and others have argued that these doses should be increased.[4; 8] A proposed alternative is to provide 2 x 200,000 IU orally to the mother shortly after birth, and 3 x 50,000 IU to the infant, one dose at each DTP (diphtheria, tetanus, pertussis) immunization visit. Where HIV prevalence is ≥10%, an additional dose of 50,000 to the infant at birth may be added to this schedule.

4. Potential Benefits for the other

What health benefits might result from vitamin A supplementation given to the nursing mother and her baby? First, maternal supplementation can dramatically reduce maternal mortality. The Nepal supplementation trial (described above in terms of its effects on milk vitamin A and infant deficiency) also assessed the impact of weekly supplementation with vitamin A or β-carotene from pre-conception to 3 months post-partum in 45,000 women.[9] The two treatment arms reduced mortality during pregnancy and lactation by around 45%. The weekly dosing used in this study was designed to mimic the effects of improved diet, but may not be feasible to implement in many places. In this regard, it is important to note that supplementation had similar effects on pregnancy mortality and post-partum mortality (measured up to 12 weeks post-partum), and that two-thirds of deaths occurred in the post-partum period. Though no study to date has measured the impact of post-partum dosing on maternal mortality, it is likely that a substantial proportion of these post-partum deaths could be prevented by the simpler post-partum intervention.

Table 3. High-dose universal distribution schedule for prevention of vitamin A deficiency

Infants < 6 months of age	
Non-breast fed infants	50,000 IU orally
Breastfed, whose mothers have not received supplemental vitamin A	50,000 IU orally
Infants 6-12 months of age	100,000 IU every 4-6 months
Children > 12 months of age	200,000 IU every 4-6 months
Mothers	200,000 IU within 8 weeks of delivery

Reprinted with permission from *Vitamin A Supplements. A guide to their use in the treatment and prevention of vitamin A deficiency and xerophthalmia.* World Health Organization, Geneva, 1997.

Second, maternal supplementation may prevent and/or cure night blindness. Night blindness has been thought to occur mainly in pregnancy, but in Nepal the prevalence of this condition during lactation was also strikingly high (8.1% prevalence in pregnant women vs. 6.2% in lactating women 0-5 months post-partum).[10] The pattern of occurrence in lactating women (2.0 % prevalence in women 0-2 months post-partum vs. 10.8% in women 3-5 months post-partum) suggests a relation to depletion of vitamin A status as breastfeeding continues. Although factors other than vitamin A also contribute to night blindness in women,[11] it is likely that post-partum vitamin A supplementation could help to prevent this condition.

Third, post-partum maternal vitamin A supplementation may reduce puerperal sepsis. In Java, Indonesia, 680 newly pregnant women were randomly allocated to receive daily doses of vitamin A, zinc, both nutrients, and neither beginning in the second trimester of pregnancy.[12] Puerperal sepsis was defined as a body temperature >38°C on at least one day between the end of the first day and the 14th day. Treatment with vitamin A reduced the incidence of this sign by 70% (from 9.3% to 2.8%). Similarly, in 1931 Green et al. randomized 550 pregnant women attending antenatal clinic in London to take daily doses of a vitamin A preparation or not, beginning 14 -30 days prior to their calculated day of delivery.[13] They defined puerperal sepsis as having 2 rises in temperature >100°F between the end of the first and the end of the 8th day after delivery (a more conservative definition than used in Java). Vitamin A reduced the incidence of this sign from 4.7% to 1.1%. When they defined cases as any puerperal fever during the 8 days post partum, vitamin A reduced the incidence from 35.6% to 20.4%.

Finally, and most speculatively, post partum maternal vitamin A supplementation may reduce the risk of sexually acquired HIV infections among women during the post partum year. The ZVITAMBO trial is a 4-armed trial of 14,110 mother-infant pairs on-going in Zimbabwe, where 1/3 of antenatal women are HIV+. Mother baby pairs are randomized to receive one of 4 treatments within 96 hours of delivery: 400,000 IU vitamin A to mother and 50,000 IU vitamin A to baby, only mother supplemented, only baby supplemented, or neither. One of the primary outcomes is whether the maternal dose can reduce sexually acquired incident HIV infections among post partum women.

The basis for this is that animal work and more limited human studies show that vaginal epithelium is highly sensitive to the earliest stages of vitamin A deficiency –

undergoing cornification and reduced vaginal integrity – which responds within days to supplementation[14-16]. So reliable and sensitive is this response that it became the basis of a biologic assay – the vaginal smear assay widely used to measure activity of various forms of the vitamin.[17] In addition we know women are often vitamin A deficient at delivery, and that reduced vaginal integrity is a risk factor for sexually acquired HIV infection. The ZVITAMBO trial completed recruitment in January of this year and plans complete follow up and break code in mid-2001.

5. Potential Benefits for the Baby

What health benefits for the baby might result from improving the vitamin A status of the nursing mother-infant dyad? First, several studies have investigated whether maternal and/or infant vitamin A supplementation can reduce mortality in the first 6 months of life. This is of great importance because 2/3 of all infant mortality occurs during these early months of life when other common child survival interventions (i.e., immunizations) have little impact. (Exclusive breast feeding, of course is the exception!) Finally, it is plausible to believe supplementation could have this effect on this age group because babies are born with meager stores and their deficient mothers provide milk with low vitamin A concentration. Nonetheless, whether supplementation is delivered to the woman during pregnancy, to the woman post-partum, or to the infant directly, the evidence that it will reduce early infant mortality has been inconsistent – with the bulk of evidence to date showing no effect on early infant mortality or morbidity.

One randomized trial demonstrated benefit. In a randomized controlled study in Indonesia (n=2000), a single 50,000 IU dose of vitamin A given to infants on the first day of life reduced mortality by 64%.[18] The effect was observed only during the first 4 months when the intervention also reduced sick clinic visits for cough and fever, suggestive of pneumonia, the major cause of death in the age group.

Three trials found no benefit of vitamin A supplementation on early infant mortality or infectious morbidity. In the first Nepal trial,[19] a 50,000 IU dose was given to children < 1 month of age, however the children were not usually reached until 3 weeks of age. No impact on mortality was observed. In the second Nepal study, weekly maternal supplementation during pregnancy and lactation had no effect on fetal wastage or infant mortality[7] — despite impressive effects on milk vitamin A and maternal mortality (discussed above). In a multicenter trial in Ghana, India and Peru,[20] mothers received 200,000 IU in the post-partum period and their babies received 25,000 IU at each of three EPI contacts before 5 months of age. No impact on infant morbidity or mortality was observed during the first 6 months of life.

A distinguishing feature of the Indonesia trial was that infants were dosed very near birth. There could be a specific immunologic window of opportunity for vitamin A to act. Two replications of the Indonesian trial currently underway are the infant mortality component of ZVITAMBO, and the VASIN trial in South India, in which infants receive 2 x 25,000 IU vitamin A on successive days within 48 hours of delivery.

While the impact of vitamin A supplementation given during the first 6 months of life on mortality remains inconclusive, vitamin A supplementation given during the second 6 months of life reduces mortality by 24%. Thus if interventions given during early infancy can optimize infant vitamin A status by 6 months of age, they might reduce mortality in the second 6 months of life--regardless of any protective effect during the first 6 months. The most important outcome of interventions targeted to young infants may be whether participating infants attain sufficient vitamin A status by 6 months of age. This brings us back to Table 2, and the possible need for higher doses of supplemental vitamin A to breastfeeding mothers and infants to achieve this goal.

In HIV-endemic areas, vitamin A supplementation may have special benefits. Low serum retinol concentrations are common in HIV infection[21-23] and are associated with faster disease progression and higher mortality.[24; 25]

Two studies suggest vitamin A supplementation of HIV-infected children can reduce their burden of opportunistic infection. In South Africa, aggressive vitamin A supplementation to children born to HIV+ mothers (50,000 IU at 1 and 3 months of age, 100,000 IU at 6 and 9 months of age, 200,000 IU at 12 and 15 months) reduced overall morbidity by 31%.[26] But, the strongest effect was seen among the children who were themselves HIV infected—vitamin A reduced diarrhea episodes in these children by 49%. In Tanzania, among 687 children 6 months to 5 years of age who were admitted to hospital for pneumonia, 400,000 IU given on admission and 4 and 8 months after discharge reduced mortality by 49%.[27] But among the HIV-infected children, mortality was reduced by 63%. These studies suggest that vitamin A supplementation may be one of the few low-cost and feasible interventions available to improve the quality of life and extend the survival of HIV-infected children in developing countries.

There is strong evidence that maternal vitamin A deficiency is a risk factor for mother-to-child transmission of HIV, including transmission occurring during breast-feeding.[28, 29] However, two controlled trials of vitamin A supplementation of HIV-infected pregnant women to test the impact on mother-to-child transmission have now been published, and the results show little or no impact. In a South African trial,[30] 728 HIV-infected women entered a randomized placebo-controlled trial in which the treatment was a daily dose of 5000 IU vitamin A and 30 mg β-carotene during the third trimester of pregnancy and 200,000 IU vitamin A at delivery. There was no difference in transmission rates at 3 months (20.3% in the treated group vs. 22.3 % in the placebo group). In Tanzania, 1083 HIV-infected women entered a 2-by-2 factorial trial in which the vitamin A treatment was similar to the South African study, but half the women in each group also received daily multivitamins.[31] There was no effect of vitamin A or multivitamins on HIV transmission or infant mortality at 6 weeks of age.

Human milk with higher sodium content or sodium:potassium (Na:K) ratio (indicators of greater mammary permeability) also contains higher concentrations of certain immune factors[32; 33] and, in HIV-infected women, HIV.[32] It has been hypothesized that vitamin A status might decrease mammary permeability and therefore help prevent HIV transmission through breastmilk.[34] Indeed, total plasma carotenoid levels were inversely associated with milk sodium content in a sample of Malawian mothers with HIV infection.[32] However, in a randomized trial of post-partum vitamin A or β-carotene supplementation of Bangladeshi lactating mothers, neither form of supplementation influenced Na:K ratio or breastmilk immune factors.[33] Similarly, in a study of rural Tanzanian women, supplementation with β-carotene-rich red palm oil did not affect the Na:K ratio.[35] Whether vitamin A influences mammary permeability in a unique way in HIV-infected women remains to be tested, but it seems less likely given these findings and the lack of vitamin A effect on vertical transmission in South Africa and Tanzania.

Whether vitamin A supplementation of the breastfeeding infant of an HIV-infected mother can reduce HIV transmission during breast feeding has not yet been tested. In other words, rather than supplementing the HIV-infected mother to make her less infectious, supplement the yet uninfected nursing infant to make him less susceptible. This is another objective of the ongoing ZVITAMBO trial.

Summary

From the evidence at hand, interventions to reduce vitamin A deficiency in the breastfeeding mother and baby are likely to confer several health benefits. These include reduced puerperal morbidity, night blindness, and mortality in the mother, and reduced morbidity and mortality in the baby beginning around 6 months of age. Additional evidence is still needed to increase our confidence in these inferences, and to clarify the best interventions for achieving these outcomes. Current evidence about the effects of improved vitamin A status on early infant morbidity or mortality is mixed. There is accumulating evidence against the hypothesis that maternal vitamin A supplementation during pregnancy will reduce mother-to-child HIV transmission. There is so far no evidence about whether improved *infant* vitamin A status can prevent mother-to-child HIV transmission. In sum, the probable benefits of improving the vitamin A status of the breastfeeding mother and her baby are substantial, but to achieve these benefits, more aggressive interventions will be needed.

References

1. G. H. Beaton, R. Martorell, K. A. L'Abbe, B. Edmonston, G. McCabe, A. C. Ross *et al.*, *Effectiveness of Vitamin A Supplementation in the Control of Young Child Morbidity and Mortality in Developing Countries* (University of Toronto, Toronto, 1992).
2. S. A. Tanumihardjo, Muherdiyantiningsih, D. Permaesih, A. M. Dahro, Muhilal, D. Karyadi *et al.*, Assessment of the vitamin A status in lactating and nonlactating, nonpregnant Indonesian women by use of the modified-relative-dose-response (MRDR) test, *Am J Clin Nutr* **60**:142-147 (1994).
3. R. J. Stoltzfus and B. A. Underwood, Breast-milk vitamin A as an indicator of the vitamin A status of women and infants, *Bull WHO* **73**:703-711 (1995).
4. J. H. Humphrey and R. E. Rice, Vitamin A supplementation of young infants, *Lancet* **356**:422-424 (2000).
5. J. H. Humphrey, T. Agoestina, A. Juliana, S. Septiana, H. Widjaja, M. C. Cerreto *et al.*, Neonatal vitamin A supplementation: effect on development and growth at 3 y of age, *Am J Clin Nutr* **68**:109-117 (1998).
6. P. J. Iliff, J. H. Humphrey, A. I. Mahomva, P. Zvandasara, M. Bonduelle, L. Malaba *et al.*, Tolerance of large doses of vitamin A given to mothers and their babies shortly after delivery, *Nutr Res* **19**:1437-1446 (1999).
7. J. Katz, K. P. West Jr, S. K. Khatry, E. K. Pradhan, S. C. LeClerq, P. Christian *et al.*, Maternal low-dose vitamin A or beta-carotene supplementation has no effect on fetal loss and early infant mortality: a randomized cluster trial in Nepal, *Am J Clin Nutr* **71**:1570-1576 (2000).
8. A. L. Rice, R. J. Stoltzfus, A. de Francisco, J. Chakraborty, C. L. Kjolhede, M. A. Wahed, Maternal vitamin A or beta-carotene supplementation in lactating Bangladeshi women benefits mothers and infants but does not prevent subclinical deficiency, *J Nutr*, **129**:356-365 (1999).
9. K. P. West Jr, J. Katz, S. K. Khatry, S. C. LeClerq, E. K. Pradhan, S. R. Shrestha *et al.*, Double blind, cluster randomised trial of low dose supplementation with vitamin A or beta carotene on mortality related to pregnancy in Nepal. The NNIPS-2 Study Group [see comments], *Brit Med J* **318**:570-575 (1999).

10. J. Katz, S. K. Khatry, K. P. West Jr, J. H. Humphrey, S. C. LeClerq, E. K. Pradhan *et al.*, Night blindness is prevalent during pregnancy and lactation in rural Nepal, *J Nutr* **125**:2122-2127 (1995).
11. P. Christian, K. P. West Jr, S. K. Khatry, J. Katz, S. LeClerq, E. K. Pradhan *et al.*, Vitamin A or beta-carotene supplementation reduces but does not eliminate maternal night blindness in Nepal, *J Nutr* **128**:1458-1463 (1998).
12. M. J. Dibley and M. Hakimi, Preliminary analysis of Zibuvita trial for Mother Care Technical Advisory Group Meeting 1998 (unpublished).
13. H. N. Green, D. Pindar, G. Davis, E. Mellanby, Diet as a prophylactic agent against puerperal sepsis with special reference to vitamin A as an anti-infective agent, *Brit Med J* **ii**:595-598 (1931).
14. H. M. Evans, The effect of inadequate vitamin A on the sexual physiology of the female, *J Biol Chem* **77**:651-654 (1928).
15. K. A. Mason and E. T. Ellison, Vaginal cornification as a criterion of vitamin A deficiency in the rat, *Anat Rec* **58**:80 (1934).
16. K. D. Blackfan and S. B. Wolbach, Vitamin A deficiency in infants, A clinical and pathological study, *J Pediatr* **3**:679-706 (1933).
17. W. K. Sietsema and H. F. DeLuca, A new vaginal smear assay for vitamin A in rats, *J Nutr* **112**:1481-1489 (1982).
18. J. H. Humphrey, T. Agoestina, L. Wu, A. Usman, M. Nurachim, D. Subardja *et al.*, Impact of neonatal vitamin A supplementation on infant morbidity and mortality, *J Pediatr* **128**:489-496 (1996).
19. K. P. West Jr, J. Katz, S. R. Shrestha, S. C. LeClerq, S. K. Khatry, E. K. Pradhan *et al.*, Mortality of infants < 6 mo of age supplemented with vitamin A: a randomized, double-masked trial in Nepal, *Am J Clin Nutr* **62**:143-148 (1995).
20. WHO/CHD Immunization-Linked Vitamin A Supplementation Study Group, Randomized trial to assess benefits and safety of vitamin A supplementation linked to immunization in early infancy, *Lancet* **352**:1257-1263 (1998).
21. R. S. Beach, E. Mantero-Atienza, G. Shor-Posner, J. J. Javier, J. Szapocznik, R. Morgan *et al.*, Specific nutrient abnormalities in asymptomatic HIV-1 infection [see comments], *AIDS* **6**:701-708 (1992).
22. J. D. Bogden, H. Baker, O. Frank, G. Perez, F. Kemp, K. Bruening *et al.*, Micronutrient status and human immunodeficiency virus (HIV) infection, *Ann NY Acad Sci* **587**:189-195 (1990).
23. D. Moodley, J. Moodley, A. Coutsoudis, H. M. Coovadia, E. Gouws, Vitamin A levels in normal and HIV-infected pregnant women, *S Afr Med J* **88**:1029-1032 (1998).
24. R. D. Semba, N. M. Graham, W. T. Caiaffa, J. B. Margolick, L. Clement, D. Vlahov, Increased mortality associated with vitamin A deficiency during human immunodeficiency virus type 1 infection, *Arch Intern Med* **153**:2149-2154 (1993).
25. R. D. Semba, W. T. Caiaffa, N. M. Graham, S. Cohn, D. Vlahov, Vitamin A deficiency and wasting as predictors of mortality in human immunodeficiency virus-infected injection drug users [see comments], *J Infect Dis* **171**:1196-1202 (1995).
26. A. Coutsoudis, R. A. Bobat, H. M. Coovadia, L. Kuhn, W. Y. Tsai, Z. A. Stein, The effects of vitamin A supplementation on the morbidity of children born to HIV-infected women [see comments], *Am J Public Health* **85**:1076-1081 (1995).
27. W. W. Fawzi, R. L. Mbise, E. Hertzmark, M. R. Fataki, M. G. Herrera, G. Ndossi *et al.*, A randomized trial of vitamin A supplements in relation to mortality among human immunodeficiency virus-infected and uninfected children in Tanzania, *Pediatr Infect.Dis J* **18**:127-133 (1999).
28. R. D. Semba, P. G. Miotti, J. D. Chiphangwi, A. J. Saah, J. K. Canner, G. A. Dallabetta *et al.*, Maternal vitamin A deficiency and mother-to-child transmission of HIV-1 [see comments], *Lancet* **343**:1593-1597 (1994).
29. R. W. Nduati, G. C. John, B. A. Richardson, J. Overbaugh, M. Welch, J. Ndinya-Achola *et al.*, Human immunodeficiency virus type 1-infected cells in breast milk: association with immunosuppression and vitamin A deficiency, *J Infect Dis* **172**:1461-1468 (1995).
30. A. Coutsoudis, K. Pillay, E. Spooner, L. Kuhn, H. M. Coovadia, Randomized trial testing the effect of vitamin A supplementation on pregnancy outcomes and early mother-to-child HIV-1 transmission in Durban, South Africa, South African Vitamin A Study Group, *AIDS* **13**:1517-1524 (1999).
31. W. W. Fawzi, G. Msamanga, D. Hunter, E. Urassa, B. Renjifo, D. Mwakagile *et al.*, Randomized trial of vitamin supplements in relation to vertical transmission of HIV-1 in Tanzania, *JAIDS* **23**:246-254 (2000).
32. R. D. Semba, N. Kumwenda, T. E. Taha, D. R. Hoover, T. C. Quinn, Y. Lan *et al.*, Mastitis and immunological factors in breast milk of human immunodeficiency virus-infected women, *J Hum Lact* **15**:301-306 (1999).

33. S. M. Filteau, A. L. Rice, J. J. Ball, J. Chakraborty, R. Stoltzfus, A. de Francisco *et al.*, Breast milk immune factors in Bangladeshi women supplemented postpartum with retinol or beta-carotene, *Am J Clin Nutr* **69**:953-958 (1999).
34. R. D. Semba and M. C. Neville, Breast-feeding, mastitis, and HIV transmission: nutritional implications, *Nutr Rev* **57**:146-153 (1999).
35. S. M. Filteau, G. Lietz, G. Mulokozi, S. Bilotta, C. J. Henry, A. M. Tomkins, Milk cytokines and subclinical breast inflammation in Tanzanian women: effects of dietary red palm oil or sunflower oil supplementation, *Immunology* **97**:595-600 (1999).

CHANGES IN FEEDING PATTERNS AFFECT GROWTH IN CHILDREN 0-24 MONTHS OF AGE LIVING IN SOCIOECONOMICALLY DIFFERENT AREAS OF LAHORE, PAKISTAN

Shakila Zaman, Fehmida Jalil, Munir A. Saleemi, Lotta Mellander, Rifat N. Ashraf and Lars Å. Hanson*

1. INTRODUCTION

The interaction of feeding practices, infections and growth in a young child manifests in early life in terms of malnutrition, increased morbidity and mortality. Several studies have shown that breastfeeding in early life protects young children against infections and promotes nutrition and growth particularly in the developing countries.[1-11] However, in situations where breastfeeding is the accepted predominant mode of feeding the young children, faltering in growth and with a high incidence of infections, particularly diarrhoeal diseases, are still observed.[7-8]

Pakistan is a country where breastfeeding is the major mode of feeding young children. Simultaneously, the prevalence of diarrhoeal illness and other infections is also very high leading to growth disturbances in young children living in less privileged areas.[12-13] The Demographic and Household Survey of Pakistan, 1990-1994 has shown that 13% of young children are suffering from severe degree of malnutrition. About 16-20% of the newborns are born with low birth weight.[14] The higher risk of infections and growth faltering for these children results in early malnutrition.

Recently, over the last 6-7 years, in light of the existing evidence, a major health intervention has been the promotion of optimal breastfeeding practices in Pakistan (Baby Friendly Hospital Initiative). This encourages mothers to initiate lactation as early as within ½ an hour after birth, to exclusively breastfeed for at least 4-6 months and to continue breastfeeding for at least two years. Introduction of weaning foods is recommended at 4-6 months of age. A greater impact on the improvement of nutritional status in young children, however, is not visible yet.

*Shakila Zaman, Fehmida Jalil, Munir A. Saleemi and Rifat Ashraf, Departments of Social and Preventive Paediatrics, King Edward Medical College, Lahore, Pakistan. Lotta Mellander, Department of Paediatrics, Queen Sylvia Children Hospital, University of Göteborg, Göteborg, Sweden. Lars Å Hanson, Department of Clinical Immunology, University of Göteborg, Göteborg, Sweden

The Department of Social and Preventive Pediatrics, Lahore, Pakistan has been engaged in a community based monthly follow up of children from birth to 24 months of age.[15] The analysis of data from children born during September 1984 to March 1987 in village Halloki, referred to as Cohort A, revealed that exclusive breastfeeding was uncommon in the group as a whole and partial breastfeeding was the predominant mode of feeding.[16]

This information became the basis for interventions like promotion of optimal breastfeeding practices and nutrition counselling. These interventions were offered to the children born during Dec. 1990 – Dec. 1992 referred here as Cohort B.

The aim of the larger follow up study was to explore the possible factors involved in the interaction between breastfeeding, infections and growth. This provided us with the opportunity to investigate the impact of the nutrition education programme focusing on promotion of optimal breastfeeding practices on the growth of the young children living in less privileged areas.

2. Material

From an initial cross-sectional survey of nearly 3000 households (a population of 15000), the pregnant and lactating mothers were identified in three socio-economically different areas in and around Lahore, Pakistan indicating the natural path of urbanization i.e., a village, a periurban slum and an urban slum inside the city. An upper middle class group (UM Class) was also identified to act as a control group for comparing growth and other parameters. All mothers were followed from the 5th month of gestation till the child was born. Every child born was followed by monthly home visits till 24 months of age. Such surveys were repeated every 5 months to identify all the pregnant mothers in the areas during the study periods.[15]

3. Study Population

The children followed during the two study periods constituted the study population. They were categorized into belonging to Cohort A (n=1476) i.e., born between Sept. 1984 till March 1987 and Cohort B (n=1264) where children born were registered between Dec. 1990 – Dec. 1992 and followed till 2 years of age, Table 1.

4. Study areas

The village Halloki was situated at a distance of 40 kilometers from Lahore, Pakistan. The characteristics of the village were those of a typical Punjabi village with a mean family size of 6.8 and income of less than Rs 1000/month in 75% of the families. Illiteracy among mothers was high (93%) and 80% of the families lived in mud houses.[15]

The periurban slum was selected from around the periphery of the main city along the railway track. The housing standards were poor with 96% living in mud houses, illiteracy among mothers was as high as 95% and the mean family size was 6.1 with 87% having an income of less than Rs. 1000/month. The added disadvantage of these people was that the closed organized community of the village life was absent here with families moving from the village looking for better life and settling down in the mud huts without any water supply and sanitation.[15]

The urban slum families enjoyed a more settled life with an average family size of 6.4, maternal illiteracy of 51% and 98% having brick houses. Only 42% of the families had an income less than Rs. 1000/month. The social fabric in this area was stronger than the periurban slum with access to tap water and closed drainage system. Further details are given elsewhere.[15]

Table 1. The number of children born and followed over the study periods in Cohort A and B from birth to 24 months of age. The number of children leaving the study due to deaths or refusal or moving from the areas is also given.

	Areas of living				
	Village	Periurban slum	Urban slum	UM class	Total
	n	n	n	n	n
Cohort A					
Live births	485	398	353	240	1476
Birth to 24 months					
Deaths	61	57	37	4	159
Refused/moved	59	91	64	75	289
Followed	365	249	253	161	1028
Cohort B					
Live births	544	250	470		1264
Birth to 24 months					
Deaths	55	31	13		99
Refused/moved	109	74	58		241
Followed	380	145	392		917

5. Methods

In cohort A, pregnant mothers were identified by the routine epidemiological survey and were followed till the outcome of pregnancy. A trained Paediatrician examined the newborns as soon as possible after birth. At the same time, anthropometric measurements were made; history of initiation of breastfeeding and prelacteals was taken. Later, they were followed monthly by home visits and anthropometric measures were taken and feeding history was registered in detail. The frequency of breastfeeding or amount of weaning foods consumed, however, was not obtained.

The anthropometric measurements were done using an infant weighing scale permitting an error of less than 100 gms. Length was measured using locally made measuring boards accepting an error of less than 1cm.

In Cohort B, the mothers were followed every two months starting from 5th month of pregnancy and on each contact were counselled on optimal breastfeeding practices, motivating the mothers using educational material and discussions. The follow up of the children was, however, done in a similar fashion as the earlier cohort.

The reference population data comes from the UM Class group which was selected according to a preset criteria: the family had a maximum of three children, owned the house, earned more than Rs. 10,000/month and the mothers had a schooling of at least ten years. The growth patterns in the children from this group were close to the NCHS standard population.[13]

Informed verbal consent was obtained from the village council and from the families included in our follow up. Ethical clearance to conduct the study was obtained from the Ethics' Committees of the Göteborg University and the Pakistan Medical Research Council.

6. Results

Main patterns of breastfeeding n Cohort A by age are depicted in Figure 1. The village, periurban slum and the urban slum are pooled together. All the newborns had received prelacteal feeds in this cohort. Only 12% of the newborns were exclusively breastfed (with prelacteals) at one month of age, rapidly declining over the next four months. The predominant pattern that emerged was that of partial breastfeeding. During the early ages, mainly breastmilk with water was fed and was gradually taken over by breastfeeding along with other milks and semisolids or animal milk feeding and semisolids at later ages.

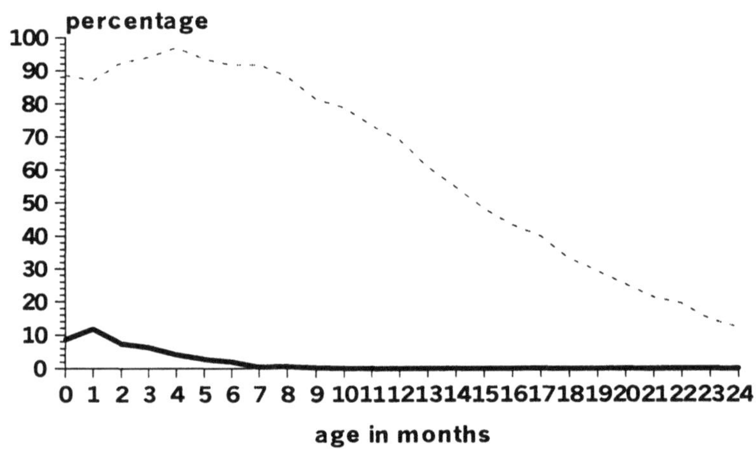

Figure 1. Exclusive breastfeeding (-) was the least common mode of feeding after birth in Cohort A. Partial breastfeeding (--) especially breastmilk and water feeding was common during the first 6 months of life.

The feeding patterns in Cohort B are shown in Figure 2. Only 48% of the infants had received any prelacteals soon after birth. Exclusive breastfeeding (without prelaceals) became a predominant mode of feeding during the first 3-4 months in the three poor areas of living with 70% of the mothers exclusively breastfeeding at 1 month of age and declining to 25% at 4 months of age. Human milk with water and other milks took over after 4-6 months and were gradually replaced by semisolids along with human milk.

Mean values for length, shown in Figure 3 were derived using the UM Class as the reference population providing the functional values for calculating the mean length of the children measured from 0-24 months of age in Cohort A belonging to the three areas of living i.e., village, Periurban slum and the urban slum. At each given age, in all these areas, 929-586 children were measured. The mean length followed parallel to the UM

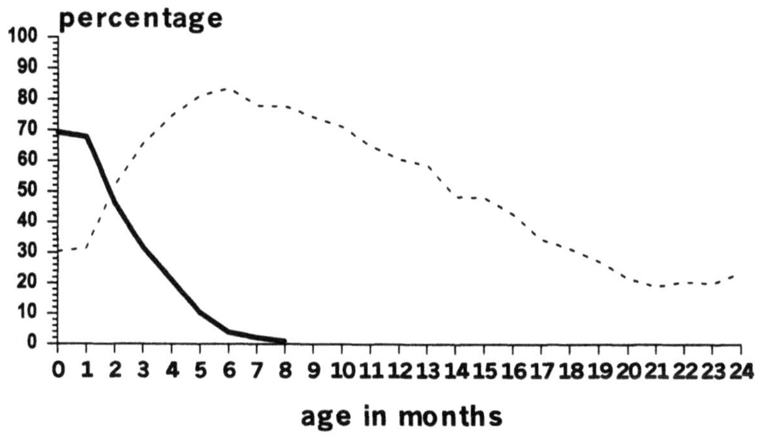

Figure 2. Exclusive breastfeeding (-), represented was the most common mode of feeding in Cohort B initially. Partial breastfeeding gradually replaced exclusive breastfeeding after 4 months of age. Partial breastfeeding (--) was common in later half of the first year.

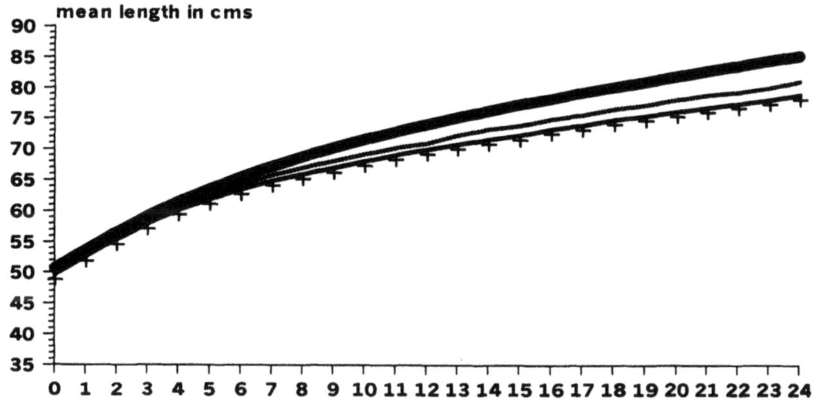

Figure 3. The Upper Middle Class, shown as the bold solid line, was used as the reference population to compare the mean length at each age in the three poor areas of living which are pooled here as Cohort A (+) and Cohort B (-).

class although at a much lower level especially deviating from it close to four months of age. The deficit enlarged as the age advanced. In Cohort B, however, the mean length is closer to the reference line at all ages and hence, the deficit appears less then Cohort A. The number of children represented at different ages in this cohort varied from 1029-222.

Absolute differences in cms are shown in Figure 4. By 24 months of age, a deficit of 6 cms is seen in Cohort A compared to the children living in the more privileged area of living. The overall deficit in height had reduced by 2.8 cms in Cohort B when the three areas were pooled together. However, when examined by area of living, a deficit of 8 cms was seen in the periurban slum, 7 cms in the village and 4 cms in the urban slum when compared to the UM class (data not shown).

This deficit had reduced by 3 cms at 24 months of age in the village and the periurban slum, while in the urban slum the deficit has further diminished another one cms in Cohort B (data not shown). The deficit in weight was also shown to improve in Cohort B compared to Cohort A by 0.5 Kgs at 24 months of age.

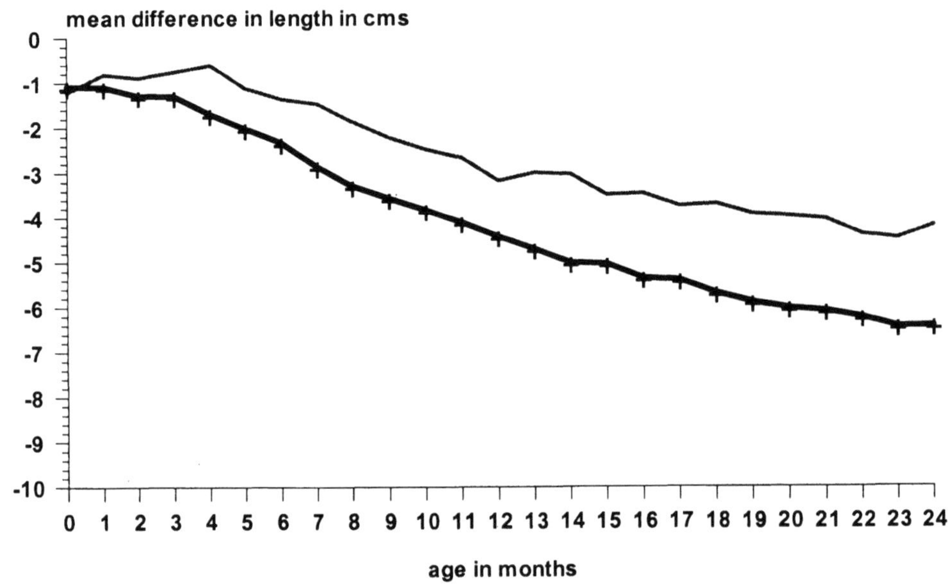

Figure 4. The mean difference in cms is shown with the Upper Middle Class as the reference population which is represented by the 0. The deficit in cms is shown to improve by 2.8 cms at 24 months of age in Cohort B (-) compared to Cohort A (+) after the introduction of optimal breastfeeding practices.

Discussion

The village and the periurban slum are the poorest areas with a high birth rate, low income, high illiteracy and poor housing standards.[15] This deficit in weight and length in Cohort A can be explained on the basis of differences in the areas of living. At the same time, the occurrence of infections and the type and amount of food given to the children cannot be ignored at all. Diarrhoea and respiratory tract infections contributed to 33% and 20% of the disease load, respectively, during the follow up. It was found that on an average each child in the village had 4.3 episodes of diarrhea per year.[12] The better rate of growth in Cohort B may be attributed to the increase in exclusive breastfeeding and reduced infections in this cohort. Thus, by providing exclusive breastfeeding to their babies for at least 4-6 month not only the growth has shown to improve but also the reduction in infections can also reduce the excessive energy loss in young children.[17-18]

It has been argued by researchers that an exclusively breastfed baby is leaner as compared to the bottle-fed child especially during the first 12 months of life because of low protein intake from the breastmilk.[19] But, at the same time, studies have shown that nutritional interventions can have an impact on reducing morbidity and improving growth.[19-20]

After the introduction of simple interventions like nutritional counseling, promotion of optimal breastfeeding practices, supplementary foods and improving maternal health, improved child health has been shown to be possible in areas which are less privileged.[9, 19-22] Thus, simple nutrition educational messages to the mothers will ensure not only her own health but that a healthy newborn will be less likely to suffer from the scourges of poverty and social deprivations if he/she is provided with breastmilk and proper supplementary foods at the appropriate ages.

Acknowledgements

The present investigation was part of a longitudinal study being carried out in Lahore, Pakistan. This was supported by the grants from SAREC (Swedish Agency for Research Cooperation with Developing Countries) and King Edward Medical College, Lahore, Pakistan.

The authors would like to express thanks to the field teams, all the mothers and their children for making this study possible without loosing a single moment of interest. We are indebted to Miss Khalida Chaudhary, who punched and cleaned the data during the quality control and made the analysis possible.

References

1. R.G. Feacham, and M.A. Koblinsky, Interventions for the control of diarrhoeal diseases among young children: promotion of breastfeeding, Bull WHO 62 (2):271-291 (1984).
2. R.G. Feachem, Preventing Diarrhoea: what are the policy options? Hlth. Policy and Planning 1(2):109-117 (1986).
3. K.H. Brown, R.E. Black, G. Lopez de Romana, and H. Creed de Kanashiro, Infant feeding practices and their relationship with diarrhoeal and other diseases in Huascar (Lima), Peru, Pediatr. 83(1):31-40 (1989).
4. S. Diaz, C. Herreros, R. Aravena, M.E. Casado, M.V. Reyes, and V. Schiappacasse. Breast feeding duration and growth of fully breast fed infants in a poor urban Chilean population, Amer J Clin Nutr 62(2):371-376 (1995).
5. K.G. Dewey, M.J. Heinig, and L.A. Nommsen-Rivers, Differences in morbidity between breastfed and formula-fed infants, J Pediatr 126(3):696-702 (1995).
6. L. Å. Hanson, I. Adlerberth, B. Carlsson, U. Dahlgren, M. Hahn-Zoric, A. Morikawa, I. Narayanan, K. Nilsson, D. Roberton, A. Wold, S. Zaman, and F. Jali,. Breastmilk's attack on microbes: is it of clinical significance?. In: *Human Lactation IV. Breastfeeing, nutrition, infection and infant growth in developed and emerging countries, edited by* SA Atkinson, LÅ Hanson, RK Chandra (ARTS Biomedical Publishers and Distributors, St John's, Newfoundland, Canada, 1990), pp 55-65.

7. F. Jalil, I. Adlerberth, R.N. Ashraf, B. Carlsson, L.Å. Hanson, J. Karlberg, K. Khalil, B.S. Lindblad, R. Nazir, S. Zaman, and S.R. Khan, Methodological problems in assessment of long-term health outcomes in breastfed versus bottlefed infants. In: *Human Lactation IV. Breastfeeing, nutrition, infection and infant growth in developed and emerging countries, edited by* SA Atkinson, LÅ Hanson, RK Chandra (ARTS Biomedical Publishers and Distributors, St John's, Newfoundland, Canada, 1990), pp 381-394.
8. J. Karlberg, S. Zaman, L.Å. Hanson, S.R. Khan, B.S. Lindblad, and F. Jalil, Aspects of infantile growth and the impact of breastfeeding: A case control study of the infants from four socioeconomically different area in Pakistan, In: *Human Lactation IV. Breastfeeing, nutrition, infection and infant growth in developed and emerging countries, edited by* SA Atkinson, LÅ Hanson, RK Chandra (ARTS Biomedical Publishers and Distributors, St John's, Newfoundland, Canada, 1990), pp 219-247.
9. R.J. Cohen, K.H. Brown, J. Canahuati, L.L. Rivera, and K.G. Dewey, Determinants of growth from birth to 12 months among breast fed Honduran infants in relation to age of introduction of complementary foods, J. Pediatr. 96 (3):504-510 (1995).
10. WHO Collaborative Study Team. Effect of breastfeeding on infant and child mortality due to infectious diseasesin less developed countries: a pooled analysis, Lancet 355:451-455 (2000).
11. R.N. Ashraf, F. Jalil, S. Zaman, J. Karlberg, S.R. Khan, B.S. Lindblad, and L.Å. Hanson, Breastfeeding and protection against sepsis in a high risk population, Arch. Dis. Childh. 66:488-490 (1991).
12. S. Zaman, F. Jalil, J. Karlberg, and L.Å. Hanson, Early child health in Lahore, Pakistan: VI. Morbidity, Acta Paediatr. (82):390 (suppl), 63-78 (1993).
13. J. Karlberg, R.N. Ashraf, M.A. Saleemi, M. Yaqoob, and F. Jalil, Early child health in Lahore, Pakistan: XI. Growth, Acta Paediatr. (82):390 (suppl), 119-149 (1993).
14. Pakistan Demographic and household survey, 1990-1994. Pakistan Institute of Population studies and Govt of Pakistan.
15. F. Jalil, B.S. Lindblad, L.Å. Hanson, S.R. Khan, R.N. Ashraf, B. Carlsson, S. Zaman, and J. Karlberg, Early child health in Lahore Pakistan: I. Study design, Acta Paediatr. (82):390 (suppl), 3-16 (1993).
16. R.N. Ashraf, F. Jalil, S.R. Khan, S. Zaman, J. Karlberg, B.S. Lindblad, and L.Å. Hanson, Early child health in Lahore, Pakistan.: V. Feeding patterns, Acta Paediatr. (82):390 (suppl), 48-62 (1993).
17. F.C. Barros, T.C. Semer, S.T. Filho, E. Tomasi, and C.G. Victora, The impact of lactation centers on breast feeding pattern, morbidity and growth: a birth cohort study, Acta Paediatr. 84:1221-1226 (1995).
18. T.H. Tulchinsky, S. El Ebweini, G.M.Y. Ginsberg, et al., Growth and nutrition patterns of infants associated with a nutrition education and supplementation program in Gaza, 1987-92, Bull WHO 72:(6), 869-875 (1994).
19. K.G. Dewey, M.J. Heinig, L.A. Nommsen, J.M. Peerson, and B. Lonnerda,. Breastfed infants are leaner than formula-fed infatns at 1 year of age: the DARLING study, Am. J. Clin. Nutr. 57(1):140-145 (1993).
20. L.H. Allen, Nutritional influences on linear growth: a general review, Eur. J. Clin. Nutr., 48:(suppl), S75-S89 (1994).
21. B.N. Tandon, Nutritional interventions through primary health care: impact of the ICDS projects in India, Bull WHO 67:(1), 77-80 (1989).
22. G.H. Beaton, and H. Ghassemi, Supplementary feeding programs for young children in developing countries, Amer. J. Clin. Nutr. 35:(4 Suppl), 863-916 (1982).

IMPACT OF VITAMIN B-12 DEFICIENCY DURING LACTATION ON MATERNAL AND INFANT HEALTH

Lindsay H. Allen[1]

1. Structure and Metabolism of Vitamin B-12

Vitamin B-12 is a "cobalamin", a term that describes structures with a corrin ring which has a central cobalt atom, phosphate, a base, and ribose. One of several groups can be attached to the cobalt: cyanide, forming vitamin B-12; 5'deoxyadenosine, forming 5' deoxyadenosylcobalamin; or a methyl group, forming methylcobalamin. Methylcobalamin is a cofactor for the conversion of homocysteine to methionine, and adenosylcobalamin is required for the conversion of methyl malonyl CoA to succinyl CoA. These cofactor functions explain why in vitamin B-12 deficiency there is elevated methylmalonic acid (MMA) in urine and serum, and plasma homocysteine (Hcy).

The digestion and absorption of vitamin B-12 from foods is relatively complex. The vitamin is released from its binding to proteins in food by gastric acid and pepsin. The free vitamin then binds to an R binder (haptocorrin) produced by the salivary glands and the stomach. In the alkaline environment of the small intestine, pancreatic proteases partially degrade the R binders, releasing vitamin B-12 that then binds to intrinsic factor produced in the stomach. The intrinsic factor-vitamin B-12 complex binds to receptors in the ileal mucosa and appears in serum bound to holotranscobalamin II (holoTC II), which delivers the vitamin to tissues. HoloTC II comprises approximately 20% of the total serum vitamin B-12 but is the biologically important component.

Because approximately half of the vitamin excreted in bile is reabsorbed on a daily basis, it has been estimated that liver stores of 1 mg will prevent deficiency symptoms from appearing for 3 years even if little is consumed in the diet. The liver of healthy younger adults contains about 2 to 3 mg of the vitamin.

[1]Department of Nutrition, University of California, Davis, California 95616, USA

Integrating Population Outcomes, Biological Mechanisms and Research Methods in the Study of Human Milk and Lactation
Edited by Davis *et al.*, Kluwer Academic/Plenum Publishers, 2002

2. Symptoms and Causes of Vitamin B-12 Deficiency

Indicators of B-12 deficiency, in descending order of sensitivity, include: elevated MMA in plasma and urine, elevated plasma Hcy, low serum vitamin B-12 (<200 ug/mL indicating severe deficiency and 200-300 ug/mL, marginal deficiency); low serum holotranscobalamin II (holoTC II), and in relatively severe deficiency, megaloblastic anemia and macrocytosis. Abnormal mental and motor function may occur through mechanisms that are not well understood. They may involve defective methylation of myelin, elevated concentrations of MMA and S-adenosyl methionine, higher rates of incorporation of odd chain fatty acids into myelin, and/or high concentrations of TNFα in spinal fluid.

There are two main causes of vitamin B-12 deficiency. One is malabsorption of the vitamin, which is commonly encountered in elderly individuals with gastric atrophy or in diseases where the ileum is damaged. Malabsorption can also occur as a result of pernicious anemia, an autoimmune disorder that affects approximately 2 - 3% of adults. In this condition, there is a lack of intrinsic factor. The other risk factor for deficiency is a low dietary intake of the vitamin, which will occur if animal product consumption is low. In the case of infants, deficiency can result from consuming breast milk from vitamin B-12 deficient mothers that contains a low concentration of the vitamin. This situation occurs in strict vegetarians and in many women and infants in developing countries. When a sub optimal dietary pattern is the cause of vitamin B-12 deficiency it is, however, likely that the intake of many other nutrients – including iron, zinc, and vitamin A - is also inadequate. Thus apparent associations between vitamin B-12 deficiency and poor functional outcomes could be explained by simultaneous deficiencies of other nutrients. Nevertheless, surprisingly similar symptoms are seen in infants fed by mothers with pernicious anemia and mothers who are vegans. This may be due to vitamin B-12 being the nutrient most lacking in vegan diets, and the vulnerability of the infant to B-12 deficiency.

3. Recommended Intakes of Vitamin B-12

The recommended intakes of vitamin B-12 for lactating women are based on an estimated average milk concentration of 0.42 ug/L, amounting to a daily secretion of 0.33 ug/d during the first 6 months of lactation and 0.25 ug/d during the second six months.[1] By adding an allowance that is sufficient to replace this B-12 secreted in milk to the RDA for non-lactating women, the RDA during lactation is 2.8 ug/d. The recommended intake (Adequate Intake) for infants is 0.4 ug/d during the first six months of life, and 0.5 ug/d during the second six months, values that are based on the estimated intake of vitamin B-12 from breast milk by healthy infants.[1]

4. Vitamin B-12 Transfer to the Fetus

Maternal vitamin B-12 absorption may become more efficient during pregnancy. HoloTC II is taken up by receptors on he maternal placenta and vitamin B-12 released into the fetal circulation.[2] In the fetus the free vitamin binds to transcobalamin II produced by fetal liver. The adequacy of current maternal intake and absorption of vitamin B-12 may have a stronger influence on the transfer of the vitamin to the fetus than does the amount in maternal stores.[3] The fetus accumulates 0.1 to 0.2 ug/d, amounting to 25 to 30 ug of stores in a well-nourished infant at birth. Assuming that infant requirements are 0.4 ug/d, normal stores at birth should last for approximately 3 months even if there is no vitamin B-12 in their diet. When the mother's intake or absorption of the vitamin during pregnancy is poor, however, infant stores at birth can be very low.[1]

5. Vitamin B-12 during Lactation

Vitamin B-12 is taken up by the mammary gland from holoTC II in maternal serum. In the mammary gland the vitamin is released from transcobalamin II and combines with haptocorrin synthesized by the mammary tissue.[4] Almost all of the vitamin B-12 in human milk is present as haptocorrin, which is stable to the proteolytic enzymes of the gastrointestinal tract and may be absorbed intact by human infants.[5]

Published values for the concentration of vitamin B-12 in breast milk vary considerably even among studies in the same country. Average reported values range from about 0.2 to 1.0 ug/L. Part of the reason for this variability is due to differences among analytical methods. Prior to analysis the vitamin B-12 must be released from its tight binding to haptocorrin in milk by a digestive enzyme such as papain, or by heat treatment. There is no obvious diurnal or within-feed variability in milk vitamin B-12 concentration, although values fluctuate considerably within a day in the same individual.[6] The vitamin B-12 content of colostrum appears to be slightly higher than that of mature milk.[6] While several investigators report that maternal serum and milk vitamin B-12 concentrations are strongly correlated,[7,8] in other studies the association was weak or non-existent.[9]

6. Effects of Maternal Deficiency on Infant Vitamin B-12 Status and Function

Table 1 summarizes information from reports on the age of onset and diagnosis of symptoms of vitamin B-12 deficiency in breastfed infants. Some were born to mothers with pernicious anemia, and others to strict vegetarians. Table 2 summarizes the symptoms of vitamin B-12 deficiency observed in these studies.

6.1. Maternal Pernicious Anemia

Maternal pernicious anemia may first be detected in the early postpartum as a result of the infant failing to develop and thrive normally Vitamin B-12 depletion of the infant in this condition is the combined result of poor maternal absorption and subsequently low infant stores at birth, and the lower concentrations of the vitamin in breast milk. Because not all of the investigators measured or reported all of the outcomes, the data in Table 1 cannot be used to assess the prevalence of each symptom. Also, diagnosed cases of deficiency would be expected to be the most severe. Symptoms of deficiency were first observed when the infants reached between 4 and 14 months of age. Both maternal and infant serum concentrations of vitamin B-12 were very low.

Megaloblastic anemia was reported in most cases, usually accompanied by neutropenia. Other clinical symptoms include lethargy, poor appetite, vomiting, delayed developmental milestones, jerkiness, poor reflexes and abnormal cardiac function.

Table 1. Case Studies Reporting Age at Onset of Symptoms of Vitamin B-12 Deficiency

Condition	n	Age at onset of symptoms (months)	Age at Diagnosis (months)	Serum B-12 (ug/mL)	First author, journal
Maternal pernicious anemia					
PA	1	2	4	- - -	Zetterstrom, Acta Paediatr. 43:379, 1954
Subclinical PA	1	4	4	- - -	Lampkin, N. Engl. J. Med. 274:1169, 1966
PA	1	3	3.5	38	Heaton, N. Engl. J. Med. 300:202, 1979
PA	1	12.5	14	46	Hoey, J. Roy. Soc. Med. 75:656, 1982
PA+hypothyroid	1	4 .5	5	23	Johnson, J. Pediatr. 100:917, 1982
PA	1	6	11		

< 50

Sadowitz, Clin. Pediatr. 25: 369, 1986

PA

1

5

5

< 50

McPhee, Arch. Dis. Child. 63:921, 1988

PA

2

4-6

7-10

28-41

Emery, Pediatrics 99:255, 1997

Subclinical PA

2

6

8-10

61-102

Danielsson, Acta Paediatr. Scand. 77:310, 1988

Maternal vegetarian diet

Lacto-ovo

1

9

10

90

Lampkin, J. Pediatr. 75:1053, 1969

Vegan

1

4

6

20

Higginbottom, N. Engl. J. Med. 299, 1978

Vegan

1

3

9

126

Wighton, Med. J. Aust. 2:1, 1979

Vegan

1

6

6

ND

Davis, Am. J. Dis. Child. 135:566, 1981

Vegan

1

NR

12

NR

Close, Br. Med. J. 286:473, 1983

Vegan

1

8

11

102

Gambon, Eur. J. Pediatr, 145:570, 1986

Vegan

1

5

7

< 100

Sklar, Clin. Pediatr. 25:219, 1986

Vegan

1

6

14

63

Stollhoff, Eur. J. Pediatr. 146:201, 1987

Vegan

1

6

9

28

Kuhne, Eur. J. Pediatr. 150:205, 1991

Vegan

1

12

14

ND

von Schenck, Arch. Dis. Child. 77:137, 1997

NR = not reported, ND = not detectable

Table 2. Case Studies Reporting Symptoms of Vitamin B-12 Deficiency in Infants

	Maternal Condition	
	Pernicious Anemia (n=9)	Vegetarian (n=10)
Anemia	5	10
Lethargy	3	8
Poor appetite	3	1
Vomiting	3	3
Late milestones	2	6
Glossitis/abnormal tongue	2	2
Jerkiness	2	2
Poor reflexes	1	6
Cardiac problem		2

The usual treatment for such infants is to provide 1 mg vitamin B-12 daily by intramuscular injection for 4 days, followed by large oral doses to replete stores. Within two days the infants are reported to be more alert and active, and to have a better appetite. Reticulocytosis is also increased rapidly. Involuntary movements disappear within 2 to 4 weeks. Some infants seem to recover completely but the majority are still developmentally delayed (by about 3 months) at age 12 months. This range of response is likely to be due to variability in the severity and duration of the deficiency but there are insufficient data to answer this question. Relatively few studies reported following up the treated infants, and especially in earlier reports the assessment was relatively superficial or even non-existent. Developmental delays have been reported to persist for years in some case reports.

6.2. Maternal Vegetarianism

Two studies report the impact of strict vegetarianism on the vitamin B-12 status of women and their infants. One of these was conducted in Boston, U.S.A.[10] The participants included 64 women, and 42 children aged 1.5 to 11.7 years (median age 3.9 years). The infants had been breastfed, and weaned onto a macrobiotic diet. Their parents had been macrobiotic for from 1 month to 30 years (median 3.5 years). Low serum vitamin 12 concentrations (<220 ug/mL) were highly prevalent in the adults, affecting about one third of those who had been vegetarian for about one year and virtually all of those who had adhered to the diet for > 5 years. Serum vitamin B-12 concentrations were inversely correlated with urinary methylmalonic acid concentrations, as would be expected. Over half of the children also had high urinary MMA, one third were stunted, and there appeared to an association between elevated MMA and stunting.

The investigators also evaluated vitamin B-12 concentrations in serum and milk of 17 of the vegetarian mothers and 6 omnivorous controls, as well as the urinary MMA of their infants.[7] Over half of the vegan mothers had a serum vitamin B-12 level indicating deficiency, and their infants had higher urinary MMA. Breast milk vitamin B-12 concentrations averaged about 0.3 ug/L in the vegan mothers and 0.5 ug/L in the omnivores. Levels of the vitamin in breast milk were strongly inversely related to the length of time that women had consumed a macrobiotic diet. In this relatively small sample, women who had been strict vegetarians for over 4 to 5 years clearly had lower breast milk content of the vitamin. This is a shorter length of time than is usually assumed to be needed for vitamin B-12 depletion to occur, and suggests that maternal and infant vitamin B-12 deficiency may occur quite rapidly if animal product intake is low.

The concentrations of the vitamin in breast milk were strongly correlated with maternal serum levels ($r = 0.79$, $P < 0.001$) and inversely correlated with maternal urinary MMA ($r = -0.830$, $P < 0.001$). In addition, infant urinary MMA started to increase below a breast milk vitamin B-12 concentration of approx. 362 pmol/L (0.5 ug/L). This is one of the few pieces of information that can be used for defining a cut-point for an inadequate concentration of vitamin B-12 in breast milk.

The other study of a group of macrobiotic infants occurred in The Netherlands.[11] This longitudinal study included a birth cohort of 53 infants whose mothers were macrobiotic, and a matched control group of 57 omnivorous infants. Virtually all were breastfed, to 14 months on average in the macrobiotic group and to 7 months of age in the controls. Complementary feeding started at about 4.8 months but complementary foods plus breast milk supplied only 0.3 ug/d of vitamin B-12 to macrobiotic infants compared to 2.9 ug/d in the omnivorous controls. Breast milk concentrations were only about 0.35 ug/L in the macrobiotics compared to 0.44 ug/L in the omnivores.

There was a high prevalence of macrocytic anemia in the macrobiotic infants. Their weight and length gain was slow, and their rate of muscle mass accretion was half that of the omnivorous controls. There was major wasting of skin and muscle in 30% of the macrobiotic group and their gross motor and language development were slower.

Serum vitamin B-12, homocysteine and MMA were later measured in 41 of the macrobiotic infants aged 10 to 20 months, and compared to values in 50 matched omnivorous controls.[12] Both homocysteine and MMA were markedly increased in the macrobiotic infants (8-fold and 2-fold, respectively), and both were inversely related to serum B-12 concentrations as expected. MMA was the strongest predictor of dietary group, followed by plasma homocysteine and serum vitamin B-12. MMA predicted dietary group with 85% sensitivity and 83% specificity, while MMA and Hcy combined increased the sensitivity of the prediction to 98%.

6.4. Long-term Consequences of Vitamin B-12 Deficiency in Early Childhood

Apart from a few case studies there are relatively few reports of the effects of vitamin B-12 repletion on the recovery of deficient infants. In cases where their strictly vegetarian parents were advised to start feeding animal products, these would have supplied other nutrients in addition to vitamin B-12. Reviews of small groups of formerly deficient infants suggest that developmental delays persist for many years.[13, 14]

The largest longitudinal study of the function of formerly vitamin B-12 deficient infants was a follow-up of the macrobiotic children in The Netherlands when they became adolescents (mean age approximately 12 years).[15, 16] The parents had been advised to include some animal products in the diets of these children animal products after the adverse effects of the macrobiotic diet became known. The introduction of these foods started at about age 6 years on average. Surprisingly the serum vitamin B-12 concentrations of the previously macrobiotic adolescents were only half those of the omnivores, and their mean red cell volumes and serum MMA were significantly higher. Importantly, on most psychological tests the adolescent formerly-macrobiotic subjects scored lower, and there was a significant association between serum B-12 and fluid intelligence. It cannot be claimed with certainty that the persistently poor vitamin B-12 status of these adolescents was caused by low intakes and stores of the vitamin during infancy, because their vitamin B-12 intakes were still only half those of omnivorous teenagers. Presumably this situation resulted from a lower consumption of animal products in their households. Nevertheless it is striking that these differences in vitamin B-12 status persisted in spite of their current consumption of some meat and milk.

7. Vitamin B-12 Deficiency in Developing Countries

Because vitamin B-12 is found only in animal products, intake is highly dependent on the consumption of these foods. For example, the vitamin B-12 intake of vegans in England was 0.7 ug/d (geometric mean, 0.3 ug/d), compared to 2 to 3 ug/d for lacto-ovo vegetarians. The average intake by young adult women in the United States is 4.8 ug/d.[1]

Evidence is accumulating that the prevalence of vitamin B-12 deficiency in developing countries is much higher than was previously suspected. In several studies in Mexico and Guatemala we have found consistently that approximately one third of preschoolers, schoolers and adults studied have deficient or marginal serum concentrations of vitamin B-12.[8, 9, 17] A recent study from Nepal reported about half of pregnant women to have a serum B-12 concentration <200 pg/mL.[18] In rural Kenya we find concentrations < 300 ug/L in approximately half of the schoolers (unpublished data).

In a subgroup of 16 Mexican women studied longitudinally from prior to conception through 8 months of lactation, serum concentrations fell during pregnancy as expected, but had returned to pre-pregnancy values at 8 months postpartum.[8] Based on a cut-off concentration of 0.470 ug/L, the breast milk concentration of the vitamin was low in 31/50 samples. In a poor suburb of Guatemala City, 113 women and their infants were recruited at 3 months of lactation with no exclusion criteria except for poor health.[9] There was a high prevalence (47%) of low or marginal vitamin B-12 concentrations in maternal serum, and low concentrations were found in 31% of breast milk samples. There was a weak but significant inverse association between urinary MMA and breast milk concentrations of the vitamin.

The cause of this high prevalence of vitamin B-12 deficiency in developing countries could be low intake of the vitamin, and/or malabsorption that would exacerbate the impact of inadequate intakes. Factors that could cause malabsorption of the vitamin in developing countries include infection with Helicobacter pylori and atrophic gastritis, bacterial overgrowth or infection with Giardia lamblia. Further research is needed to confirm the relative importance of these potential risk factors for vitamin B-12 deficiency.

Conclusions

Maternal vitamin B-12 deficiency occurs as a result of a low intake of the vitamin (when animal product consumption is low) or its malabsorption (in pernicious anemia, and possibly as a result of intestinal infections especially in developing countries). The prevalence of deficiency has been reported to be high in some developing countries, and there is a strong risk of deficiency symptoms developing in infants born to mothers who consumed a strict vegetarian diet, even if this occurred for a relatively short period prior to conception, and throughout the pregnancy. Most reports show that maternal vitamin B-12 deficiency reduces the amount of the vitamin in breast milk, which can result in anemia, failure to thrive, major developmental delays and other clinical symptoms in the infant. These first appear at around 4 to 6 months after birth, probably sooner in the case of infants born to women with pernicious anemia because their stores at birth will also be low. Treatment with the vitamin apparently results in full recovery in some cases but not others. Because of the high prevalence of vitamin B-12 deficiency in some countries it is urgent that the causes and consequences of this problem receive more attention.

References

1. Institute of Medicine. *Dietary Reference Intakes for Thiamin, Riboflavin, Niacin, Vitamin B-6, Folate, Vitamin B-12, Pantothenic Acid, Biotin and Choline* (National Academy Press, Washington D.C., 1998).
2. L.H. Allen. Vitamin B-12 metabolism and status during pregnancy, lactation and infancy, in: *Nutrient Regulation during Pregnancy, Lactation and Infant Growth,* edited by L.Allen, J.King and B. Lonnerdal, pp. 173-186, (Plenum Press, New York, 1994).
3. A.L. Luhby, J.M. Cooperman, A.M. Donnenfeld, J.M. Herrero, D.N. Teller, and J.B. Wenig. Observations on transfer of vitamin B-12 from other to fetus and newborn, Am. J. Dis. Child. 96:532-533.
4. Y. Adkins and B. Lonnerdal. High affinity binding of the transcobalamin II-cobalamin complex and expression of haptocorrin by human mammary epithelial cells, Biochim. Biophys. Acta (submitted).
5. Y. Adkins. The Role of Haptocorrin in Vitamin B-12 Nutrition in Infancy, PhD Thesis, Department of Nutrition, University of California, Davis, 2000.
6. N. Trugo and F. Sardinha. Cobalamin and cobalamin-binding capacity in human milk, Nutr. Res. 14:22-33, 1994.
7. B.L. Specker, A. Black, L. Allen, and F. Morrow. Vitamin B-12: low milk concentrations are related to low serum concentrations in vegetarian women and to methylmalonic aciduria in their infants, Am. J. Clin. Nutr. 52:1073-1076, 1990.
8. A.K. Black, L.H. Allen, G.H. Pelto, M.P. de Mata, and A. Chavez. Iron, vitamin B-12 and folate status in Mexico: associated factors in men and women and during pregnancy and lactation, Am. J. Clin. Nutr. 124:1179-1188, 1994.
9. J.E. Casterline, L.H. Allen, and M.T. Ruel. Vitamin B-12 deficiency is very prevalent in lactating Guatemalan women and their infants at three months postpartum, *J. Nutr.* 127:1966-1972, 1997.
10. D.R. Miller, B.L. Specker, M.L. Hop, and E.J. Norman. Vitamin B-12 status in a macrobiotic community, Am. J. Clin. Nutr. 53:524-529 (1991).
11. S.M. Graham, O.M. Arvela, and G.A. Wise. Long-term neurologic consequences of nutritional vitamin B-12 deficiency in infants, J. Pediatr. 121:710-714.
12. U. von Schenck, C. Bender-Gotze, and B. Koletzko. Persistence of neurological damage induced by dietary vitamin B-12 deficiency in infancy, Arch. Dis. Child. 77:139, 1997.
13. P.C. Dagnelie, W.A. F.J.V.R.A. Vergote, J. Burema, M.A. van't Hof, J.D. van Klaveren, and J.G.A.J. Hautvast. Nutritional status of infants aged 4 to 18 months on macrobiotic diets and matched omnivorous control infants: A population-based mixed-longitudinal study, Eur. J. Clin. Nutr. 43:325-338, 1989.
14. J. Schneede, P.C. Dagnelie, W.A. van Staveren, S.E. Vollset, H. Refsum, and P.M. Ueland. Methylmalonic acid and homocysteine in plasma as indicators of functional cobalamin deficiency in infants on macrobiotic diets, Pediatr. Res. 36:194-201, 1994.
15. M. Van Dusseldorp, J. Schneede, H. Refsum, P.M. Ueland, C.M.G. Thomas, and E. de Boer. Risk of persistent cobalamin deficiency in adolescents fed a macrobiotic diet in early life, Am. J. Clin. Nutr. 69:664-671, 1999.

16. M.W.J. Louwman, M. van Dusseldorp, F.J.R. van de Vijver, C.M.G. Thomas, J. Schneede, P.M. Ueland, H. Refsum, and W.A. van Staveren. Signs of impaired cognitive function in adolescents with marginal cobalamin status, Am. J. Clin. Nutr. 72:762-769, 2000.
17. L.H. Allen, J.L. Rosado, J.E. Casterline, H. Martinez, P. Lopez, E. Munoz, and A.K. Black. Vitamin B-12 deficiency and malabsorption are highly prevalent in rural Mexican communities, Am. J. Clin. Nutr. 62:1013-1019, 1995.
18. G.T. Bondevik, B. Eskeland, R.J. Ulvik, M. Ulstein, R.T. Lie, J. Schneede, and G. Kvale. Anemia in pregnancy: possible causes and risk factors in Nepali women, Eur. J. Clin. Nutr. 54:3-8, 2000.

ZINC AND BREASTFED INFANTS: IF AND WHEN IS THERE A RISK OF DEFICIENCY?

Nancy F. Krebs and Jamie Westcott

1. INTRODUCTION

Recent results of randomized controlled zinc supplementation trials in young children have confirmed that globally, zinc deficiency is a major public health problem contributing to significant morbidity and mortality from infectious diseases.[1] On the basis of a pooled analysis of the zinc supplementation trials, it is estimated that assuring adequate zinc status may have preventative effects on diarrhea exceeding the effects of clean water and sanitation, as well as those of promotion of breastfeeding. Similarly, the estimated preventive effect of zinc supplementation on pneumonia is similar to that estimated for breastfeeding. Thus there seems little question that zinc deficiency occurs in young infants, including those who are breastfed, but the circumstances are not yet well characterized.

This paper will consider risk of zinc deficiency for breastfed infants, with attention to the conditions and circumstances that affect such risk. Because of the very dynamic nature of zinc in human milk, a brief review of longitudinal changes in zinc concentrations in human milk and zinc intake of the breastfed infants sets the stage for such a discussion. Similarly, a brief review of the major processes of zinc homeostasis also is appropriate.

2. Review of Zinc in Human Milk and Zinc Homeostasis

Milk zinc concentrations decline sharply from approximately 4 μg/ml at 2 wk post-partum to approximately 1.5 μg/ml at 3 mo, and decline more slowly thereafter. By 6 mo, average milk zinc concentration is 1.0 μg/ml.[2] Although volume of milk intake increases in the early weeks post-partum, this does not supersede the decline in

*Nancy F. Krebs, University of Colorado, School of Medicine, 4200 E. 9th Avenue/Box C225, Denver, CO 80262

Integrating Population Outcomes, Biological Mechanisms and Research Methods in the Study of Human Milk and Lactation
Edited by Davis *et al.*, Kluwer Academic/Plenum Publishers, 2002

concentrations. Thus, overall zinc intake from human milk also declines from birth through 7 months for the exclusively breastfed infant.[3]

Zinc is a component of hundreds of enzymes, and it functions in regulation of gene transcription and expression. The zinc finger proteins constitute one of the largest classes of transcription factors, and as such impact synthesis of a wide range of compounds. Zinc also has a role in stabilization of membranes. These fundamental biological roles translate to critical functions in growth, immune function, and neurologic development and maturation.[4] For infants and young children, because of their relatively high zinc requirement, even a mild deficiency can have significant adverse effects on growth, immune function, and development and activity.

Zinc homeostasis is maintained primarily via the gastrointestinal tract through the processes of absorption of dietary zinc and through endogenous secretion of zinc into the intestinal lumen, some of which is reabsorbed, and some of which is excreted.[5] Although zinc is also excreted through the kidneys, this route is a relatively minor factor in conservation of zinc, except in situations of moderate to severe zinc restriction. In studies of zinc homeostasis in infants, because of the difficulties of collecting complete and uncontaminated urine specimens, zinc excretion from this route is generally not included. The term 'net absorption of zinc' thus is used to refer to absorbed zinc (dietary zinc x fractional absorption) minus the endogenous zinc in feces. In the growing infant, net absorption is ideally sufficiently positive to account for unmeasured losses in urine, skin and sweat, and to allow zinc retention for accretion of lean tissue.

3. Young Term Infants (Birth to Approximately 6 Mo)

The favorable bioavailability of the zinc in human milk has long been appreciated. For example, prior to the recognition that the inherited condition of acrodermatitis enteropathica was due to a defect in zinc transport, the therapeutic benefit of human milk was exploited. In studies employing stable zinc isotopes in exclusively breastfed and formula fed infants, we have found the absorption efficiency of zinc in human milk to average 0.55,[6] slightly more than two fold that from standard cow milk based formula.[7] These differences are likely due to several compositional features of human milk, including the modest concentrations of zinc and other minerals, the protein matrix, and the presence of low molecular weight compounds.[8]

In addition to the efficient absorption of zinc from human milk, the breastfed infant also excretes modest amounts of endogenous zinc in the feces. Under normal circumstances, the endogenous zinc in feces is positively correlated to the amount of absorbed zinc (Figure 1). This relationship has been observed both in infants and in adults.[6, 7, 9]

The combination of efficient absorption and conservation of endogenous zinc results in positive net absorption of zinc of approximately 0.3 mg/d in 2-5 month old breastfed infants.[6] Compared to estimated needs for zinc retention in relation to growth rates, this amount seems marginally adequate, particularly for the 4-5 mo old infant.[10, 11]

Is there evidence of zinc deficiency in healthy breastfed infants under 6 mo of age? Data are limited, but studies which have longitudinally examined zinc status in breastfed infants have found serum/plasma zinc levels to be within normal limits at this age.[12, 13, 14] In our longitudinal study of 71 breastfed infants, the mean plasma zinc was

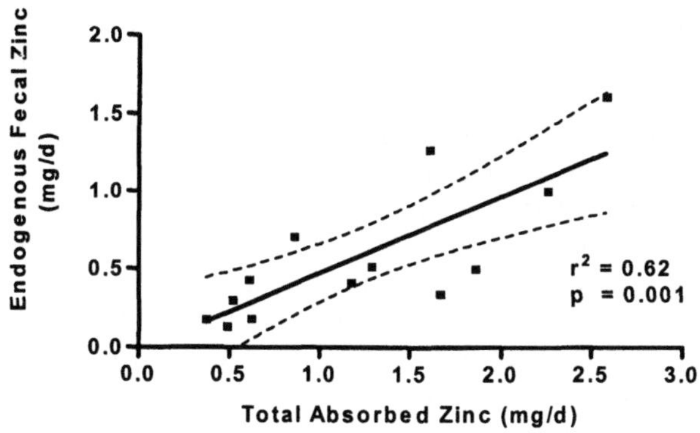

Figure 1. Relationship between amount of absorbed zinc and fecal excretion of endogenous zinc in 3-5 mo old breastfed (n=7) and formula fed infants (n=6).

also normal (78 µg/dl at 6 months.[3] Two studies have found an inverse relationship between growth and plasma zinc levels, however, suggesting a relative depletion of metabolically active zinc in those who were growing more rapidly.[3, 14] Breastfed infants in India were found to have higher plasma and leukocyte zinc levels in the early months of life compared to formula fed infants, but levels were lowest at 4 to 6 months and improved to normal levels by nine months, following weaning.[15] Although plasma zinc is not an optimally sensitive index of zinc status, mean plasma zinc level has been found to be predictive of a growth response to zinc supplementation in research studies. Thus the findings of low plasma zinc concentrations are likely to be associated with suboptimal zinc status.[16]

The gold standard for assessment of zinc status continues to be the randomized controlled supplementation trial. On this basis, there is mixed evidence of benefit from a zinc supplement before 6 mo of age. Two supplementation trials initiated a zinc supplement in exclusively breastfed infants at 4 mo and followed infants' growth for 3-6 mo after. In one, there was no effect on growth (weight gain or linear growth), morbidity, or development,[17] whereas in the other the females showed a modest but significant growth response and there was biochemical evidence of a benefit of the zinc supplement.[18] In another randomized supplementation trial, breastfed infants over a wider age range but including 4 mo olds, were supplemented with zinc and showed significantly greater weight gain and linear growth.[19]

Low birth weight infants receiving human milk exclusively may be particularly vulnerable to zinc deficiency. This is likely due to lower body stores of zinc, relatively higher requirements for growth, and possibly less efficient conservation of endogenous zinc. Present evidence does not suggest that pre-term human milk is higher in zinc content, so the physiologic decline in milk zinc concentrations may predispose to inadequate intake in the early weeks post-partum in the face of relatively high requirements.[11]

In combination, these data suggest that for breastfed infants approximately 6 mo of age, zinc may become growth limiting. Depending on milk zinc levels, milk intake, and growth rate, healthy infants are likely to need an additional source of zinc beyond that in human milk approximately mid-way through the first year of life. Furthermore, the impacts of non-exclusive breastfeeding and the use of non-milk liquids, early introduction of complementary foods, and/or single nutrient supplements on zinc bioavailability have not been well characterized.

It is also important to recognize that infants in less protected environments, especially those with a high infectious burden, may have a relatively increased zinc requirement. This is in part due to the potential for impaired fractional absorption and for excessive endogenous losses via the gastrointestinal tract, but may also be associated with relatively increased requirements for repeated immune system responses to acute infections, and with reduced intake during infections. Requirements may also then be relatively increased to support periods of catch-up growth.

4. The Older Infant (> 6 Mo)

By 7 mo of age, the intake of zinc from human milk averages 0.5 mg/d; typical intake from complementary foods at this age in infants from the U.S. is an additional 0.5 mg/d.[20,3] Calculations by Gibson et al.[21] demonstrate that meeting zinc and iron needs of the older infant is particularly dependent on complementary foods. The zinc content of early complementary foods in the U.S. is quite low, e.g. iron-fortified cereals, pureed fruits and vegetables. Poultry, meat, and vegetable mixtures are often not introduced until 8-9 mo or later, and these can vary substantially in their zinc content. In developing countries, legume and cereal mixtures are often the primary complementary food staples. Such choices are not only relatively low in zinc but often have quite high phytate-to-zinc ratios, which is likely to impair the bioavailability of zinc, as well as of iron and calcium.[21]

The effect of meat as a complementary food in older infants has been shown to enhance nonheme iron absorption and possibly to benefit iron status.[22, 23] We have investigated the effect of using meat or cereal as an early complementary food for breastfed infants. Exclusively breastfed infants were enrolled at 4 mo and randomized to receive either iron fortified rice cereal or pureed beef as their first complementary food at 5-6 mo of age. These infants were followed through 12 mo of age to assess effects of the food assignment on growth, biochemical indices, and cognitive development. Zinc and protein intakes were significantly higher in the meat group at 7 mo, whereas iron intake was significantly lower. Thereafter, nutrient intakes were similar between groups. There were modest differences in growth and no differences in biochemical iron and zinc status or development between groups.[24]

Zinc absorption was studied in a subgroup of infants from each group by application of stable isotope methods. The zinc fractional absorption was slightly higher from the meat, but because of the much higher zinc content of the meat test meal (0.4 vs

0.03 mg), there was a highly significant difference in absorbed zinc from the meat compared to the cereal test meal. We also examined the effect of the addition of human milk to the complementary food. For both the meat and the cereal groups, fractional absorption was lower on the second day when human milk was added to the complementary food. The size of the exchangeable zinc pools (EZP), which reflect metabolically active zinc, was also measured by stable isotope techniques in this subgroup of infants. For both groups, there was a significant positive correlation between zinc intake from complementary foods and the EZP. Our interpretation of these results is that increasing zinc intake in the older breastfed infant will have a positive effect on metabolically available zinc.[25]

Is there evidence that zinc deficiency occurs in the older breastfed infant? In addition to the theoretical risk as calculated from intake vs requirements, studies of zinc status in older breastfed infants also suggest they may become at risk for zinc deficiency. Michaelson, et al.[13] examined zinc status longitudinally in healthy Danish infants and found the lowest levels at 9 mo of age, but this was not related to feeding status, i.e., being partially breastfed or weaned. In this study, serum zinc was positively associated with growth velocity. In Swedish infants followed from birth to 12 mo, as many as one-third were estimated to be zinc depleted on the basis of serum zinc concentrations.[26] As noted above, recent zinc supplementation trials have demonstrated a significant positive effect on morbidity and mortality from diarrhea and pneumonia in developing countries where infants would be expected to be at least partially breastfed. Unfortunately detailed dietary data were generally not reported for these populations so specific information on extent of breastfeeding is not available.

A recent supplementation trial in Ethiopia, however, found a benefit of zinc supplementation for breastfed infants from 6-12 mo of age.[27] In this study, stunted and non-stunted breastfed infants were pair matched for age, sex and anthropometry, and were then randomly and blindly assigned to receive 10 mg/d of zinc for 6 months. All infants were from extremely impoverished living conditions. Positive effects were found on growth (weight gain and linear growth), morbidity, and possibly on appetite. The effects were greatest in the zinc supplemented stunted infants.[27] This is the strongest evidence to date that zinc deficiency occurs in breastfed infants in the second half of the first year of life, at least in the setting of a developing country. Reasons for the differences between the results of this trial and of those undertaken in the U.S. include the timing of the intervention, the presumed lower bioavailability of the complementary foods for the Ethiopian group, as well as the higher infectious burden in the latter group.

5. Summary and Conclusions

Infancy is a time of relatively high zinc requirements. Human milk provides an excellent source of highly bioavailable zinc and generally meets the needs of the healthy young exclusively breastfed infants for the first several months of life. Investigations of exclusively breastfed infants less than 6 mo of age have generally found zinc homeostasis and status to be adequate, although there are indications that zinc intake from human milk alone may become limiting by around 6 mo of age. Exceptions may be small for gestational age and low birth weight infants, who may well benefit from increased zinc intake before 6 mo of age. The older infant clearly becomes dependent on non-human milk sources of zinc, i.e., from complementary foods. Traditional early complementary foods, such as cereals, fruits, and vegetables provide very modest amounts of zinc, and for those high in phytic acid, bioavailability may be low. Introduction of animal products or zinc supplementation may be important to meet the older infant's zinc requirements. This is likely to be particularly important in less protected environments with a high infectious burden and limited dietary options.

References

1. Z.A Bhutta, R.E. Black, and K.H. Brown, et al., Prevention of diarrhea and pneumonia by zinc supplementation in children in developing countries: pooled analysis of randomized controlled trials. Zinc Investigators' Collaborative Group, J Pediatr. 135:689-697 (1999).
2. N. F. Krebs, C.J. Reidinger, S. Hartley, A. D. Robertson, and K. M. Hambidge, Zinc supplementation during lactation: effects on maternal status and milk zinc concentrations, Am. J. Clin. Nutr. 61:1030-1036 (1995).
3. N. F. Krebs, C. J. Reidinger, A. D. Robertson, and K. M. Hambidge, Growth and intakes of energy and zinc in infants fed human milk, J. Pediatr. 124:32-39 (1994).
4. R. J. Cousins. A role of zinc in the regulation of gene expression, Proc. Nutr. Soc. 57:307-11 (1998).
5. J. C. King, D. M. Shames, and L. R. Woodhouse, Zinc homeostasis in humans, J. Nutr. 130:1360S-1366S (2000).
6. N. F. Krebs, C. J. Reidinger, L. V. Miller, and K. M. Hambidge, Zinc homeostasis in breast-fed infants, Pediatr. Res. 39:661-665 (1996).
7. N. F. Krebs, C. J. Reidinger, L. V. Miller, and M. W. Borschel, Zinc homeostasis in healthy infants fed a casein hydrolysate formula [see comments], J. Pediatr. Gastroenterol. Nutr. 30:29-33 (2000).
8. B. Lonnerdal, Dietary factors influencing zinc absorption, J. Nutr. 130:378S-383S (2000).
9. L. Sian, X. Mingyan, L. V. Miller, L. Tong, N. F. Krebs, and K. M. Hambidge, Zinc absorption and intestinal losses of endogenous zinc in young Chinese women with marginal zinc intakes, *Am. J. Clin.* Nutr. 63:348-353 (1996).
10. N. F. Krebs, and K. M., Hambidge, Zinc requirements and zinc intakes of breast-fed infants, Am. J. Clin. Nutr. 43:288-292 (1986).
11. N. F. Krebs, Zinc transfer to the breastfed infant, J. Mammary Gland. Biol. Neoplasia 4:259-268 (1999).
12. K. M. Hambidge, P. A. Walravens, C. E. Casey, R. M. Brown, and C. Bender, Plasma zinc concentrations of breast-fed infants, J. Pediatr. 94:07-08 (1979).
13. N. F. Michaelsen, G. Samuelson, T. W. Graham, and B. Lonnerdal, Zinc intake, zinc status and growth in a longitudinal study of healthy Danish infants, Acta Paediatr. 83:1115-1121 (1994).
14. L. Salmenpera, J. Perheentupa, V. Nanto, and M. A. Siimes, Low zinc intake during exclusive breast-feeding does not impair growth, J. Pediatr. Gastroenterol. Nutr. 18:361-370 (1994).
15. P. Hemalatha, P. Bhaskaram, P. A. Kumar, M. M. Khan, and M. A. Islam, Zinc status of breastfed and formula-fed infants of different gestational ages, J. Trop. Pediatr. 43:52-54 (1997).
16. K. H. Brown, Effect of infections on plasma zinc concentration and implications for zinc status assessment in low-income countries, Am. J. Clin. Nutr. 68:425S-429S (1998).
17. M. J. Heining, K. H. Brown, B. Lonnerdal, and K. G. Dewey, Zinc supplementation does not affect growth, morbidity, or motor development of U.S. breastfed infants at 4-10 months, FASEB J. 12:A970 (1998).
18. N. F. Krebs, J. E. Westcott, and N. Butler-Simon, Effect of zinc supplement on growth of normal breastfed infants, FASEB J. 10:A230 (1996).
19. P. A. Walravens, A. Chakar, R. Mokni, J. Denise, and D. Lemonnier, Zinc supplements in breastfed infants [see comments], Lancet 340:683-685 (1992).
20. S. A .Abrams, K.O. O'Brien, J. Wen, L. K. Liang, and J. E. Stuff, Absorption by 1-year-old children of an iron supplement given with cow's milk or juice, Pediatr. Res. 41:171-175 (1997).
21. R. S. Gibson, E. L. Ferguson, and J. Lehrfeld, Complementary foods for infant feeding in developing countries: their nutrient adequacy and improvement, Eur. J. Clin. Nutr. 52:764-770(1998).

22. M. D. Engelmann, B. Sandstrom, and K. F. Michaelsen, Meat intake and iron status in late infancy: an intervention study, J. Pediatr. Gastroenterol. Nutr. 26:26-33 (1998).
23. M. D. Engelmann, B. Sandstrom, T. Walczyk, R. F. Hurrell, and K. F. Michaelsen KF., The influence ofmeat on nonheme iron absorption in infants, Pediatr. Res. 43:768-773 (1998).
24. J. Westcott, N. B. Simon, and N. F. Krebs, Growth, zinc and iron status, and development of exclusively breastfed infants fed meat vs cereal as a first weaning food, FASEB J 12:A487 (1998).
25. S. Jalla, J. Westcott, M. Steirn, L. V. Miller, and N. F. Krebs. Zinc absorption and exchangeable zinc pool sizes in breastfed infants fed meat or cereal as first complementary food, J. Pediatr. Gastroenterol. Nutr. Submitted for publication (2000).
26. L. A. Persson, M. Lundstrom, B. Lonnerdal, and O. Hernell, Are weaning foods causing impaired iron and zinc status in 1-year-old Swedish infants? A cohort study, Acta Paediatr. 87:618-622 (1998).
27. M. Umeta, C. E. West, J. Haidar, P. Deurenberg, and J. G. Hautvast, Zinc supplementation and stunted infants in Ethiopia: a randomised controlled trial, Lancet 355:2021-2026 (2000).

BREAST FEEDING AND COGNITIVE OUTCOME IN CHILDREN BORN PREMATURELY

Ruth Morley*

Children who were breast fed generally perform better in tests of development or cognition, verbal ability or school performance. However, there are two major sources of confounding.

- The intimacy of breast feeding may be important for infant development.
- Mothers who choose to breast feed are different from mothers who choose not to. In studies conducted on subjects born in the last few decades the findings are confounded by the fact that breast fed children come from a higher socio-economic group than those fed formula Thus any differences may reflect, for example, higher parental socio-economic status and educational level, better nutrition and housing, a more stimulating environment, better access to pre-school education or a more positive attitude to education.

Many studies have adjusted for a range of potentially confounding factors and in some, but not all, the advantage for breast fed children has remained.[1, 2]

Two studies were conducted on subjects born at a time when artificially fed children came from the most socio-economically advantaged stratum of society. Hoeffer studied children born in the USA between 1915 and 1921 and showed that breast fed children performed better than those who were artificially fed, though children who were exclusively breast fed beyond 9 months did less well.[3] A study of elderly subjects born in the UK in the 1930's showed that despite lower socio-economic status those who were breast fed had higher unadjusted IQ.[4]

Infants born preterm are a high risk group and generally have lower neurodevelopmental scores that children born at term. It is possible that early post-natal nutrition is more influential in such children, in terms of their neurodevelopment.

*Senior Research Fellow, University of Melbourne Dept. of Paediatrics, Royal Children's Hospital, Flemington Road, Parkville, Victoria 3052, Australia

We conducted large multicentre randomised trials of nutrition and later development in preterm infants. In 1982 the milks routinely fed to preterm infants differed substantially in nutrient content, ranging from donor breast milk to a standard term or a nutrient enriched preterm formula.

A summary of the findings of our multicentre randomised outcome trials of early diet in preterm infants[5, 6] is presented here. Altogether 926 infants, born weighing under 1850g in five centres in the UK in 1982 to 1984 were randomly allocated their early enteral diet. The randomization was as shown in the figure.

STUDY DESIGN

STUDY 1, comparing banked donor milk with preterm formula

Mother asked if she wished to provide her own expressed breast milk for her baby

↓	↓
No	Yes
↓	↓
Randomise to:	Randomise to:
Banked donor milk vs. Preterm formula	Banked donor milk vs. Preterm formula
as sole diets	*as supplements to mother's milk*
n=159	n=343

STUDY 2, comparing term with preterm formula

Mother asked if she wished to provide her own expressed breast milk for her baby

↓	↓
No	Yes
↓	↓
Randomise to:	Randomise to:
Term formula vs. Preterm formula	Term formula vs. Preterm formula
as sole diets	*as supplements to mother's milk*
n=160	n=264

Study 1 compared banked donor milk (collected by breast feeding mothers in the community as it dripped from one breast while their baby fed from the other) with a preterm formula, either as sole diets or as a supplement to mother's expressed breast milk. Study 2 compared term with preterm formula as sole diets or as a supplement to mother's milk.

Milk was fed to the infants via a nasogastric tube in volumes of up to 180 ml/kg per day and until the infant attained 2000 g in weight or was discharged from the neonatal unit, whichever was sooner. The major nutrient constituents of the trial milks are as shown in table 1.

Table 1

Major nutrient constituents of the trial milks. Values for human milk are mean values from pooled samples.

Per 100 ml	Preterm formula	Term formula	Mother's expressed breast milk	Banked donor breast milk
Protein (g)	2.0	1.5	1.5	1.1
Fat (g)	4.9	3.8	3.0	1.7
Carbohydrate (g)	7.0	7.0	7.0	7.1
KCals	80	68	62	46
Na (mg)	45	19	23	16
Ca (mg)	70	35	35	35
P (mg)35	29	15	15	

In terms of nutrient content the banked donor milk was the poorest diet, with approximately half the protein and total energy content of the preterm formula. We therefore hypothesized that if early diet is important for developmental outcome, the most disadvantaged group would be those fed donor breast milk.

Surviving children from both studies were assessed at 18 months post term using the mental and motor scales of the Bayley Scales of Infant Development. This was completed first in study 2, the comparison of term with preterm formula. In the group fed the milks as their sole diet there was a significant 14.7 point advantage in Bayley Psychomotor Development Index (PDI) for children fed the preterm formula.[5]

In sharp contrast to the findings from study 2, and to our surprise, there was little difference in PDI between the groups fed donor milk versus preterm formula, in study 1.[6]

These findings led us to consider the possibility that human milk contains a factor or factors that promote neurodevelopment and ameliorated or offset the adverse effects of poor nutrition in the group fed donor breast milk. We found that among over 700 children without evidence of neuromotor or neurosensory impairment, those children whose mothers had chosen to provide their own breast milk were performing better at both 18 months and 7.5 to 8 years than those whose mothers had decided not to do so. Even after adjusting for a range of social and demographic differences between the two groups a significant developmental advantage remained for children of mothers choosing to provide breast milk[7] (*and unpublished data*). It is possible that we did not identify or adequately measure all relevant differences between the groups, so failing to adjust out the difference. For example, the mother's choice to express her milk may have reflected a deeper commitment to parenting, with better stimulation of her child's development during infancy and childhood and we had no measure of parenting or home environment.

In support of a causal link between human milk feeding and higher neurodevelopmental scores, we also found evidence that there was a dose-response relationship between intake of maternal milk and performance at 7.5 to 8 years. However, this relationship too could be confounded, because mothers with the greatest commitment to their infant's well being may have worked harder to provide sufficient milk to supply their needs.

Our findings from the randomized trials are more convincing. Among children fed the assigned milks as their sole diets, (those whose mothers who had all chosen not to provide their milk), there was a significant advantage for those fed preterm rather than term formula but no advantage for those fed preterm formula rather than donor milk. This was despite the poor nutrient content of the latter. These findings support the hypothesis that human milk contains factors promoting brain growth or maturation.

An advantage for very preterm infants fed breast milk was also seen by Hagan et al. in Perth,[8] who found that children who were fed breast milk had higher IQ and lower distractibility / hyperactivity scores (on the Parenting Stress Index) than those who were not. However, Doyle et al. in Melbourne reported findings from 181 children with lower mean birthweight and shorter mean gestation than in our study (mean birthweight 1094 g versus 1410 g in our study and mean gestation 28.8 versus 31.2 weeks respectively).[9] They found that children who were actually breast fed had significantly higher scores than those who were not, both before and after adjustment for potentially confounding factors. Conversely they found no benefit for children fed expressed breast milk versus formula in the neonatal unit and suggest that factors associated with the act of breast-feeding are more important than the breast milk itself. These data contrast with ours. More detailed comparisons of data from different cohorts may help resolve the issue of whether the developmental advantage for breast-fed infants is due to factors in human milk or confounding.

If human milk itself does indeed promote neurodevelopment in children, what might be the important factors? The brain and retina are rich in DHA and there is considerable interest in whether the DHA available to the infant from human milk might be the explanation, at least in part, for any developmental advantage associated with breast feeding. Human milk contains a range of long chain polyunsaturated fatty acids (LCPUFA's), notably docosahexaenoic acid (DHA, 22:6ω3), whereas conventional formulas contain a negligible amount. DHA and arachidonic acid (AA) uptake by the brain and retina increase substantially during the last trimester of pregnancy and for several months after birth. The use of conventional formulas has been shown to result in deterioration in LCPUFA status, particularly in infants born preterm.

Whether these long chain polyunsaturated fatty acids are important for later development is a question best answered by randomized trials comparing conventional formulas with those supplemented with DHA and AA. Meta-analyses of randomized trials to date, in both term and preterm infants, have been unable to confirm that this is the case.[10, 11]

The level of DHA in human milk varies greatly and depends on maternal dietary intake. Recent data from a small study by Gibson et al.[12] suggest that supplementing lactating mothers with DHA increases the level in their milk and in their infants' red cell membrane DHA levels. Preliminary data suggest that the infants' developmental performance is positively related to their red cell DHA levels. This is an important area for further research.

Other possible candidate factors include oligosaccharides and the large range of biologically active peptides in milk, (including for nerve growth factor and insulin-like growth factors). The extent to which the latter reach target tissues in the infant and whether they are important for development is unclear.

However, despite potential benefits, human milk may fail to meet preterm infants' nutrient requirements for growth. We tested the hypothesis that fortified breast milk, fed alone or with preterm formula, would improve neurodevelopment at 18 months post-term. Altogether 275 preterm infants from two centres, with birth weight < 1850 g and mean gestation 29.8 +/- 2.7 wk), whose mothers chose to provide breast milk were randomly assigned to receive for a mean of 39 days a multinutrient fortifier or control supplement containing phosphate and vitamins. Developmental scores at 18 months were slightly but not significantly higher in the fortified group. Beneficial effects of breast milk fortification on development remain unproven.[13]

While research should be undertaken to improve infant formula milks, for the sake of children whose mothers choose not to breast feed, mothers should be strongly encouraged to provide their own milk for their infants, whether they are born before or at term. If human milk does indeed promote development we are still a long way from understanding the reason, and it may well be multifactorial. No formula can approach the complexity or suitability of human milk.

References

1. Golding J, Rogers IS, and Emmett PM. Association between breast feeding, child development and behaviour, Early Hum. Dev. 49 Suppl:S175-S184 (1997).
2. Malloy MH and Berendes H. Does breast-feeding influence intelligence quotients at 9 and 10 years of age? Early Hum. Dev. 50:209-217 (1998).
3. Hoefer C, and Hardy MC. Later development of breast fed and artificially fed infants. JAMA 92:615-619 (1929).
4. Gale CR and Martyn CN. Breast feeding, dummy use and adult intelligence, Lancet 347:1072-1075 (1996).
5. Lucas A, Morley R, Cole TJ et al. Early diet in preterm babies and developmental status at 18 months, Lancet 335: 1477-1481 (1990).
6. Lucas A, Morley R, Cole TJ, and Gore SM. A randomized multicentre study of human milk versus formula and later development in preterm infants, Arch. Dis. Child. 70: F141-146 (1994).
7. Lucas A and Morley R. Breast milk and subsequent intelligence quotient in children born preterm, Lancet 339: 261-264 (1991).
8. Hagan R, French N, Evans S et al. Breast feeding, distractibility and IQ in very preterm infants (abstract), Pediatr. Res. 39:266A (1996).
9. Doyle LW, Rickards AL, Kelly EA, Ford GW, and Callanan C. Breastfeeding and intelligencem Lancet, 339:744-745 (1992).
10. Simmer K. Longchain polyunsaturated fatty acid supplementation in preterm infants, Cochrane Database Syst. Rev. (2) (2000).

11. Simmer K. Longchain polyunsaturated fatty acid supplementation in infants born at term, Cochrane Database Syst. Rev. (2) (2000).
12. Gibson RA and Neumann MA. Makrides M Effect of increasing breast milk docosahexaenoic acid on plasma and erythrocyte phospholipid fatty acids and neural indices of exclusively breast fed infants, Eur. J. Clin. Nutr. 51:578-845 (1997).
13. Kuschel CA and Harding JE. Multicomponent fortified human milk for promoting growth in preterm infants, Cochrane Database Syst. Rev. (2) (2000).

THE ROLE OF HUMAN MILK IN NECROTIZING ENTEROCOLITIS

Michael S. Caplan, Michael Amer and Tamas Jilling*

1. INTRODUCTION

Necrotizing enterocolitis (NEC) is the most common gastrointestinal emergency of premature infants, and accounts for significant morbidity and mortality worldwide (Kliegman et al., 1984; Uauy et al., 1991). Although the etiology of intestinal inflammation and necrosis that characterizes this disease remains poorly understood, studies have suggested that human milk feedings reduce its' incidence significantly (Lucas et al., 1990). As compared to neonatal formula, mother's milk has a myriad of bioactive components that provide immunoprotection and host defense that presumably contribute to the beneficial effects (Goldman, 2000; Goldman et al., 1990; Hanson, 1999). One key component of breast milk may be PAF-acetylhydrolase (PAF-AH), a specific PAF degrading enzyme that is deficient in human newborns, and has been shown in animal models to markedly alter the development of neonatal NEC (Caplan et al., 1990a; Caplan et al., 1997b; Farr et al., 1983). Since data strongly support the role of endogenous PAF in the initiation of intestinal injury (Hsueh et al., 1994; Hsueh et al., 1987), the presence or absence of PAF-AH may be a critical factor in the pathophysiology of the neonatal disease.

2. Effect of Human Milk on NEC

Although most neonatologists assume that human milk feedings reduce the incidence of NEC, there are only a few scientific reports addressing this clinical effect. Over twenty years ago, there were two retrospective analyses that examined the effectiveness of human milk supplementation (Kliegman et al., 1979; Moriartey et al., 1979). In one study, 109 cases of NEC were examined, and the findings identified

*Department of Pediatrics, Evanston Northwestern Healthcare Research Institute, Northwestern University Medical School, Evanston, IL 60201

Integrating Population Outcomes, Biological Mechanisms and Research Methods in the Study of Human Milk and Lactation
Edited by Davis *et al.*, Kluwer Academic/Plenum Publishers, 2002

several cases fed fresh human milk exclusively, and were unable to identify specific risk factors peculiar to any of the three feeding classes that included human milk, milk and formula, and formula alone (Kliegman et al., 1979). In another study, many fewer cases of NEC were analyzed according to feeding type, and the authors concluded (based on soft data) that frozen human milk was beneficial to all infants, but that it did not fully protect against the disease in the highest risk, lowest birth weight groups (Moriartey et al., 1979). To better assess the beneficial effect of human milk, Lucas and Cole conducted a multicenter, prospective, controlled trial that was randomized for some infants. For the randomized patients, donor milk reduced the incidence of NEC compared to preterm formula, although inadequate power precluded statistical significance (1% vs 5% confirmed cases). Nonetheless, in the non-randomized groups, donor milk reduced the incidence of NEC in all birth weight subcategories (Lucas et al., 1990). Despite the flaws in study design, these data suggested a beneficial role for human milk on the development of NEC. While physicians empirically suggest that fresh breast milk may provide greater benefit than frozen or pasteurized preparations, specific randomized trials have not confirmed this supposition.

3. Bioactive Factors in Human Milk

Experimental laboratory evidence strongly supports the beneficial role of milk on intestinal injury. Barlow and colleagues studied the effect of formula feeding and asphyxia in a neonatal rat model in the early 1970's, and subsequently identified the importance of mother's milk and specifically milk leukocytes as a key protective factor against experimental NEC (Barlow et al., 1975; Pitt et al., 1977). Nonetheless, there have been many bioactive factors described in human milk that contribute to intestinal health by strengthening mucosal integrity, supporting immuno-host defense, or modulating intestinal inflammation (Avery et al., 1991; Walker, 1997). These bioactive factors span multiple components of milk (Donovan et al., 1994; Goldman, 2000; Goldman et al., 1990; Hanson, 1999; Pickering et al., 1998; Shulman et al., 1998), including cytokines, chemokines, enzymes, growth factors, immunoglobulins and a number of other bioactive molecules (see table 1). Nonetheless, it remains undetermined what specific role each of these compounds has on maintaining intestinal health of the newborn infant, and more importantly as it pertains to NEC, in the premature neonate.

Table 1. Examples of bioactive molecules in human milk

Cytokines and Chemokines	Enzymes, growth factors and other proteins	Other bioactive molecules
IL-6, IL-10, IL-11, IL-12, TNF	Catalase, lysozyme, PAF-acetylhydrolase, glutathione peroxidase EGF, insulin, IGF-I, IGF-II, NGF, relaxin, TGF-α, TGF-β1 and TGF-β2, GMCSF, lactoferrin, IgA, IgG, IgM	Prostaglandins, PAF, neuropeptides, nucleotides, polyunsaturated fatty acids, oligosaccharides, β-carotene, ascorbate, vitamin E, α-tocopherol

4. Role of PAF and the Inflammatory Cascade in the Pathophysiology of NEC

Platelet activating factor (PAF) is a potent, endogenous phospholipid mediator that is present in most body tissues and fluids (Benveniste, 1988; Snyder, 1990). PAF is synthesized from phosphatidylcholine precursors under the influence of the PAF-specific enzyme phospholipase A_2-II and subsequent acetylation by acetyltransferase (Figure 1) (Bussolino et al., 1995). The bioactive compound PAF activates multiple cell types via the G protein-coupled PAF receptor that appears to be expressed in the highest amounts in intestinal homogenate (Wang et al., 1997). PAF maintains an extremely short half-life due to the presence of the PAF-degrading enzyme PAF-acetylhydrolase (PAF-AH). This 43 kD protein is present in human milk, and has been shown to be deficient in the circulation of newborn infants, and lower in patients who develop classic NEC (Caplan et al., 1990a; Caplan et al., 1990b; Moya et al., 1994). In addition, circulating PAF activity is elevated in patients with NEC compared to age-matched controls (Caplan et al., 1990b). Furthermore, supplementation of infant formula with PAF-AH has been shown to reduce the incidence of NEC in a neonatal rat model (Caplan et al., 1997b).

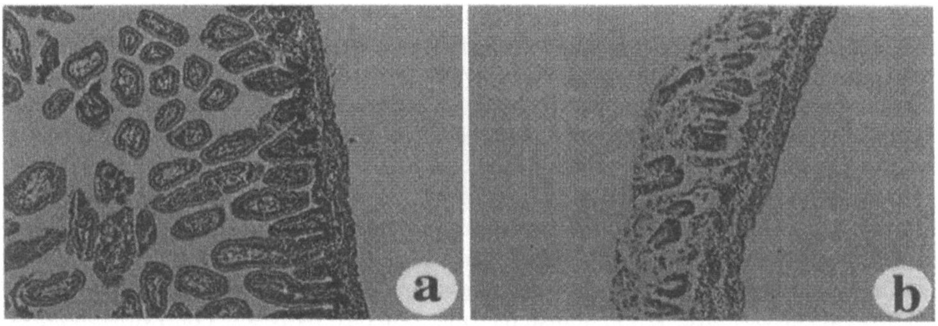

Figure 1. The effect of formula feeding and cold/asphyxia stressing on gut histology. Shown are photomicrographs of H&E stained intestinal sections from neonatal rats reared by their mothers (a), and from neonatal rats delivered with cesarean section and subjected to formula feeding and cold/asphyxia stressing (b).

Accumulating evidence suggests that the well-known risk factors for NEC (i.e. formula feeding, hypoxia/ischemia, and bacterial colonization) stimulate the inflammatory cascade, and that activation of inflammatory mediators promotes the development of intestinal injury (Caplan et al., 1993; Kliegman, 1990). Bacterial endotoxin activates macrophages and other cells to synthesize and secrete cytokines, PAF, prostaglandin metabolites, nitric oxide, and cell adhesion molecules, which subsequently stimulate additional second messengers of inflammation (Arbibe et al., 1997; Christman et al., 1993; Hsueh et al., 1990; Natanson et al., 1994; Vos et al., 1997). Using adult rats, Hsueh and coworkers have shown that endotoxin stimulates endogenous, intestinal PAF production, that exogenous PAF causes ischemic intestinal necrosis, and that PAF receptor antagonists can prevent NEC following endotoxin and TNF challenge (Gonzalez-Crussi et al., 1983; Hsueh et al., 1986a; Hsueh et al., 1987; Hsueh et al., 1986b; Hsueh et al., 1989). Recent studies have shown that PAF receptor blockade can prevent intestinal injury following hypoxia/endotoxin, nitric oxide inhibition/hypoxia, ischemia-reperfusion in adult rats, and formula feeding/asphyxia in the newborn model (Caplan et al., 1997a; Caplan et al., 1994b; Caplan et al., 1992; Caplan et al., 1990c; Mozes et al., 1989). Although the data strongly support the role of PAF in intestinal injury and NEC, the specific cellular changes responsible for these events have not been elucidated.

5. Etiology of paf-induced intestinal injury: hypothesis

At the cellular level PAF elicits a complex signaling cascade involving multiple second messengers and effector molecules. Binding of PAF to its G-protein linked receptor activates a variety of signaling pathways, including MEK and MAP kinase (Shimizu et al., 1996), MAP kinase 3, p38 MAP kinase (Nick et al., 1997), P42/p44 MAP kinase (MEK kinase) and protein kinase C (Bonaccorsi et al., 1997). These signaling mechanisms elicit a series of biochemical changes that result in accentuated inflammation with clinical changes including capillary leak, pulmonary vasoconstriction, myocardial depression, bronchoconstriction, neutropenia, and thrombocytopenia (Benveniste, 1988; Snyder, 1990). Studies in animal models have shown that PAF causes altered intestinal mucosal permeability, mesenteric artery vasoconstriction, and intestinal leukocyte activation and chemotaxis, but these experiments fail to clarify the roles of specific cell types and molecular mechanisms that are responsible for these mucosal responses leading to tissue necorsis (Kubes et al., 1991; Musemeche et al., 1991).

Studies from the neonatal rat in our laboratory confirm the importance of asphyxia and formula feeding together in the development of NEC (Caplan et al., 1994a). As illustrated in figure 1, formula feeding and cold asphyxia stressing together result in histological changes in the neonatal rat intestine that bear strong resemblance to the histological changes observed in human intestines affected by NEC. Following 96 hrs of formula feeding and cold/asphyxia stressing the deterioration of villus structure is observed in approximately 75% of the neonatal rats subjected to formula feeding and cold/asphyxia stressing. In contrast, such changes were not observed in mother-fed pups even if they were subjected to the same cold/asphyxia stressing protocol as the formula fed animals. Interestingly, similar histological alterations were found, albeit in a smaller percentage (38%) in intestinal sections of formula fed animals even without cold/asphyxia stressing.

Furthermore, we found that formula feeding and asphyxia stressing together stimulated phospholipase A_2-II and PAF receptor expression (Figure 2). Correlating with the histological findings, mother-fed pups exhibited unaltered PLA2-II and PAF receptor expression even when they were subjected to cold/asphyxia stressing. These findings together underlie the protective effect of milk against altered PAF metabolism and experimental NEC in this model. As mentioned above, enteral supplementation of PAF-AH, a potential key protective factor in human milk markedly reduced the incidence of NEC in this model. Additional studies showed that PAF-AH retained functional activity in the distal intestine, was identifiable in intestinal epithelial cells using immunohistochemical staining, and was not measurable in the systemic circulation (Caplan et al., 1997b). Of interest, intraperitoneal dosing of PAF-AH did not confer protection against NEC using similar dosing to the enteral route (Table 2). These findings further support the theory that PAF acts on the mucusal surface of the intestine rather than via the systemic circulation as previously surmised.

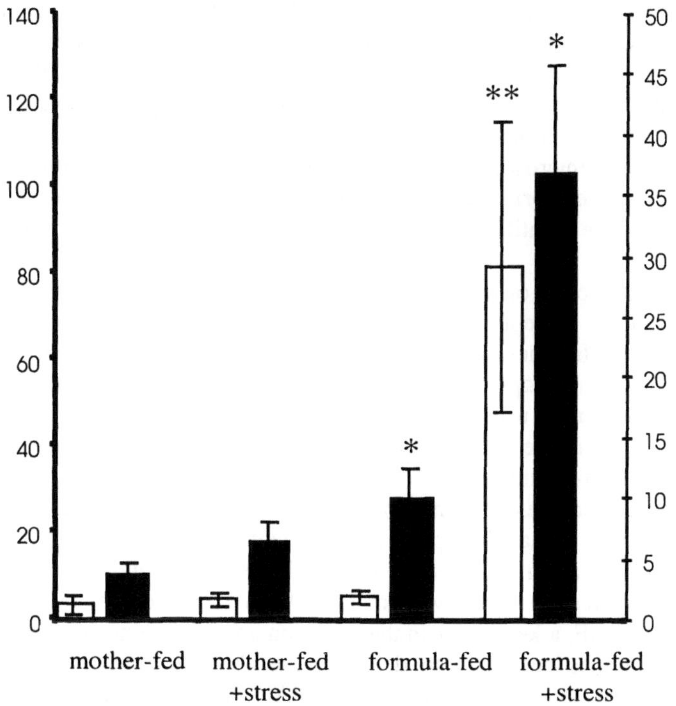

Figure 2. Changes in PLA2-II and PAFR expression as a consequence of formula feeding and cold/asphyxia stressing in a neonatal rat model of NEC. Neonatal rat pups were treated as indicated, then intestines were harvested, mRNA was isolated and PAFR and PLA_2-II transcripts were quantified using quantitative RTPCR. Results shown are mean ± S.E.M.

Table 2. The effect of rPAF-AH supplementation on experimental NEC

	Death	NEC
Formula	12/17	10/17
Formula+ rPAF-AH (enteral)	7/14*	3/14**
Formula+ rPAF-AH (IP)	8/11	6/11

* p = 0.10 and ** p = 0.12 (two tailed Fisher's Exact Test)

We have recently initiated a series of experiments to delineate the specific effects of PAF on the mucosal epithelium (Claud et al, submitted 2000). Using intestinal epithelial cell monolayers in an Ussing chamber system, we have found that PAF potently stimulates chloride transport in a dose-dependent manner to a degree similar to a cell-permeant cAMP analog, the strongest agonist known to produce this short-circuit current. Additional studies to delineate sidedness of this response have confirmed that mucosal PAF (and not serosal PAF) results in this chloride-transport response. Confocal immunofluorescence microscopy and image analysis supports this finding and shows that PAF receptor is present primarily on the mucosal surface of intestinal epithelium, and RT-PCR studies have identified mRNA expression of PAF receptor in various intestinal epithelial cells.

In another series of intestinal cell monolayer studies, we have found that PAF stimulates apoptosis of epithelial cells in a time and dose-dependent manner, as evidenced by DNA fragmentation and elevated caspase 2 and 3 activity (Jilling et al, submitted 2000). This apoptosis appears to be inhibited by PKC activation, and can be independently augmented using PKC blockade. These studies are of particular interest because experiments in the neonatal rat have shown apoptosis to precede necrosis using our asphyxia/formula-feeding model (Jilling et al., submitted). As such, mother's milk feeding markedly attenuates intestinal epithelial apoptosis in this model.

In summary, human milk plays an important role in maintaining intestinal health in premature infants, and it may reduce the incidence of neonatal NEC. Although milk contains many interesting bioactive factors, PAF-AH is one that down-regulates the bioavailability of PAF, a key factor in the initiation and pathogenesis of intestinal injury. Additional research is necessary to delineate the absolute benefit of human milk in neonatal NEC, and to evaluate the role of PAF-AH supplementation in premature formula and/or breast milk to prevent this complex disease.

Acknowledgement

This work was supported in part by NIH grant HD37581, a grant from the March of Dimes (#6-FY99-278) and by the Jessica Jacobi Golder endowment.

References

Arbibe, L., Vial, D., Rosinski-Chupin, I., Havet, N., Huerre, M., Vargaftig, B.B., and Touqui, L., Endotoxin induces expression of type II phospholipase A2 in macrophages during acute lung injury in guinea pigs: involvement of TNF-alpha in lipopolysaccharide-induced type II phospholipase A2 synthesis. Journal of Immunology 159(1):391-400 (1997).

Avery, V.M., and Gordon, D.L., Antibacterial properties of breast milk: requirements for surface phagocytosis and chemiluminescence. Eur. J. Clin. Microbiol. Infect. Dis. 10(12):1034-1039 (1991).

Barlow, B., and Santulli, T.V., Importance of multiple episodes of hypoxia or cold stress on the development of enterocolitis in an animal model. Surgery 77(5):687-690 (1975).

Benveniste, J., Paf-acether, an ether phospho-lipid with biological activity. Progress in Clinical & Biological Research 282:73-85 (1988).

Bonaccorsi, L., Luconi, M., Maggi, M., Muratori, M., Forti, G., Serio, M., and Baldi, E., Protein tyrosine kinase, mitogen-activated protein kinase and protein kinase C are involved in the mitogenic signaling of platelet-activating factor (PAF) in HEC-1A cells. Biochim. Biophys. Acta 1355(2):155-166 (1997).

Bussolino, F., and Camussi, G., Platelet-activating factor produced by endothelial cells. A molecule with autocrine and paracrine properties. [Review] [115 refs]. European Journal of Biochemistry 229(2):327-337 (1995).

Caplan, M., Hsueh, W., Kelly, A., and Donovan, M., Serum PAF acetylhydrolase increases during neonatal maturation. Prostaglandins 39(6):705-714 (1990a).

Caplan, M.S., Hedlund, E., Adler, L., and Hsueh, W., Role of asphyxia and feeding in a neonatal rat model of necrotizing enterocolitis. Pediatric Pathology 14(6):1017-1028 (1994a).

Caplan, M.S., Hedlund, E., Adler, L., Lickerman, M., and Hsueh, W., The platelet-activating factor receptor antagonist WEB 2170 prevents neonatal necrotizing enterocolitis in rats. Journal of Pediatric Gastroenterology & Nutrition 24(3):296-301 (1997a).

Caplan, M.S., Hedlund, E., Hill, N., and MacKendrick, W., The role of endogenous nitric oxide and platelet-activating factor in hypoxia-induced intestinal injury in rats. Gastroenterology 106(2):346-352 (1994b).

Caplan, M.S., Kelly, A., and Hsueh, W., Endotoxin and hypoxia-induced intestinal necrosis in rats: the role of platelet activating factor. Pediatric Research 31(5):428-434 (1992).

Caplan, M.S., Lickerman, M., Adler, L., Dietsch, G.N., and Yu, A., The role of recombinant platelet-activating factor acetylhydrolase in a neonatal rat model of necrotizing enterocolitis. Pediatric Research 42(6):779-783 (1997b).

Caplan, M.S., and MacKendrick, W., Necrotizing enterocolitis: a review of pathogenetic mechanisms and implications for prevention. Pediatr. Pathol. 13(3):357-369 (1993).

Caplan, M.S., Sun, X.M., Hseuh, W., and Hageman, J.R., Role of platelet activating factor and tumor necrosis factor-alpha in neonatal necrotizing enterocolitis. Journal of Pediatrics 116(6):960-964 (1990b).

Caplan, M.S., Sun, X.M., and Hsueh, W., Hypoxia causes ischemic bowel necrosis in rats: the role of platelet-activating factor (PAF-acether). Gastroenterology 99(4):979-986 (1990c).

Christman, B.W., Christman, J.W., Dworski, R., Blair, I.A., and Prakash, C., Prostaglandin E2 limits arachidonic acid availability and inhibits leukotriene B4 synthesis in rat alveolar macrophages by a nonphospholipase A2 mechanism. J. Immunol. 151(4):2096-2104 (1993).

Donovan, S.M., and Odle, J., Growth factors in milk as mediators of infant development. Annual Review of Nutrition 14:147-167 (1994).

Farr, R.S., Wardlow, M.L., Cox, C.P., Meng, K.E., and Greene, D.E., Human serum acid-labile factor is an acylhydrolase that inactivates platelet-activating factor. Fed. Proc. 42(14):3120-3122 (1983).

Goldman, A.S., Modulation of the gastrointestinal tract of infants by human milk. Interfaces and interactions. An evolutionary perspective. J. Nutr. 130(2S Suppl):426S-431S (2000).

Goldman, A.S., Goldblum, R.M., and Hanson, L.A., Anti-inflammatory systems in human milk. Adv. Exp. Med. Biol. 262:69-76 (1990).

Gonzalez-Crussi, F., and Hsueh, W., Experimental model of ischemic bowel necrosis. The role of platelet-activating factor and endotoxin. American Journal of Pathology 112(1):127-135 (1983).

Hanson, L.A., Human milk and host defense: immediate and long-term effects. Acta Paediatr. Suppl. 88(430):42-46 (1999).

Hsueh, W., Caplan, M.S., Sun, X., Tan, X., MacKendrick, W., and Gonzalez-Crussi, F., Platelet-activating factor, tumor necrosis factor, hypoxia and necrotizing enterocolitis. [Review] [55 refs]. Acta Paediatrica. Supplement. 396:11-17 (1994).

Hsueh, W., Gonzalez-Crussi, F., and Arroyave, J.L., Platelet-activating factor-induced ischemic bowel necrosis. An investigation of secondary mediators in its pathogenesis. American Journal of Pathology 122(2):231-239 (1986a).

Hsueh, W., Gonzalez-Crussi, F., and Arroyave, J.L., Platelet-activating factor: an endogenous mediator for bowel necrosis in endotoxemia. FASEB Journal 1(5):403-405 (1987).

Hsueh, W., Gonzalez-Crussi, F., Arroyave, J.L., Anderson, R.C., Lee, M.L., and Houlihan, W.J., Platelet activating factor-induced ischemic bowel necrosis: the effect of PAF antagonists. European Journal of Pharmacology 123(1):79-83 (1986b).

Hsueh, W., Sun, X., Rioja, L.N., and Gonzalez-Crussi, F., The role of the complement system in shock and tissue injury induced by tumor necrosis factor and endotoxin. Immunology 70(3):309-314 (1990).

Hsueh, W., and Sun, X.M., Tumor necrosis factor-induced bowel necrosis: the role of platelet-activating factor. Advances in Prostaglandin, Thromboxane, & Leukotriene Research 19:363-366 (1989).

Kliegman, R.M., Models of the pathogenesis of necrotizing enterocolitis. [Review] [45 refs]. Journal of Pediatrics 117(1 Pt 2) (1990).

Kliegman, R.M., and Fanaroff, A.A., Necrotizing enterocolitis. [Review] [165 refs]. New England Journal of Medicine 310(17), 1093-1103 (1984).

Kliegman, R.M., Pittard, W.B., and Fanaroff, A.A., Necrotizing enterocolitis in neonates fed human milk. J. Pediatr. 95(3):450-453 (1979).
Kubes, P., Arfors, K.E., and Granger, D.N., Platelet-activating factor-induced mucosal dysfunction: role of oxidants and granulocytes. Am. J. Physiol. 260(6 Pt 1), G965-971 (1991).
Lucas, A., and Cole, T.J., Breast milk and neonatal necrotising enterocolitis [see comments]. Lancet 336(8730):1519-1523 (1990).
Moriartey, R.R., Finer, N.N., Cox, S.F., Phillips, H.J., Theman, A., Stewart, A.R., and Ulan, O.A., Necrotizing enterocolitis and human milk. J. Pediatr. 94(2):295-296 (1979).
Moya, F.R., Eguchi, H., Zhao, B., Furukawa, M., Sfeir, J., Osorio, M., Ogawa, Y., and Johnston, J.M., Platelet-activating factor acetylhydrolase in term and preterm human milk: a preliminary report. Journal of Pediatric Gastroenterology & Nutrition 19(2), 236-239 (1994).
Mozes, T., Braquet, P., and Filep, J., Platelet-activating factor: an endogenous mediator of mesenteric ischemia-reperfusion-induced shock. American Journal of Physiology 257(4 Pt 2) (1989).
Musemeche, C., Caplan, M., Hsueh, W., Sun, X., and Kelly, A., Experimental necrotizing enterocolitis: the role of polymorphonuclear neutrophils. Journal of Pediatric Surgery 26(9):1047-1049 (1991).
Natanson, C., Hoffman, W.D., Suffredini, A.F., Eichacker, P.Q., and Danner, R.L., Selected treatment strategies for septic shock based on proposed mechanisms of pathogenesis. Annals of Internal Medicine 120(9):771-783 (1994).
Nick, J.A., Avdi, N.J., Young, S.K., Knall, C., Gerwins, P., Johnson, G.L., and Worthen, G.S., Common and distinct intracellular signaling pathways in human neutrophils utilized by platelet activating factor and FMLP. J. Clin. Invest. 99(5):975-986 (1997).
Pickering, L.K., Granoff, D.M., Erickson, J.R., Masor, M.L., Cordle, C.T., Schaller, J.P., Winship, T.R., Paule, C.L., and Hilty, M.D., Modulation of the immune system by human milk and infant formula containing nucleotides. Pediatrics 101(2):242-249 (1998).
Pitt, J., Barlow, B., and Heird, W.C., Protection against experimental necrotizing enterocolitis by maternal milk. I. Role of milk leukocytes. Pediatr. Res. 11(8):906-909 (1977).
Shimizu, T., Mori, M., Bito, H., Sakanaka, C., Tabuchi, S., Aihara, M., and Kume, K., Platelet-activating factor and somatostatin activate mitogen-activated protein kinase (MAP kinase) and arachidonate release. J. Lipid Mediat. Cell Signal 14(1-3):103-108 (1996).
Shulman, R.J., Schanler, R.J., Lau, C., Heitkemper, M., Ou, C.N., and Smith, E.O., Early feeding, antenatal glucocorticoids, and human milk decrease intestinal permeability in preterm infants. Pediatr Res 44(4):519-523 (1988).
Snyder, F., Platelet-activating factor and related acetylated lipids as potent biologically active cellular mediators. [Review] [141 refs]. American Journal of Physiology 259(5 Pt 1) (1990).
Uauy, R.D., Fanaroff, A.A., Korones, S.B., Phillips, E.A., Phillips, J.B., and Wright, L.L., Necrotizing enterocolitis in very low birth weight infants: biodemographic and clinical correlates. National Institute of Child Health and Human Development Neonatal Research Network. Journal of Pediatrics 119(4):630-638 (1991).
Vos, T.A., Gouw, A.S., Klok, P.A., Havinga, R., van Goor, H., Huitema, S., Roelofsen, H., Kuipers, F., Jansen, P.L., and Moshage, H., Differential effects of nitric oxide synthase inhibitors on endotoxin-induced liver damage in rats [see comments]. Gastroenterology 113(4):1323-1333 (1997).
Walker, W.A., Breast milk and the prevention of neonatal and preterm gastrointestinal disease states: a new perspective. Chung Hua Min Kuo Hsiao Erh Ko I Hsueh Hui Tsa Chih 38(5):321-331 (1997).
Wang, H., Tan, X., Chang, H., Gonzalez-Crussi, F., Remick, D.G., and Hsueh, W., Regulation of platelet-activating factor receptor gene expression in vivo by endotoxin, platelet-activating factor and endogenous tumor necrosis factor. Biochem. J. 322(Pt 2):603-608 (1997).

THE EFFECT OF STRESS ON LACTATION - ITS SIGNIFICANCE FOR THE PRETERM INFANT

Chantal Lau

The population of infants born prematurely has been increasing steadily over the last 2 decades (Schanler 1996). Inasmuch as these infants are born at progressively shorter gestational ages, the length of their hospitalization is increasing. Among the numerous medical advances that have benefited these infants, the advantages offered by mother's milk, more specifically, the presence of growth and host-defense factors which cannot be replicated in formulae, have been demonstrated unequivocally (Gartner et al., 1997). Consequently, the provision of mother's milk for the preterm neonate has become an important aspect of neonatal care. Unfortunately, for reasons not fully understood, mothers of preterm infants often cannot meet the nutritional needs of their infants.

Stress is often proposed as the culprit when mothers have difficulty providing milk for their infants. This notion is based primarily on animal studies demonstrating that stress can interfere with lactation (Grosvenor & Mena, 1967; Lau, 1992). In the particular case of mothers with preterm infants, studies have shown significant maternal psychological distress (Singer et al., 1999; Meyer et al., 1995). However, no study has monitored concurrently lactation performance and stress levels in mothers of preterm infants. Therefore, one may query whether the lactation insufficiency frequently observed in these mothers can be attributed solely to the stress resulting from their infant's fragility. Poor lactation may ensue from other causes, such as immature mammary development, hormonal inadequacy, and poor infant's sucking skills. Understanding if and how stress affects lactation would greatly benefit the preterm infant.

The Nursing Dyad

The important aspects of mother-infant interactions are presented in Figure 1. This model distinguishes 3 major interplays existing between the members of the dyad. Mother and/or infant can influence the outcome of each interplay. Interplay I addresses our primary concern, namely the provision of mother's milk. At this level, the better an infant can feed, the more sustained lactation will be. In return, the more appropriate the

*Department of Pediatrics/Neonatology, Baylor College of Medicine, Houston, TX

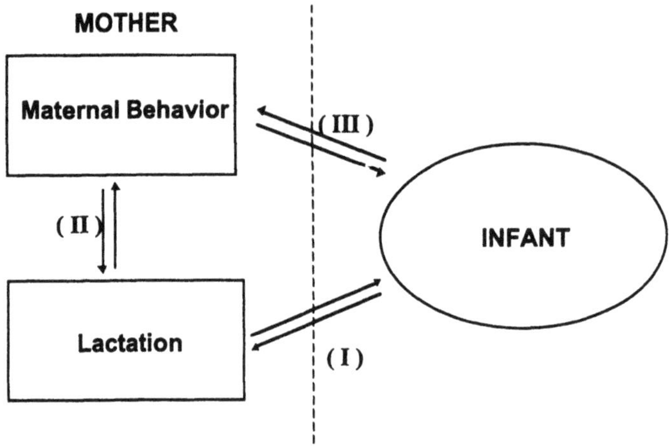

Figure 1. Model of the mother-infant nursing dyad. Interplay I relates to the interaction existing between the infant and lactation *per se*. Interplay II addresses the interaction between maternal behavior and lactation. Interplay III pertains to the interaction existing between the infant and maternal behavior.

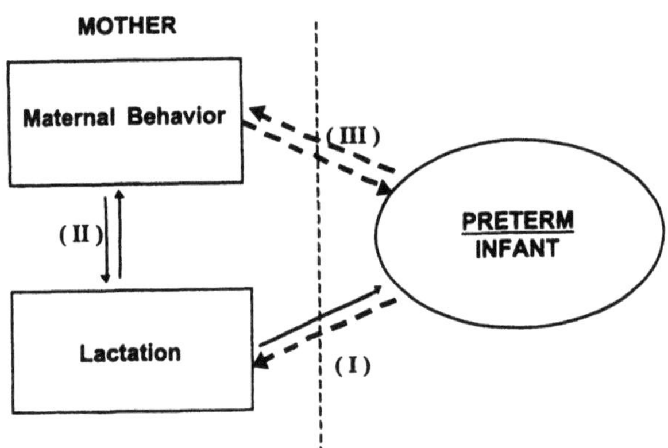

Figure 2. Model of the mother-preterm infant nursing dyad. Broken arrows show the interactions presumed altered as a result of the infant's prematurity.

lactation, the better an infant will feed. Interplay II relates to the expression of maternal behavior that leads to the appropriate drive to breastfeed or express milk. On the one hand, the stronger the maternal behavior, the stronger her drive will be to provide milk, and thus the more likely she will meet her infant's need. On the other hand, poor lactation may decrease the intensity of maternal behavior, thereby leading to a decreased drive or interest to breastfeed/express milk. Interplay III addresses the concept of mother-infant bonding (Tessier et al., 1998). The more interaction or contact a mother has with her infant, the closer she will feel towards her infant, potentially leading to a greater desire to breastfeed/express milk. The longer a mother is kept away from her infant, the more rapidly this desire will dissipate. With a healthy term nursing dyad, when all these interplays are balanced, proper lactation will be expected. Alterations in any of these interplays will require the ability of one of the partners to compensate for the imbalance in order to preserve the integrity of the nursing dyad.

Maternal Attributes

To optimize the growth and development of the infant, the primary function of maternal behavior is to nurture and provide milk. Maternal attitudes, health, education, lifestyle and psychosocial support contribute to such behavior. A caring attitude and recognition of infant's needs are positive elements in the promotion of the infant's physical and psychological development. Similarly, an interest to breast-feed or provide milk will facilitate lactation. However, the latter can be affected by other factors such as inadequate ductal and alveolar development of the mammary gland. There is little information on whether adequate mammary development is reached at the time of a premature delivery and/or whether "catch-up" maturation can occur following precocious parturition. Anatomical components such as types of maternal nipples may interfere with the normal latching-on process of infants, particularly those who are sick or born prematurely. This may hamper feeding, and subsequently lactation (Lau and Hurst, 1999). The primary lactogenic hormones necessary for adequate milk synthesis are prolactin, insulin, glucocorticoids and for milk ejection, oxytocin. Lactation performance will be hampered if any of these hormonal levels are inappropriate. In addition, other hormones have been demonstrated to indirectly affect milk synthesis/ejection. For instance, leptin has been implicated in mammary development (Laud et al.,1999) and opiates, elevated during the suckling period, can affect lactogenic hormones (Riskind et al., 1984; Lau, 1992; Merchenthaler, 1994).

Infant Attributes

Although infants may play a less active role than mothers, they must be capable of providing appropriate feedback if the integrity of the nursing dyad is to be preserved. Mothers appear to become more attentive to their infant's needs following infant's suckling or mere touching of the maternal areolar region immediately after birth (Widstrom et al., 1990). It is interesting to note that the mutual feedback experienced by mother and infant during breastfeeding does not seem to be reproduced to the same extent with bottle feeding (De Andraca et al., 1995). Lactation is maintained if an infant can readily latch on to the breast, has appropriate sucking skills, coordinated suck-swallow-breathe and endurance so as to complete feedings with no adverse effect. Conditions such as prematurity, congenital anomalies, sickness or prolonged hospitalization can affect these abilities. Figure 2 shows the imbalance that occurs with a preterm nursing dyad. Interplays I and III are disrupted as a result of the infant's inadequate oral motor skills and the physical separation of mother and infant, respectively. Under these circumstances, the mother is the one most likely capable of compensating for the imbalance. In fact, the preterm infant, due to his/her immaturity, poor oral motor skills, and fragility, can be regarded as a passive participant in the exchange. The prolonged and forced separation of mother and infant negatively affects both mother and infant. Mothers may become more detached from their infant and may experience stress through guilt and feeling of incompetence in the care of their child. If the mother takes on the task of compensating for the inadequacy of her infant, the pressure to preserve the integrity of their dyad may lead to additional stress. Infant feeding behavior may be affected as well. Indeed, studies have demonstrated how social and physical isolation can lead to growth and developmental stunting or psychosocial dwarfism (Bowden &

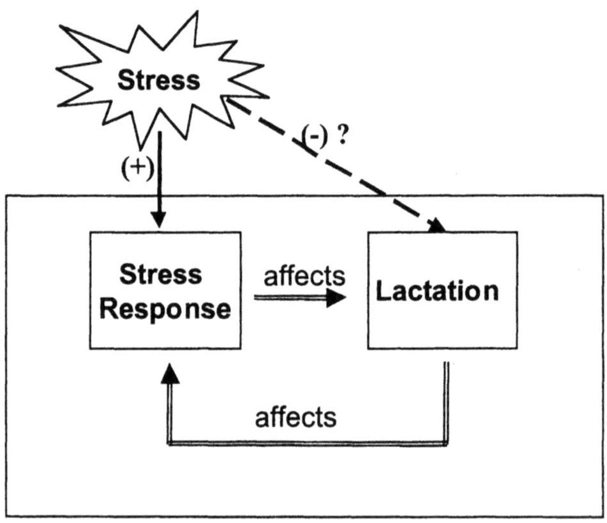

Figure 3. Model of the effect of stress on lactation. (-): inhibitory effect; (+): stimulatory effect.

Hopwood, 1982). Schanberg et al (1984) demonstrated in rats that physical contact between mother and pups stimulated ornithine decarboxylase activity, an enzyme implicated in the synthesis of DNA and RNA in the young. Ronca et al (1996) advanced that perinatal sensory input facilitates the postpartum adaptive function of the young.

Interventions

There are interventions used to correct lactation insufficiency (see rev Lau and Hurst, 1999). Increased frequency of breastfeeding/milk expression ($\geq$ 6 times per day) and complete breast emptying have shown beneficial effects. The decreased lactation that routinely follows incomplete breast emptying has been attributed in part to the elevated levels of a protein in the residual milk, feedback inhibitor of lactation, believed to act at an autocrine level on mammary epithelial cells (Blatchford et al., 1998). The use of metoclopramide, a dopamine antagonist, offers transient improvement for some women. The practice of skin-to-skin holding has been used also to ameliorate lactation. However, the lack of consistent improvement following these interventions emphasizes the existence of multifactorial causes responsible for lactation insufficiency.

The Effect of Stress on Lactation

Figure 3 depicts potential means by which stress can affect lactation. It may directly suppress the hormones implicated in milk synthesis/ejection and/or indirectly through its effect on other physiologic responses or maternal behavior. It is important to remember that if stress affects lactation, in return, lactation affects how a lactator responds to stress. Research, in fact, has focused significantly more on the latter aspect, i.e., how stress response differs between lactators and non-lactators, than on the former, namely, if and how stress alters lactation. Teleologically, the existence of such negative feedback may be viewed as benefiting the survival of the young.

Clinical studies investigating the effect of stress on lactation have been limited. Ueda et al. (1994) measured the level of circulating oxytocin in women subjected to a psychological stress consisting of mental calculation or noise. Compared to non stressed counterparts, these mothers demonstrated a decrease in oxytocin level, but no change in prolactin or milk yield. Dewey et al (1994) observed no adverse effect following aerobic exercise (4-5 times/week) beginning 6-8 weeks postpartum on milk volume, composition, and plasma levels of prolactin. Although these studies would support the notion that not all stressors hamper lactation, careful interpretation of the results is needed. First, these studies examined the effect of acute and intermittent stress. It is well recognized that responses to acute vs. chronic stressors are different (Natelson, 1988). Indeed, prolonged exposure to specific stimuli has been shown to lead to physiologic adaptation. This is best illustrated by the extensive literature available on stress-induced analgesia (see rev. Vaccarino et al. 1999). Second, varied effects are observed depending upon the nature or type of stress administered. Based on our own studies (Lau, 1992), we are speculating that a stimulus may be more stressful when it is directed towards the young or threatens the integrity of the nursing dyad, than when directed towards the mother herself. Further studies are necessary to verify such hypothesis. Third, the intensity of the stressor is a variable that also comes into play. Threshold sensitivity to a particular stimulus will vary depending upon subjects and experimental conditions. It is known that in response to a particular stress, physiological responses may differ between strains, gender, and physiological states (Olsen et al., 1997; Cruz et al., 1996; Hard and Hansen, 1985). Of particular interest lactators exhibit increased aggression towards intruders and increased pain threshold when compared to non-lactators. It was mentioned earlier that a number of factors can affect the expression of maternal behavior, e.g., psychological make-up of the mother, her own health, lifestyle (e.g., work), education, perception of support. Therefore, if all the above elements were taken together, studies on the effect of stress on the lactation of mothers with preterm infants need to take into consideration both maternal as well as stress characteristics. The relative importance of these 2 categories of factors would explain why certain interventions to improve lactation, such as increased breastfeeding frequency, use of metoclopramide, or skin-to-skin holding, work for certain mothers and not for others.

In summary, this presentation aims to emphasize the importance of considering not only stress-specific factors, but also maternal elements when working with a preterm nursing dyad. A stimulus is only stressful if it is perceived as such by the recipient. For this reason and as a first step, it is suggested that there is a greater urgency to understand maternal characteristics in order to identify the stimuli that are stressful to individual mothers rather than to study the effect of stress *per se* on lactation.

References

D.R. Blatchford, K.A. Hendry, and C.J. Wilde, Autocrine regulation of protein secretion in mouse mammary epithelial cells, Biochem. Biophys. Res. Commun. 248:761-766 (1998).

M.L. Bowden, and N.J. Hopwood, Psychosocial dwarfism: identification, intervention and planning, Soc. Work Health Care 7:15-36 (1982).

Y. Cruz, M. Martinez-Gomez, J. Manzo, R. Hudson, and P. Pacheco, Changes in pain threshold during the reproductive cycle of the female rat, Physiol. Behav. 59:543-547 (1996).

I. De Andraca, and R. Uauy, Breastfeeding for optimal mental development. In: Simopoulos AP, Dutra de Oliveira JE, Desai ID (eds). Behavioral and metabolic aspects of breastfeeding. World Rev Nutr Diet. Basel: Karger; 1995 pp 1-27.

K.G. Dewey, C.A. Lovelady, L.A. Nommsen-Rivers, M.A. McCrory, B. Mand, and A. Lonnerdal, Randomized study of the effects of aerobic exercise by lactating women on breast-milk volume and composition, N. E. J. Med. 330:449-453 (1994).

L.M. Gartner, L.S. Black, A.P. Eaton, R.A. Lawrence, A.J. Naylor, M.E. Neifert, D. Ohare, R.J. Schanler, M. Georgieff, Y. Piovanetti, and J. Queenan, Breastfeeding and the use of human milk, Pediatrics 100:1035-1039 (1997).

C.E. Grosvenor, and F. Mena, Effect of auditory, olfactory and optic stimuli upon milk release and suckling-induced release of prolactin in lactating rats, Endocrinology 80:840-848 (1967).

E. Hard, and S. Hansen, Reduced fearfulness in the lactating rat, Physiol. Behav. 35:641-644 (1985).

C. Lau, and N. Hurst, Oral feeding in infants, Curr. Probl. Pediatr. 29:101-128 (1999).

C. Lau, Effects of various stressors on milk release in the rat, Physiol. Behav. 51:1157-1163 (1992).

K. Laud, I. Gourdou, L. Belair, D.H. Keisler, and J. Djiane, Detection and regulation of leptin receptor mRNA in ovine mammary epithelial cells during pregnancy and lactation, FEBS Lett. 463:194-198 (1999).

I. Merchenthaler, Induction of enkephalin in tuberoindundibular dopaminergic neurons of pregnant, pseudopregnant, lactating and aged female rats, Neuroendorcinology 60:185-193 (1994).

E.C. Meyer, C.T. Garcia Coll, R. Seifer, A. Ramos, E. Kilis, and W. Oh, Psychological distress in mothers of preterm infants, J. Dev. Behav. Pediatr. 16:412-417 (1995).

B.H. Natelson, J.E. Ottenweller, J.A. Cook, D. Pitman, R. McCarty, and W.N. Tapp, Effect of stressor intensity on habituation of the adrenocortical stress response, Physiol. Behav. 43:41-46 (1988).

G.A. Olson, R.D. Olson, A.L. Vaccarino, and A.J. Kastin, Endogenous opiates: 1997. Peptides 1998; 19:1791-1843.

P.N. Riskind, W.J. Millard, and J.B. Martin, Opiate modulation of the anterior pituitary hormone response during suckling in the rat, Endocrinology 114:1232-1237 (1984).

A.E. Ronca, R.A. Abel, and Alberts, Jr., Perinatal stimulation and adaptation of the neonate, Acta Paediatr. Suppl. 416:8-15 (1996).

S.M. Schanberg, G. Evonivk, and C.M. Kuhn, Tactile and nutritional aspects of maternal care: specific regulators of neuroendocrine function and cellular development, Proc. Soc. Exp. Biol. Med. 175:135- 146 (1984).

R.J. Schanler, The low birth weight infant. In: Walker WA, Watkins JB, eds. Nutrition in pediatrics: basis science and clinical applications. 2nd ed. Hamilton (ON): BC Decker Inc; 1996; pp 392-412.

L.T. Singer, A. Salvador, S. Guo, M. Collin, L. Lilien, and J. Baley, Maternal psychological distress and parenting stress after the birth of a very low-birth-weight infant, JAMA 281:799-805 (1999).

R. Tessier, M. Cristo, S. Velez, M. Giron, Z.F. de Calume, J.G. Ruiz-Palaez, Y. Charpak, and N. Charpak, Kangaroo mother care and the bonding hypothesis, Pediatrics 102:e17 (1998).

T. Ueda, Y. Yokoyama, M. Irahara, and T. Aono, Influence of psychological stress on suckling-induced pulsatile oxytocin release, Obstet. Gynecol. 84:259-262 (1994).

A.L. Vaccarino, G.A. Olson, R.D. Olson, and A.J. Kastin, Endogenous opiates: 1998. Peptides 1999; 20: 1527-574.

A.M. Widstrom, V. Wahlberg, S.A. Werner, and J. Winberg, Short-term effects of early suckling and touch of the nipple on maternal behavior, Early Hum. Dev. 21:153-163 (1990).

IMMUNE SYSTEM MODULATION BY HUMAN MILK

Lars Å. Hanson, Sven-Arne Silfverdal, Marina Korotkova, Valdemar Erling, Louise Strömbeck, Per Olcén, Marina Ulanova, Mirjana Hahn-Zoric, Shakila Zaman, Rifat Ashraf and Esbjörn Telemo*

1. INTRODUCTION

The mother's immune system seems to have a much more active role for the development and protection of her fetus and newborn than previously realized. We have recently learned that various cytokines of the mother's immune response to her fetus play a major role in actively monitoring pregnancy, such as mediating the implantation in the decidua of the fertilized egg and the growth of the placenta. Breastfeeding we have mainly thought of as a process providing passive protection of the infant by transferring various ready made protective factors like secretory IgA antibodies, lactoferrin, oligosaccharides functioning as receptor analogues etc. Such protection is well proven against numerous infections. Now we realize that breastfeeding also brings various active stimuli for the immune system of the infant providing possibilities for enhanced protection against infections also after the termination of breastfeeding.

2. Suggestions of a Long Term Improved Protection against Infections in Breastfed Infants

Saarinen reported already in 1982 that breastfeeding for >4 months decreased the risk of attracting otitis media up to the age of 3 years.[1] Howie et. al. gave evidence that breastfeeding beyond 13 weeks of age resulted in enhanced protection against diarrhoea for another 3 years.[2] The study of Howie et al. was extended and it was found that children who had been exclusively breastfed for 15 weeks, without any solid food being introduced during that period remained better protected against respiratory tract

*Lars Å.Hanson, Valdemar Erling, Louise Strömbeck, Marina Ulanova, Mirjana Hahn-Zoric, Esbjörn Telemo, Department of Clinical Immunology, Göteborg University. Marina Korotkova also Department of Pediatrics, Göteborg University, Göteborg. Sven-Arne Silfverdal, Per Olcén, Departments of Pediatrics and Microbiology and Immunology, Örebro Medical Centre Hospital, Örebro, Sweden. Shakila Zaman and Rifat Ashraf, Department of Social and Preventive Pediatris, King Edward Medical College, Lahore, Pakistan

Integrating Population Outcomes, Biological Mechanisms and Research Methods in the Study of Human Milk and Lactation
Edited by Davis *et al.*, Kluwer Academic/Plenum Publishers, 2002

infections for some 7 years compared to those who were not breastfed.[3] (See further Howie et al. in this publication).

There is also evidence that protection against wheezing illness may be enhanced by breastfeeding and remain so for up to 6-7 years after breastfeeding has ended, compared with non-breastfeeding.[3, 4] However, this improved protection was mainly seen among non-atopic children and it is assumed that the effect mostly stems from reduction i.a. of the acute viral infections which initiate 80 - 85% of asthma attacks among school children.[5] Interestingly it was noted that the protection was further improved for each additional week of breastfeeding.

Significant prevention of wheezing bronchitis up to the age of 18 months by exclusive breastfeeding was reported in another study.[6] They also found that those breastfed for < 3 months and with > 5 upper respiratory infections before 18 months showed a relative risk of wheezing bronchitis of 10.7.

Long term prevention of acute appendicitis and recurrent tonsillitis has been claimed.[7, 8] It will be interesting to see whether this will be confirmed in further investigations.

Two important studies, one with a case-control design, and one ecologic study, have been published on the long term protective effects of exclusive breastfeeding on the risk of attracting invasive *Haemophilus influenzae* type b infections.[9, 10] Infants exclusively breastfed 13 weeks or more were compared with those breastfed less than 13 weeks or more were compared with those breastfed less than 13 weeks. The longer breastfeeding gave significantly better protection. This association was stronger for children older than 12 months. The protection was enhanced for each additional week of lactation. The ecologic study indicated a protective effect at a population level and this effect was still present 5 and 10 years, but not 15 years, later.

In developing countries where the rate of infection usually is very high among the many poor children, the immediate protective role of breastfeeding is of fundamental importance. If, in addition, a long-term enhanced protection would remain after the termination of breastfeeding it would be important.

We have tried to search for a prolonged effect by breastfeeding on diarrhoea in our previous study material from Pakistan.[11] To avoid the many confounding factors involved with breastfeeding and its effects we have applied a newly developed statistical method where propesity schores are used to try to overcome these problems.[12] With this method we could not confirm any prolonged protective effect against diarrhoeal infections.[13] Considering the much heavier microbial exposure in Pakistan than in Industrialized countries where a long term protection has been shown[2] it might be that the enhanced immunity induced by breastfeeding is insufficient.

There might be additional reasons that breastfeeding in Pakistan may not lead to the same immuno-stimulatory effects as in the West. There is good evidence of high secretory IgA antibody contents in the milk also of often undernourished women in developing countries against numerous micro-organisms which is both effective and of many specificities.[14-16] In contrast, little if anything is known about the possible effect of undernutrition on many other immune-related components in the milk like the cells, the cytokines and the many various protective factors like lactoferrin etc. It may well be that inadequate nutrition of the mother could affect some of these many factors.

Recently we have studied the effect of reduction of the intake of essential fatty acids (EFA) in rat dams and their pups on their serum leptin.[17] The EFA deficient diet was given during the last ten days of gestation and throughout lactation. Firstly, low serum leptin levels were noted in the control rat dams during the whole lactation period. For those rat dams on the EFA deficient diet the serum leptin levels were even lower. The pups of the dams on the deficient diet had markedly low serum leptin levels although they via the milk seemingly drained their mothers of leptin (Korotkova et al., unpublished observation). Since the hormone leptin has the structure of a cytokine and has a strong stimulatory activity on the T helper type 1 side of the immune system[18] it is possible that this nutritional deficiency of EFA also leads to an immune deficiency of mothers and offspring.

Enhanced responses to vaccines during or after breastfeeding might also be taken as support for the thesis that breastfeeding may actively stimulate the immune response of the infant. Several studies have shown such an effect with vaccines against tetanus and diphtheria toxoids, BCG, *Haemophilus influenzae* type b (Hib) and live oral poliovirus (OPV).[19-21] Increased responses of serum antibodies as well as secretory IgM in the stool and secretory IgA in the saliva have been reported.[20]

Pickering et al.[22] showed that giving a formula enriched with nucleotides, in similarity to human milk resulted in higher antibody responses to Hib and diphtheria, but not to OPV and tetanus, than those breastfed and still higher than those given an identical formula without nucleotides. Breastfeeding was accompanied by increased titres against OPV compared to the other groups – even months after the termination of breastfeeding.

In other studies a better response was not found among the breastfed infants. In some of those studies live vaccines have been used and it cannot always be excluded that the vaccine has been neutralized by a breastfeed with all its maternal milk secretory IgA antibodies.[23] But studies with killed vaccines against Hib and tetanus toxoid have also failed to induce a better response among the breastfed.[24, 25] This may be related to the definition of breastfeeding used, the duration of breastfeeding and the immunological status of the mother. Recently we could show that idiotypic and anti-idiotypic antibodies given to newborn rats could have priming or tolerogenic effects on the immune response of the neonate, due to the dose used. Furthermore these effects could be noted in the next generation.[26]

Breastfeeding has also been reported to significantly enhance the size of the thymus in the infant; exclusive breastfeeding more than partial.[27] However, these findings have been debated.[28], 29 A further study supporting the initial finding has been presented

3. Suggestions of Down-modulating Effects of Immunological Diseases by Breastfeeding

Single studies suggest that breastfeeding may reduce the risk of attracting inflammatory bowel diseases like ulcerous colitis and Crohn's disease.[30, 31] There is more solid evidence that breastfeeding may diminish the risk of developing coeliac disease.[32] See the review in this volume by Persson et al.

The reports claiming protection against a few autoimmune diseases, like diabetes have caused much debate. This is further discussed by Dosch et al. in the present volume.

The debate on whether or not breastfeeding protects against allergic diseases has been going on for a long time and this problem has been repeatedly reviewed.[33, 34] One interesting new contribution in this field is the long-term study by Wright et al. which showed that children whose mothers had with severe asthma may increase their risk of attracting the disease with longer duration of breastfeeding.[35] Such a connection may have confused previous studies. The protective effect noted by breastfeeding in Wright's study was seen in non-atopic children with recurrent wheeze as also reported previously by other researchers.[3, 4] However, the study by Saarinen and Kajosaari[36] who followed 150 children for 17 years noted a significant protective effect by breastfeeding during the first 1-3 years against food allergy, later to be followed by protection against inhalant allergy. It should be noted that there were few cases of asthma in Saarinen-Kajosaari's study and asthma is a more complex disease than other inhalant allergies.

Another recent study from Australia of impressive size (n=2187) and a design with careful control of confounders and good statistics showed that exclusive breastfeeding <4 months significantly increased the risk of asthma, wheezing and positive skin tests to one common aeroallergen.[37]

4. Possible Mechanisms behind the Immunomodulating Effects of Human Milk

It has been shown that neonatal mice can have their immune system primed by anti-idiotypic antibodies.[26, 38] Such effects were also obtained if the anti-idiotypic antibodies reached the neonate via the milk.[38, 39] We have shown the presence of anti-idiotypic antibodies also in human milk.[40]

Already Beer and Billingham[41] indicated the possibility that viable leukocytes may be transferred from mother to offspring and that this might have positive, as well as negative effects. Since then numerous studies have demonstrated that such a transfer of milk cells takes place in several species including man and that it also includes lymphocytes.[42-46] In two studies transfer of enhanced specific immunity has been demonstrated.[46, 47] This topic is further discussed by Dr. Tuboly in the present volume.

A most striking effect of the milk cell uptake by the breastfed infant is that tolerance seems to appear in the offspring to the maternal HLA.[48-50] This was confirmed by foster feeding in rats.[51] Furthermore breastfed individuals had a lower frequency of precursors of cytotoxic T cells reacting with maternal HLA than non breastfed. As a consequence renal transplantation with the mother as a donor and the breastfed offspring as a recipient is more successful than if the father is the donor or if the offspring has not been breastfed.[48]

These observations agree with previous work showing that a tuberculin-positive mother may, at least temporarily, make her infant tuberculin-positive by breastfeeding.[52, 53] However, another study did not confirm this.[54]

An example of successful lymphocyte transfer from mother to offspring and was shown in B-cell deficient mice who could accept the maternal milk B-cells and survive thanks to their antibody production.[55]

Human milk contains numerous cytokines.[56] However their functions in the offspring are largely unknown, although it has been shown for TGF-β that foster feeding rescued TGF-β gene-disrupted newborn mice to perinatal survival and normal development.[57] The fact that the milk is relatively rich in IFN-γ has been considered to possibly counteract the risk of developing allergic diseases because of its capacity to depress the T helper cell promoted responses, which include production of IgE antibodies. Further, it has recently been discussed that allergic and non-allergic mothers may differ as to their milk cytokine content.[58]

Breastfed infants have been reported to respond with more IFN-γ production to respiratory syncytial virus than non-breastfed.[59] In agreement, measles-mumps-rubella vaccination in breastfed infants resulted in IFN-γ synthesis, not seen among non-breastfed.[60]

We recently found that children who attracted invasive Hib infections showed different antibody responses related to how long they had been breastfed previously.[61] Thus, among children 18 months or older a significantly higher IgG2 anti-Hib response was noted for those exclusively breastfed for > 13 weeks (mean 19.3 weeks) compared to those exclusively breastfed < 13 weeks (mean 5,4 weeks). It has already been shown that breastfeeding significantly enhances the protection against Hib for up to 10 years after the termination of breastfeeding.[9, 10] The new data suggests that one factor behind this enhanced protection could be the increased production of protective IgG2 antibodies against Hib. This isotope is known to be dependent especially of IFN-γ for its production.[62, 63] This IFN-γ may come from the milk originally.

Milk of rats contains leptin and it can be reduced in the serum of the offspring by a deficient intake of essential fatty acids of the rat dams during late gestation and lactation.[17] Since the hormone leptin is a cytokine by structure, which can stimulate T helper type1 immune responses (e.g. delayed type hypersensitivity) it might in adequate concentrations counteract T helper 2 dependent IgE production decreasing the risk for development of allergy.

Whereas it thus seems that breastfeeding may stimulate the immune system to enhanced protection, one might also consider that down-regulation may result. This could possibly be useful by decreasing the risk of developing immunologic diseases such as inflammatory bowel disease, or allergy. In a preliminary study in rats we put rat pups and dams either in a cage with ovalbumin (OA) in the food or in a cage without. Moving the dams from one cage to another we noted that a decreased immune response to OA was seen also among the pups in the cage without OA, but after suckling a dam who had eaten OA in the other cage. Bystander tolerance was not seen to another unrelated antigen suggesting that the down regulation was not mediated by a non specifically immunosuppressive cytokine as TGF-β, as we have seen in the gut of orally tolerized rats.[64]

We investigated how restriction of dietary EFA intake in rat dams influenced the development of oral tolerance to the food antigen ovalbumin (OA) in their pups. During lactation the dams were either given OA orally, or not. At 3 weeks of age all pups were immunized s.c. with OA. The T cell and antibody response to OA were similar in pups of the mothers fed the control diet either given OA orally or not. In the pups from the deficient dams given OA orally, the T cell and antibody response to OA were significantly decreased and the TGF-β mRNA levels was significantly higher in the draining lymph nodes compared to those from of immunized pups of dams not given OA. Our data indicate that the restriction of dietary EFA intake promotes the development of oral tolerance in the pups via the milk. Thus non-optimal feeding may have effects on development of immunological tolerance to dietary antigen ingested by the mother.[65] These early observations obviously need to be confirmed.

References

1. U.M. Saarinen, Prolonged breast feeding as prophylaxis for recurrent otitis media, Acta Paediatr. Scand. 71:567-571 (1982).
2. P.W. Howie, J.S. Forsyth, S.A. Ogston, A. Clark and C. du Florey, Protective effect of breast feeding against infection, BMJ 300:11-16 (1990).
3. A.C. Wilson, J.S. Forsyth, S.A. Greene, L. Irvine, C. Hau and P.W. Howie, Relation of infant diet to childhood health: seven year follow up of cohort of children in Dundee infant feeding study, BMJ 316:21-25 (1998).
4. M.L. Burr, E.S. Limb, M.J. Maguire, L. Amarah, B.A. Eldridge, J.C. Layzell, and T.G. Merrett, Infant feeding, wheezing, and allergy: a prospective study, Arch. Dis. Child. 68:724-728 (1993).
5. S.L. Johnston, P.K. Pattemore, G. Sanderson, S. Smith, F. Lampe, L. Josephs, P. Symington, S. O'Toole, S.H. Myint, D.A. Tyrrell, and et al., Community study of role of viral infections in exacerbations of asthma in 9-11 year old children, BMJ 310:1225-1229 (1995).
6. E. Rylander, G. Pershagen, M. Eriksson, and L. Nordvall, Parental smoking and other risk factors for wheezing bronchitis in children, Eur. J. Epidemiol. 9:517-526 (1993).
7. A. Pisacane, U. de Luca, N. Impagliazzo, M. Russo, C. De Caprio, and G. Caracciolo, Breast feeding and acute appendicitis, BMJ 310:836-837 (1995).
8. A. Pisacane, N. Impagliazzo, C. De Caprio, L. Criscuolo, A. Inglese, and M.C. Pereira de Silva, Breast feeding and tonsillectomy, BMJ 312:746-747 (1996).
9. S.A. Silfverdal, L. Bodin, S. Hugosson, O. Garpenholt, B. Werner, E. Esbjörner, B. Lindquist, and P. Olcén, Protective effect of breastfeeding on invasive Haemophilus influenzae infection: a case-control study in Swedish preschool children, Int. J. Epidemiol. 26:443-450 (1997).
10. S.A. Silfverdal, L. Bodin, and P. Olcén, Protective effect of breastfeeding: an ecologic study of Haemophilus influenzae meningitis and breastfeeding in a Swedish population, Int. J. Epidemiol. 28:152-156 (1999).
11. R.N. Ashraf, F. Jalil, S.R. Khan, S. Zaman, J. Karlberg, B.S. Lindblad, and L.Å. Hanson, Early child health in Lahore, Pakistan: V. Feeding patterns, Acta Paediatr. Suppl. 82 Suppl 390:47-61 (1993).
12. A. Carlquist, V. Erling, and M. Frisén, Longitudinal methods for analysis of the influence of breastfeeding on early child health in Pakistan. Research Report, Dept. Statistics, Göteborg University, 11 (1999).
13. V. Erling, A. Carlquist, M. Frisén, S. Zaman, and L.Å. Hanson, Seasonal impact on early childhealth in relation to short and long-term effect of breastfeeding in Lahore, Pakistan (in manuscript).
14. R.I. Glass, A.M. Svennerholm, B.J. Stoll, M.R. Khan, K.M. Hossain, M.I. Huq, and J. Holmgren, Protection against cholera in breast-fed children by antibodies in breast milk, N. Engl. J. Med. 308:1389-1392 (1983).
15. J.R. Cruz, L. Gil, F. Cano, P. Caceres, and G. Pareja, Breast milk anti-Escherichia coli heat-labile toxin IgA antibodies protect against toxin-induced infantile diarrhea, Acta Paediatr. Scand. 77:658-662 (1988).
16. G.M. Ruiz-Palacios, J.J. Calva, L.K. Pickering, Y. Lopez-Vidal, P. Volkow, H. Pezzarossi, and M.S. West, Protection of breast-fed infants against Campylobacter diarrhea by antibodies in human milk, J. Pediatr. 116:707-713 (1990).
17. M. Korotkova, B. Gabrielsson, L.Å. Hanson, and B. Strandvik, Maternal essential fatty acid deficiency depresses serum leptin levels in suckling rat pups, J. Lipid Res. 42:359-365 (2001).
18. G.M. Lord, G. Matarese, J.K. Howard, R.J. Baker, S.R. Bloom, and R.I. Lechler, Leptin modulates the T-cell immune response and reverses starvation-induced immunosuppression, Nature 394:897-901 (1998).
19. H.F. Pabst, J. Godel, M. Grace, H. Cho, and D.W. Spady, Effect of breast-feeding on immune response to BCG vaccination, Lancet 1:295-297 (1989).
20. M. Hahn-Zoric, F. Fulconis, I. Minoli, G. Moro, B. Carlsson, M. Böttiger, N. Räihä, and L.Å. Hanson, Antibody responses to parenteral and oral vaccines are impaired by conventional and low protein formulas as compared to breast-feeding, Acta Paediatr. Scand. 79:1137-1142 (1990).
21. D.P. Greenberg, C.M. Vadheim, S. Partridge, S.J. Chang, C.Y. Chiu, and J.I. Ward, Immunogenicity of Haemophilus influenzae type b tetanus toxoid conjugate vaccine in young infants, J. Infect. Dis. 170:76-81 (1994).
22. L.K. Pickering, D.M. Granoff, J.R. Erickson, M.L. Masor, C.T. Cordle, J.P. Schaller, T.R. Winship, C.L. Paule,and M.D. Hilty, Modulation of the immune system by human milk and infant formula containing nucleotides, Pediatrics 101:242-249 (1998).

23. M.E. Pichichero, Effect of breast-feeding on oral rhesus rotavirus vaccine seroconversion: a metaanalysis, J. Infect. Dis. 162:753-755 (1990).
24. N. Watemberg, R. Dagan, Y. Arbelli, I. Belmaker, A. Morag, L. Hessel, B. Fritzell, A. Bajard, and L. Peyron, Safety and immunogenicity of Haemophilus type b-tetanus protein conjugate vaccine, mixed in the same syringe with diphtheria-tetanus-pertussis vaccine in young infants, Pediatr. Infect. Dis. J. 10:758-763 (1991).
25. S. Stephens, C.R. Kennedy, P.K. Lakhani, and M.K. Brenner, In-vivo immune responses of breast- and bottle-fed infants to tetanus toxoid antigen and to normal gut flora, Acta Paediatr. Scand. 73:426-432 (1984).
26. B.S. Lundin, A. Dahlman-Höglund, I. Pettersson, U.I.H. Dahlgren, L.Å. Hanson, and E. Telemo, Antibodies given orally in the neonatal period can affect the immune response for two generations: Evidence for active maternal influence on the newborn's immune system, Scand. J. Immunol. 50:651-656 (1999).
27. H. Hasselbalch, M.D. Engelmann, A.K. Ersboll, D.L. Jeppesen, and K. Fleischer-Michaelsen, Breast-feeding influences thymic size in late infancy, Eur. J. Pediatr. 158:964-967 (1999).
28. J.M. Thompson, D.M. Becroft, and E.A. Mitchell, Previous breastfeeding does not alter thymic size in infants dying of sudden infant death syndrome, Acta Paediatr. 89:112-114 (2000).
29. A.M. Prentice, and A.C. Collinson, Does breastfeeding increase thymus size?, Acta Paediatr. 89:8-12 (2000).
30. S. Koletzko, P. Sherman, M. Corey, A. Griffiths, and C. Smith, Role of infant feeding practices in development of Crohn's disease in childhood, BMJ, 298:1617-1618 (1989).
31. S. Koletzko, A. Griffiths, M. Corey, C. Smith, and P. Sherman, Infant feeding practices and ulcerative colitis in childhood, BMJ 302:1580-1581 (1991).
32. S. Auricchio, D. Follo, G. de Ritis, A. Giunta, D. Marzorati, L. Prampolini, N. Ansaldi, P. Levi, D. Dall'Olio, and A. Bossi, Does breast feeding protect against the development of clinical symptoms of celiac disease in children, J. Pediatr. Gastroenterol. Nutr. 2:428-433 (1983).
33. M.S. Kramer, Does breast feeding help protect against atopic disease? Biology, methodology, and a golden jubilee of controversy, J. Pediatr. 112:181-190 (1988).
34. L.Å. Hanson, Breastfeeding provides passive and likely long-lasting active immunity, Ann. Allergy Asthma Immunol. 81:523-533 (1998).
35. A.L. Wright, C.J. Holberg, M. Taussig, and F.D. Martinez, Maternal asthma status alters the relation of infant feeding to asthma and recurrent wheeze in childhood, Thorax. 56:192-197 (2001).
36. U.M. Saarinen, and M. Kajosaari, Breastfeeding as prophylaxis against atopic disease: prospective follow-up study until 17 years old, Lancet, 346:1065-1069 (1995).
37. W.H. Oddy, P.G. Holt, P.D. Sly, A.W. Read, L.I. Landau, F.J. Stanley, G.E. Kendall, and P.R. Burton, Association between breast feeding and asthma in 6 year old children: findings of a prospective birth cohort study, BMJ, 319:815-819 (1999).
38. K.E. Stein and T. Söderström, Neonatal administration of idiotype or antiidiotype primes for protection against Escherichia coli K13 infection in mice, J. Exp. Med. 160:1001-1011 (1984).
39. Y. Okamoto, H. Tsutsumi, N. Kumar, and P. Ogra, Effect of breast feeding on the development of anti-idiotype antibody response to F glycoprotein of respiratory syncytial virus in infant mice after post-partum maternal immunization, J. Immunol. 142:2507-2512 (1989).
40. M. Hahn-Zoric, B. Carlsson, S. Jeansson, O. Ekre, A.D. Osterhaus, D.M. Roberton, and L.Å. Hanson, Anti-idiotypic antibodies to poliovirus in commercial immunoglobulin, human serum and human milk, Pediatr Res. 33:475-480 (1993).
41. A.E. Beer, and R.E. Billingham, Immunologic benefits and hazards of milk in maternal-perinatal relationship, Ann. Intern. Med. 83:865-871 (1975).
42. I.J. Weiler, W. Hickler and R. Sprenger, Demonstration that milk cells invade the suckling neonatal mouse, Am. J. Reprod. Immunol. 4:95-98 (1983).
43. R.F. Sheldrake, and A.J. Husband, Intestinal uptake of intact maternal lymphocytes by neonatal rats and lambs, Res. Vet. Sci. 39:10-15 (1985).
44. L. Jain, D. Vidyasagar, M. Xantou, V. Ghai, S. Shimada, and M. Blend, Distribution of human milk leukocytes after ingestion by newborn baboons, Arch. Dis. Child, 80:291-299 (1989).
45. C. Siafakas, W. Anderson, A. Walker, and M. Xanthou, Breast milk cells and their interaction with intestinal mucosa. (Elsevier, Amsterdam, 1997).
46. S. Tuboly, S. Bernath, R. Glavits, A. Kovacs, and Z. Megyeri, Intestinal absorption of colostral lymphocytes in newborn lambs and their role in the development of immune status, Acta Vet. Hung, 43:105-115 (1995).

47. S.N. Kumar, G.L. Stewart, W.M. Steven, and L.L. Seelig, Maternal to neonatal transmission of T-cell mediated immunity to Trichinella spiralis during lactation, Immunology, 68:87-92 (1989).
48. D.A. Campbell, Jr., M.I. Lorber, J.C. Sweeton, J.G. Turcotte, J.E. Niederhuber, and A.E. Beer, Breast feeding and maternal-donor renal allografts. Possibly the original donor-specific transfusion, Transplantation, 37:340-344 (1984).
49. W.E. Kois, D.A. Campbell, Jr., M.I. Lorber, J.C. Sweeton, and D.C. Dafoe, Influence of breast feeding on subsequent reactivity to a related renal allograft, J. Surg. Res. 37:89-93 (1984).
50. L. Zhang, S. van Bree, J.J. van Rood, and F.H. Claas, Influence of breast feeding on the cytotoxic T cell allorepertoire in man, Transplantation, 52:914-916 (1991).
51. A. Deroche, I. Nepomnaschy, S. Torello, A. Goldman, and I. Piazzon, Regulation of parental alloreactivity by reciprocal F1 hybrids. The role of lactation, J. Reprod. Immunol. 23:235-245 (1993).
52. J.J. Schlesinger, and H.D. Covelli, Evidence for transmission of lymphocyte responses to tuberculin by breast-feeding, Lancet 2:529-532 (1977).
53. S.S. Ogra, D. Weintraub, and P.L. Ogra, Immunologic aspects of human colostrum and milk. III. Fate and absorption of cellular and soluble components in the gastrointestinal tract of the newborn, J. Immunol. 119:245-248 (1977).
54. M.A. Keller, A.L. Rodgriguez, S. Alvarez, N.C. Wheeler, and D. Reisinger, Transfer of tuberculin immunity from mother to infant, Pediatr. Res. 22, 277-281 (1987).
55. E. Gustafsson, M. Arvola, L. Svensson, A. Mattsson, and R. Mattssson, Postnatally transmitted maternal B cells can cause long-term maintenance of serum IgG in B cell-deficient mice nursed by phenotypically normal dam (in manuscript),
56. A.S. Goldman, S. Chheda, R. Garofalo, and F.C. Schmalstieg, Cytokines in human milk: properties and potential effects upon the mammary gland and the neonate, J. Mammary Gland. Biol. Neoplasia, 1:251-258 (1996).
57. J.J. Letterio, A.G. Geiser, A.B. Kulkarni, N.S. Roche, M.B. Sporn, and A.B. Roberts, Maternal rescue of transforming growth factor-beta 1 null mice, Science 264:1936-1938 (1994).
58. M. Böttcher, Cytokines, and chemokines in breast milk from allergic and nonallergic mothers, ACI Internat. 12:153-160 (2000).
59. Y. Chiba, T. Minagawa, K. Mito, A. Nakane, K. Suga, T. Honjo, and T. Nakao, Effect of breast feeding on responses of systemic interferon and virus-specific lymphocyte transformation in infants with respiratory syncytial virus infection, J. Med. Virol. 21:7-14 (1987).
60. H.F. Pabst, D.W. Spady, L.M. Pilarski, M.M. Carson, J.A. Beeler, and M.P. Krezolek, Differential modulation of the immune response by breast- or formula-feeding of infants, Acta Paediatr. 86:1291-1297 (1997).
61. S.A. Silfverdal, L. Bodin, M. Ulanova, M. Hahn-Zoric, L.Å. Hanson, and P. Olcén, Antibody response to Hib in children with invasive Hib infection in relation to duration of exclusive breastfeeding (submitted),
62. R. Inoue, N. Kondo, Y. Kobayashi, O. Fukutomi, and T. Orii, IgG2 deficiency associated with defects in production of interferon-gamma; comparison with common variable immunodeficiency, Scand. J. Immunol. 41:130-134 (1995).
63. Y. Kawano, and T. Noma, Role of interleukin-2 and interferon-gamma in inducing production of IgG subclasses in lymphocytes of human newborns, Immunology 88:40-48 (1996).
64. M.R. Karlsson, H. Kahu, L.Å. Hanson, E. Telemo, and U.I. Dahlgren, Tolerance and bystander suppression, with involvment of CD25-positive cells, is induced in rats receiving serum from ovalbumin-fed donors, Immunology 100:26-33 (2000).
65. M. Korotkova, E. Telemo, L.Å. Hanson, and B. Strandvik, Modulation of neonatal oral tolerance to ovalbumin by maternal dietary essential fatty acids intake in rats. In manuscript.

INTESTINAL ABSORPTION OF COLOSTRAL LYMPHOID CELLS IN NEWBORN ANIMALS

S. Tuboly and S. Bernáth

1. INTRODUCTION

In species with an epitheliochorial placenta, colostrum-derived maternal immunoglobulins play a decisive role in the passive protection of newborn animals (Porter et al., 1970; McDowell 1973; Bourne and Curtis 1973; Brown et al., 1975; Newby and Bourne 1977; Butler et al., 1981; Banks 1982; Newby et al., 1982). At the end of pregnancy, numerous lymphoid cells from the mother's common mucosal system" (Bienenstock 1974; Solmon 1987) also accumulate in the mammary gland. These cells include T- and B-lymphocytes (Parmely and Beer 1977; Roux et al., 1977; Manning and Parmely 1980; Seelig 1980) which are transferred into the digestive tract of newborn animal with the colostrum. The newborn receives a significant number of cells trough mammary secretions, for example piglets ingest about 500-700 million viable maternal cells daily (Le Jan 1996). In rats and sheep, Sheldrake and Husband (1985) as well as Seelig and Head (1987) demonstrated that syngeneic and allogeneic maternal peripheral lymphocytes may be absorbed from the digestive tract.

There is a diversity of opinion concerning the role of these cells. In vitro studies showed, that they had antimicrobial, cytotoxic, immunoglobulin- and interferon producing properties (Riedel-Caspari and Schmidt 1991). According to some authors, they exert a protective effect in the digestive tract, while others suggest that through their mediator substances they may result in a stimulating effect on the immune system of newborn (Beer et al., 1975; Head et al., 1977), or likely have a role in long-lasting active immunity (Hanson 1998).

The ability of colostral lymphoid cells to absorb from the intestinal tract of newborn animals and their role in the development of the immune status were studied in 3 model experiments of swine and sheep.

*Tuboly, S. Szent István University Faculty of Veterinary Science, Budapest, 1078. István u. 2, Hungary. Tel.: 36-1-251-99-00 Fax: 36-1-251-92-60, e-mail: tubolys@novell.vmri.hu, Bernáth, S. Institute for Veterinary Medicinal Products, Szállás u. 8. 1107 Budapest, Hungary

Integrating Population Outcomes, Biological Mechanisms and Research Methods in the Study of Human Milk and Lactation
Edited by Davis *et al.*, Kluwer Academic/Plenum Publishers, 2002

2. Materials and Methods

Isolation of lymphoid cells. The lymphoid cells were isolated from the colostrum and blood samples with Ficoll-Paque, washed twice in phosphate-buffered saline (PBS), and the cell density was adjusted to $3x10^6$/ ml (sows), or $5x10^6$/ml (ewes).

Radiolabelling of cells. Lymphocytes were labelled with $Na^{99m}TcO_4$, in isotonic NaCl solution (Barth and Gillespie 1974).

In accordance with the objective of these experiments, some of the labelled colostral cell samples were subjected to a temperature of 56°C in a water bath for 20 min.

Histological examination and electron-microscopy. Examinations were done according to the methods described in manuals.

Radio Immuno Assay (RIA). The tetanus antibody titre was measured by RIA, the sensitivity of the method was 0.004 IU/ml (Bernáth and Habermann 1974).

Autoradiography. The sections, the lymph and blood smears were examined by autoradiography, by dipping the plates into Ilford Nuclear G emulsion. This was followed by exposure at + 4°C for 6 days in a dark room. The plates were developed in ORWO-A 49 solution, and stained with haematoxylin-eosin, and evaluated by light microscopy.

Animal experiments. The intestinal absorption of colostral lymphoid cells was studied in 3 model experiments (Tables 1, 2 and 3). The labelled lymphoid cells were administered into the digestive tract of newborn animals between postpartum hours 7-10. The animals got tea with sugar during this period. The lambs in the third model experiment got colostrum.

The intestinal absorption of radioisotope labelled colostral and blood lymphoid cells were studied in 23 piglets of four sows (A, B, C, D). B* and C*: heat-treated lymphoid cells.

The labelled cells were administered into the digestive tract of the piglets between postpartum hours 7 and 10. The piglets of sows A and B were subjected to laparotomy and the cells were injected directly into the stomach and duodenum, respectively, as indicated by the Table 1. The cell suspension was administered to the piglets of sows C and D through a naso-oesophageal tube. The animals in the different groups received cells from colostrum or blood of their own mother, or of a sow other than their mother. Three of the piglets of sows B and C received colostral cells from their own mother inactivated at 56°C.

Eight hours after they received the labelled cells, the piglets were killed by bleeding under ketamine-HCl (16 mg/kg) anaesthesia. Samples were taken from different segments of the digestive tract (duodenum and jejunum) and from the mesenteric lymph nodes, from the mesenteric lymph duct and from the blood. Some of the samples were embedded according to Saínte-Marie (1962), while cryostat sections were prepared from others.

Table 1
Porcine model experiment

Designation of sow	Designation of piglets	Lymphoid cells, used for treating piglets (10^7 cells)	Route of application
	A1	Col. A	
	A2		
A	A3	Col. B	after laparotomy into the stomach
	A4		
	A5	Blood A	
	A6		
	B1	Col. B	
	B2		
B	B3	Blood B	after laparotomy into the duodeum
	B4		
	B5	Col. B*	
	C1	Col. C	
	C2		
	C3	Col. B	
	C4		
C	C5	Blood C	Oseophageal tube
	C6		
	C7	Col. C*	
	C8		
	D1		
D	D2	Col. E	Oesophageal tube
	D3		
	D4		

Twenty ewes (group A) were treated with 3 ml tetanus anatoxin twice, while the remaining animals (group B) were left uninoculated. Lambs of group A2 were separated from their dams immediately after birth, then were administered, through an oesophageal tube, 10 ml of a suspension of lymphoid cells (cell density: $5x10^6$/ml) separated from the own maternal colostrum. Subsequently, the lambs were interchanged with lambs of nonimmunized ewes of group B1, namely they where put out to nursing. The lambs of groups A1 and B2 remained with their own dams.

At three days of age each lambs were inoculated with 3 ml tetanus anatoxin, then blood samples were taken from them 5 times in a period of 27 days for comparative examination of the humoral immune reactions.

3. Results

In the pigs model experiments we established that the lymphoid cells of the colostrum get absorbed from the digestive tract of the newborns, and by autoradiography revealed the labelled cells intercellularly in the epithelium of duodenal and jejunal

Table 2
Sheep model experiment

The intestinal absorption of colostral lymphoid cells in 17 lambs of 14 merino ewes was studied with the above described methods

Group	No. of lambs	Syngeneic cells	Allogeneic cells	Blood cells	Route of application
I.	1a, 2, 3, 4, 5a, 6	5 x 10^7	—	—	laparotomy, into the duodeum
	7a, 8, 9	5 x 10^7	—	—	oesophageal tube
II.	1b, 5b, 7b	—	5 x 10^7	—	duodenum
III.	10, 11, 12	—	—	3 x 10^7	duodenum
IV.	13, 14	—	—	—	duodenum 10 ml PBS

Explanation: lambs designated 1b, 5b, 7b (group II) are twins of lambs designated 1a, 5a, 7a (group I)

sections of piglets that had received syngenic colostral cells from their own mother (piglets A1, A2, Bl, B2, C1 and C2). By electron microscopic examination the lymphocytes were demonstrable among the epithelial cells of the jejunal mucosa in the intercellular spaces (Fig.1). Cells of maternal origin were present in smears made from the lymph of the mesenteric lymph ducts and in the cortical parts of the mesenteric lymph nodes of these piglets.

In those piglets that had received colostral allogenic cells of sows other than their own mother (piglets A3, A4, C3, C4, D1, D2, D3 and D4) the labelled cells were demonstrable in the epithelium of the intestinal wall but they were not absorbed.

In piglets that received lymphoid cells isolated from their own mother's blood (piglets A5, A6, B3, B4, C5 and C6), absorption of these lymphocytes was not observed either in the duodenum or in the jejunum.

No absorption of lymphocytes took place in piglets B5, C7 and C8, which had received inactivated colostral cells from their own mothers.

The sheep model experiments resulted in similar findings.

The third model experiment (Table 3) revealed that in lambs treated with colostral cells from immunised ewes (group A2) antibodies appeared on days 17-27 after

Table 3
Third model experiment
The effect of absorbed colostral lymphoid cells on the development of immune status of the newborn lambs was studied in a sheep model experiment involving 40 lambs of 40 ewes

EWES	Group	A		B	
	No.	20		20	
	Tetanus anatoxin	3 ml twice		—	
LAMBS	Group	A1	A2	B1	B2
	No.	10	10	10	10
	Colostral cells	—	5×10^7	—	—
	Tetanus anatoxin	3 ml	3 ml	3 ml	3 ml
Blood sampling after vaccination: days 0, 3, 6, 10, 17 and 27					

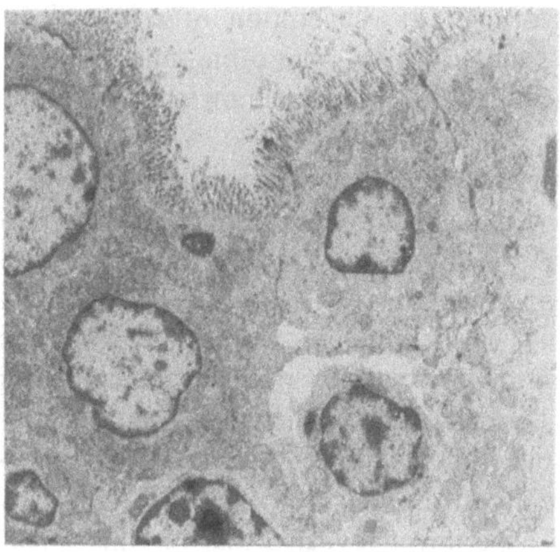

Figure 1. A lymphocyte in the dilated intercellular space between the epithelial cells. Electron micrograph, x 9600

vaccination in higher titers (0.018 and 0.038 IU/ml) than in group B2 (0.015 and 0.016 IU/ml). This allowed us to conclude that in group A2 cells carrying immunological memory were transferred with the colostral lymphocytes.

The average antibody titers in A1 and B1 groups were 6 IU and 5 IU at 3 days of age at the time of vaccination and 2 IU and 1 IU at 27 days after vaccination in the same groups. After the vaccination could not be detected any arisement of antibody titers. The results obtained showed, that maternal immunoglobulins taken up with the colostrum inhibited the development of the lambs' active immune response through a "feedback" effect.

4. Discussion

The results demonstrated that in the pigs and lambs the syngeneic colostral lymphoid cells were exclusively absorbed from the digestive tract and reached the circulation of newborn. Our results support the view of Kmetz et al. (1970), Seelig and Billingham (1981) and Weiler et al. (1983) that lymphoid cells can cross the gastrointestinal barrier. In contrast, other authors (Silvers and Poole 1975; Miller 1981) could not observe such absorption, perhaps due to inadequate sensitivity of the methods applied. In this study, colostral cells were labelled with the radioisotope technetium-99m. The advantages of technetium-99m over other radionucleides in labelling cells for *in vitro* immunological tests were demonstrated by Barth et al. (1974). The main advantages are, that the technetium-99m labelling does not influence the protein and DNA synthesis of the cells in a detectable measure. Furthermore, the technetium-99m, which was released from the labelled cells did not appear to be reutilizable for other cells.

The reason that cells from allogeneic maternal colostrum fail to cross the gastrointestinal barrier is presumably the MHC incompatibility.

Though lymphoid cells derived from the blood of syngeneic mothers could be observed in the mucosal epithelium, they failed to reach the mesenteric lymph nodes. These observations suggest that the absorption of colostral lymphoid cells may be regulated by mechanisms similar to those controlling migration into, and accumulation in the mammary gland as well as transepithelial transfer (Seelig and Beer 1987; Manning and Parmelly 1980).

As to the mode of absorption, several possibilities have been proposed, like the immunoglobulin molecules, the cells in question may get into the epithelium through the microvilli. Other authors are of the opinion that this process takes place through the intercellular spaces of the epithelium (Ogra et al., 1977; Seelig and Billingham 1981). Electron-microscopy studies on lymphocyte absorption have, however, shown that this process took place through the intercellular spaces of the epithelium. In our opinion, this finding can be attributed, besides the above-mentioned selectivity, to the activity of intact lymphoid cells, as in our experiments the heat-treated, inactivated colostral cells failed to become absorbed from the digestive tract.

The third model experiment revealed that in lambs treated with colostral cells from immunized ewes antibodies appeared in higher titres, than in the lambs of unimmunized ewes. This allows us to conclude that immunological memory transferred with the colostral lymphocytes, resulting in a secondary type immune response to tetanus anatoxin. The results obtained for groups A1 and B1 in the third model experiment demonstrated that maternal immunoglobulins taken up with the colostrum inhibited the development of the lambs' active immune response through a "feed-back" effect.

The results of this study indicated that lymphocytes from syngeneic maternal colostrum reach the newborn lamb's lymph circulation, remain immunologically active and may have transferred immune memory. Thus, they may play a role in the development of the newborn's immune status.

Summary

Intestinal absorption of colostral lymphoid cells was studied in 23 piglets of four sows and 17 lambs of ewes.

From the colostrum and blood of the dams the lymphoid cells were isolated with Ficoll-Paque and labelled with technetium ($Na^{99m}TcO_4$). In the 7th hour after birth, 10 ml volume of the cell suspensions (piglets: 10^7 cells, lambs: 5x 10^7 cells) were injected, following laparotomy, directly into the stomach or into the jejunum, or through a naso-oesophageal tube. Cryostat sections of duodenum, jejunum and lymph node samples of animals were examined by autoradiography.

It was found that lymphoid cells present in the colostrum of a piglet and lamb of their own mother were absorbed from the digestive tract and, via the lymphatic vessel, were transported to the mesenteric lypmph nodes. Electron microscopy revealed that absorption takes place intercellularly. Colostral cells of sows other than a piglet's own mother (allogeneic cells), the lymphoid cells isolated from the blood and heat-treated colostral lymphoid cells were not absorbed.

The immunization of ewes and their lambs by tetanus anatoxin demonstrate, that the absorbed lymphoid cells remain immunologically active, and may transfer immune information to the lambs.

References

K.L. Banks, Host defense in the newborn animal, J. Am.Vet. Med. Assoc. 181:1053-1056 (1982).

R.F. Barth, and G.Y. Gillespie, The use of technetium-99m as a radioisotopic label to assess cell-mediated immunity in vitro, Cell. Immunul. 10:38-39 (1974).

A.E. Beer, R.E. Billingham, and J.R. Haad, Natural transplantation during suckling, Transplant. Proc. 7:399-404 (1975).

S. Bernáth, and E. Habermann, Solid phase radioimmunoassay in antibody coated tubes for the quantitative determination of tetanus antibodies, Med. Microbiol. Immunol. 160:47-51 (1974).

J. Bienenstock, The physiology of the local immune response and the gastrointestinal tract. In: Brent, I. and Halborow, J. (eds) Progr. Immunol. Vol. II. 4. North Holland Publishing, p. 197 (1974).

F.J. Bourne, and J. Curtis, The transfer of immunoglobulin IgG, IgA and IgM from serum to colostrum and milk in the sow, Immunol. 24:157-162 (1973).

P.J. Brown, F.J. Boume, and H.R. Denny, Immunoglobulin containing cells in the pig mammary gland, J. Anat. 120:329-335 (1975).

J.E. Butler, F. Klobasa, and E. Werham, The differential localisation of IgA, IgM and IgG in the gut of suckled neonatal piglets, Vet. Immunol. Immunopathol. 2:53-65 (1981).

L.A. Hanson, Brestfeeding provides passive and likely long-lasting active immunity, Ann. Allergy, Asthma Immunol. 81:523-533 (1998).

J.R. Head, A.E. Beer, and R.E. Billingham, Significance of the cellular component of the maternal immunologic endowment in milk, Transplant. Proc. 9:1465-1471 (1977).

M. Kmetz, H.W. Dunne, and R.D. Schultz, Leukocytes as carriers in the transmission of bovine leukemia, Am. J. Vet. Res. 31:637-641 (1970).
C. Le Jan, Cellular components of mammary secretions and neonatal immunity: a rewiew, Vet. Res. 27:403-417 (1996).
L.S. Manning, and M.J. Parmely, Cellular determinants of mammary cell mediated immunity in the rat. I. The migration on radioisotopically labelled T lymphocytes, J. Immunol. 125:2508-2514 (1980).
G.H. McDowell, Local antigenic stimulation of guinea pig mammary gland, Aust. J. Exp. Biol. Med. Sci. 51:237-245 (1973).
S.C. Miller, Failure to demonstrate morphologically the presence of colostral or milk cells in the wall of the gastrointestinal tract of the suckling neonatal mouse, J. Reprod. Immunol. 3:187-194 (1981).
T.J. Newby, and F.J. Bourne, The nature of the local immune system of the bovine mammary gland, J. Immunol. 118:461-465 (1977).
T.J. Newby, C.R. Stockes, and F.J. Boume, Immunological activities of milk, Vet. Immunol. Immunopathol 3:67-94 (1982).
S.S. Ogra, D. Weintraub, and P.L. Ogra, Immunologic aspects of human colostrum milk. III. Fate and absorption of cellular and soluble components in the gastrointestinal tract of newborn, J. Immunol. 119:245-248 (1977).
M.J. Parmely, and A.E. Beer, Colostral cell mediated immunity and the concept of common secretory immune system, J. Dairy Sci. 60:655-661 (1977).
P. Porter, D.E. Nookes, and D.W. Allen, Secretory IgA and antibodies to Escherichia coli in porcine colostrum and milk and their significance in the alimentary canal of the young pig, Immunol. 18:245-257 (1970).
G. Riedel-Caspari, and F.W. Schmidt, The influence of colostral leucocytes on the immune system of the neonatal calf. I. Effects on lymphocyte responses, Dtsch. Tierärztl. Wschr. 98:325-364 (1991).
M.E. Roux, M. McWilliams, J.M. Phillips Quagliata, M. Weisz Carrington, and M.E. Lamm, Origin of IgA secreting plasma cells in mammary gland, J. Exp. Med. 146:1311- 1332 (1977).
G. Sainte Marie, A paraffin embedding technique for studies employing immunofluorescence, J.Histochem. 10:250-256 (1962).
L.L. Seelig, Jr., Dynamics of leucocytes in rat mammary epithelium during pregnancy and lactation, Biol. Reprod. 22:1211-1217 (1980).
L.L. Seelig, Jr., and A.E. Beer, Transepithelial migration of leucocytes in mammary gland of lactating rats, Biol. Reprod. 18:736-741 (1987).
L.L. Seelig, and R.E. Billingham, Capacity of "transplanted" lymphocytes to traverse in intestinal epithelium of adult rats, Transplant. 32:208-314 (1981).
L.L. Seelig, and J.R. Head, Uptake of lymphocytes fed to suckling rats. An autoradiographic study of the transit of labelled cells through the neonatal gastric mucosa, J. Reprod. Immunol. 10:258-297. (1987).
R.F. Sheldrake, and A.J. Husband, Intestinal uptake of intact maternal lymphocytes by neonatal rats and lambs, Res. Vet. Sci. 39:10-15 (1985).
W.K. Silvers, and T.W. Poole, The influence of foster nursing on the survival and immunologic competence of mice and rats, J. Immunol. 115:1117-1121 (1975).
H. Solmon, The intestinal and mammary immune system in pigs, *Vet. Immunol. Immunopathol.* 17:367-388 (1987).
J. Weiler, W. Hickler, and R. Sprenger, Demonstration that milk cells invade the suckling neonatal mouse, Am. J. Reprod. Immunol. 4:95-98 (1983).

BREAST-FEEDING PROTECTS AGAINST CELIAC DISEASE IN CHILDHOOD – EPIDEMIOLOGICAL EVIDENCE

L. Å. Persson[a,c], A. Ivarsson[a,b], and O. Hernell[b]

1. INTRODUCTION

Celiac disease, or permanent gluten sensitive enteropathy, is now recognized as a common disease in many populations. Genetic susceptibility is a prerequisite, as is dietary exposure to wheat gluten or related proteins in rye and barley. Once initiated, the disease is considered to be life-long, although transient cases have been reported. However, on exclusion of wheat gluten and related proteins from the diet, the enteropathy and symptoms almost always resolve.

In the early 1970s, the highest occurrence rate ever reported for celiac disease came from Galway, Ireland, with 1 case in 597 births (1.7 per 1000), or possibly as high as 1 in 303 (3 per 1000) if adult cases were also included.[1] In the 1980s, the occurrence in the same population decreased to 1 case in 2500 births (0.4 per 1000).[2] At that time, a prevalence of 3.8 per 1000 was reported from a screening of Swedish blood donors,[3, 4] although only 1 case per 1000 was diagnosed in clinical practice.[5, 6]

Thereafter, screening studies conducted mainly in a number of European countries and the USA, both among blood donors and in population based settings, revealed many previously undiagnosed cases.

The prevalence varies among different populations, as illustrated by Estonia with no cases identified in a village of 1461 (95% CI 0-2.5),[7] Sardinia with 11 per 1000 (95% CI 6.2-17) in childhood, and recently, Algeria with a prevalence of 56 per 1000 (95% CI 42-71) reported in children of Saharawis.[9] Cases of celiac disease have now been reported from almost all continents.[9, 10, 11-15]

The prerequisite of exposure to dietary gluten proteins for the development of celiac disease is generally accepted. Whether or not other environmental exposures also contribute to celiac disease development is a controversial issue. It has been proposed that environmental factors, besides the presence of gluten in the diet, merely modify the

[a]Department of Public Health and Clinical Medicine, Epidemiology, and [b]Department of Clinical Sciences, Pediatrics , Umeå University , Sweden, and [c]ICDDR,B: Centre for Health and Population Research, Dhaka, Bangladesh

symptoms but do not influence the development of enteropathy. [16-18] Others suggest that all individuals with a genetic susceptibility will eventually develop celiac disease, and the impact of environmental factors is mainly to delay the onset.[19-21]

However, immunological processes seem to be crucial in the pathogenesis, and the concept of oral tolerance, or the absence of such, might be important. If so, environmental exposures, varying through life, are likely to play a role in determining whether or not enteropathy develops. Only very few of all potential causal environmental exposures have thus far been explored.

2. Infant Feeding and Celiac Disease

In the 1970s, a decline in the incidence of celiac disease was reported from England, Scotland, and Ireland. Changes in infant feeding practices, in Britain promoted by new recommendations,[22] were discussed as a possible explanation for the decline.[2, 23-25] These new recommendations advocated breast-feeding for a minimum of two weeks and preferably for four to six months, use of adapted infant formulas, avoidance of solids before the age of four months, and that cereals should not be added to the milk in bottle feeding.[22] A delayed introduction of dietary gluten was also generally recommended in many other countries.[26] As a consequence of the decline in incidence, there were discussions as to whether celiac disease was disappearing,[2, 23-25] or if age at onset was increasing and was being accompanied by a shift to more vague symptoms.[27, 28] However, at that time changes in infant feeding practices, comparable to those in Britain, were also adopted in Sweden with no observed change in incidence.[29, 30]

It was suggested as early as the 1950s that breast-fed infants have a later onset of celiac disease.[31] From the 1980s and onwards, case-referent studies demonstrated that children with celiac disease had been breast-fed for a significantly shorter duration than the referents.[32-34] Furthermore, there were discussions as to whether breast-feeding really protects from celiac disease, or merely reduces the symptoms of the enteropathy.[32] Moreover, it was emphasized that the effect of breast-feeding might be indirect, i.e. as a consequence of postponed introduction of infant formulas,[32] or reduced amount of dietary gluten consumed.[34]

During the 1990s, a larger consumption of wheat gluten was reported for healthy infants in Sweden and Italy compared to Finland, Denmark and Estonia, and the former countries reported a higher occurrence of celiac disease than the latter.[35-38] Furthermore, earlier experimental studies, as mentioned above, had suggested a dose-related response to gluten proteins by the small intestinal mucosa.[39] Together, these findings indicated that quantity of gluten might be an exposure of importance for celiac disease risk.

3. Evidence from the Swedish Epidemic of Celiac Disease

In the mid 1980s pediatricians throughout Sweden diagnosed an increasing number of children with celiac disease. Most of these cases were infants and young children with symptoms as severe as those that had been observed earlier.[40, 41] It was suggested that changes in infant feeding practices were the cause of the increasing trend.

The Swedish Pediatric Association appointed a Working Group for Celiac Disease consisting of members representing all regions of the country, with the commission to investigate the question further and, if possible, to find the underlying cause(s).

A national retrospective study confirmed the clinical impression; within a few years there had been a three-fold increase in occurrence of celiac disease in children.[41] Moreover, based on results from a European multi-center study, it became obvious that Sweden had a considerably higher occurrence of childhood celiac disease than any other country.[28] Therefore, a Swedish multi-center study was initiated in 1990 that comprised the following main parts: *i)* a prospective incidence register of celiac disease in children, and *ii)* an incident case-referent study to identify environmental risk factors for celiac disease and relate these to genetic susceptibility. Results in this paper are based on retrospective reporting 1973 - 1990 from 5 pediatric departments, as well as the prospective register 1991 - 1999 based on reporting from 14 pediatric departments.

The incidence rate of celiac disease in Swedish children was analyzed for the twenty-five-year period from 1973 to 1997. In the mid-1980s, the incidence rate in children below two years of age increased three-fold within a few years, and after a ten-year period of high incidence, it equally rapidly returned to the previous level (figure 1). During the whole study period, the incidence in children between two and fifteen years of

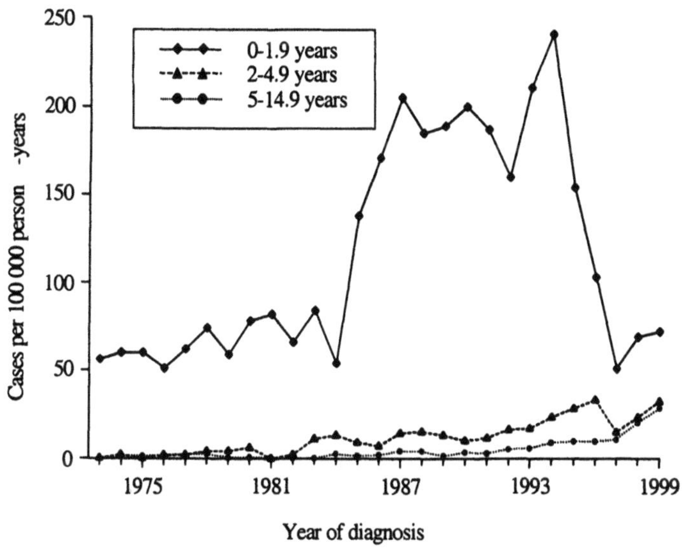

Figure 1. Annual incidence rates of celiac disease in different age groups of children from 1973 to 1999.

age increased, although it still remained at a considerably lower level than in the younger age group. According to a preliminary follow-up for the years 1998 and 1999 the incidence rate in children below two years of age has leveled off, while in the older children it continues to increase slowly (figure 1).

Throughout the study period a consistent pattern for all birth cohorts was that the cumulative incidence increased most rapidly during the first two years of life. However, the level reached at two years of age was considerably higher during the epidemic period compared to both the preceding years and the years that followed afterwards. All cohorts also had a gradual increase in cumulative incidence after two years of age. Interestingly, this was true also for those cohorts that had already reached a high level at two years of age.

4. Breast-Feeding, Dose of Gluten and Risk for Celiac Disease

An ecological study design was used to explore any temporal relationship between the epidemic incidence curve of celiac disease in children below two years of age and changes in infant dietary patterns.[42]

The rise in incidence in the middle of the 1980s was preceded by an increase in the amount of gluten consumed and a postponed introduction of dietary gluten from four until six months of age in the majority of infants, while breast-feeding habits were largely unchanged.[42] The steep decline in incidence in the middle of the 1990s coincided with a reduction in the amount of gluten consumed, although this was less pronounced than the earlier increase, and an ongoing increase in breast-feeding duration. From the incident case-referent study it was evident that celiac disease cases diagnosed before two years of age had a significantly shorter breast-feeding duration than their referents.[43]

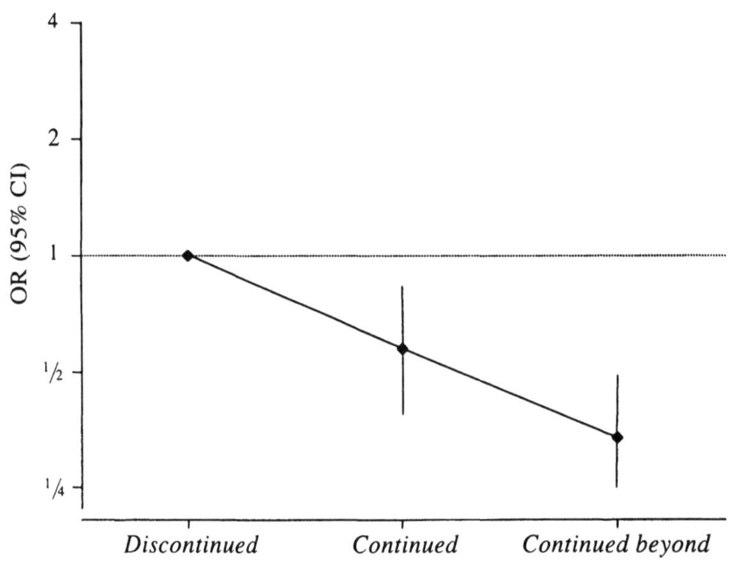

Figure 2. Breast-feeding status (BF) at introduction of gluten-containing flour into the diet and risk (OR, 95% CI) for celiac disease before two years of age.

Further analyses revealed a reduced risk for celiac disease before two years of age if the infant was still breast-fed when gluten-containing foods were introduced into the diet, an effect that was even more pronounced if breast-feeding also continued beyond that period (figure 2). Risk estimates were adjusted for the age of the infant when flour was introduced into the diet, and the amount that was then given. Large amounts of gluten at introduction increased the risk (OR=1.5, 95% CI 1.1-2.1) compared to small or medium amounts.[43] The type of food that provided the gluten was of no importance when adjusting for the amount of flour it contributed. Findings were not conclusive regarding age at introduction of gluten and disease risk. In older children these risk factors were of no importance, or at least not of large enough importance to be detectable.

From the ecological study described above, it is evident that the changes over time in infant feeding patterns influenced the proportion of infants introduced to dietary gluten by the most favorable pattern, according to the findings in the case-referent study, i.e. a gradual introduction of gluten while the child is still breast-fed. Thus, Swedish infant feeding practices have shifted over time from a favorable, to an unfavorable, and back to a favorable pattern with respect to celiac disease risk. Considering the timing of these changes, they are likely to have contributed to the rise in incidence in children below two years of age, and the later decrease in incidence in the same age group.

Based on the population attributable fraction, calculated from the results in the case-referent study, about half of the celiac disease cases during the epidemic might have been avoided if the infants had been introduced to gluten in small amounts while still being breast-fed.[43]

5. Environmental Exposure, Gender, and Age at Diagnosis

The risk for celiac disease before two years of age was significantly higher in children born during the summer compared to those born during the winter (RR=1.4, 95% CI 1.2-1.7). No such seasonality concerning births was found among celiac disease cases diagnosed between two and fifteen years of age.

These findings suggest that environmental exposure(s) with a seasonal pattern have a causal effect in the younger age group. A temporal relationship indicates that this might be due to a causal effect of infections during fetal life and/or an interaction between infections and introduction of gluten into the diet. However, other causal exposures might also contribute to seasonality in celiac disease risk.

Considering all types of infectious episodes, 36% of the cases and 28% of the referents ($p=0.007$) had experienced three episodes or more before six months of age. Excluding gastroenteritis, the difference between cases and referents decreased (34% versus 27%) but was still statistically significant ($p=0.03$). Conditional logistic regression was performed with 370 matched sets of cases and referents. The bivariate analysis resulted in an odds ratio (OR) of 1.4 (95% CI=1.1-1.9) when there were three or more infectious episodes (excluding gastroenteritis), compared to fewer number of episodes, before six months of age. Multivariate analyses resulted in the same risk estimate, but with a slightly modified confidence interval (OR=1.4, 95% CI 1.0-1.9), when adjusting for whether breast-feeding was ongoing or not when dietary gluten was introduced, amount of gluten that was then given, and the socioeconomic group to which the child's family belonged. The risk for celiac disease increased considerably if, in addition to having had three or more infections, dietary gluten was introduced in large amounts.

The relative risk (RR) for celiac disease was twice as high in girls as compared to boys (RR=1.9, 95% CI 1.7-2.1), and this did not vary significantly in spite of the epidemic incidence pattern in children below two years of age. In absolute numbers the gender gap in incidence rates was larger during the epidemic period compared to both the preceding years and the years that followed. During the pre-epidemic period the gender gap in incidence rates was 60 cases per 100,000 person-years; for boys 40 cases, and for girls 100 cases per 100,000 person-years. This gender gap had increased to 110 cases per 100,000 person-years during the epidemic period; for boys 140 cases, and for girls 250 cases per 100,000 person-years.

The families of the celiac disease children, more often than the referents, belonged to the lower socioeconomic level of society. This was true only in the group below two years of age, where 49% of the cases (n=370) and 38% of the referents (n=573) belonged to this group (p=0.001).

Conditional logistic regression in bivariate analysis resulted in an odds ratio (OR) of 1.5 (95% CI 1.2-2.0) when the family belonged to the lower as compared to the middle and upper socioeconomic strata. The risk estimate was only slightly reduced (OR=1.4, 95% CI 1.0-1.8) when adjusting for whether breast-feeding was ongoing or not when dietary gluten was introduced, amount of gluten then given, and number of infectious episodes before six months of age.

6. Scrutinizing the Evidence

In consideration of the increased knowledge concerning the immunological importance of breast-milk,[44, 45] it seems plausible that introduction of a dietary antigen while the child is still breast-fed increases the likelihood of developing oral tolerance. Whether or not this relationship holds true for dietary gluten and the risk for development of celiac disease is not clear.

Based on observations of celiac disease patients, it was suggested as early as in the 1950s that breast-feeding delays onset of the disease,[31] a view later supported by other similar studies.[46, 47] Furthermore, an increase in breast-feeding was suggested as a possible contributing factor to the decline in incidence of celiac disease in the early 1970s in England, Scotland and Ireland.[24, 48, 49]

In the 1980s, Italian case-referent studies by Auricchio *et al*[32] (216 cases/289 referents) and Greco *et al*[33] (201 cases/1949 siblings) demonstrated that the celiac disease cases were breast-fed for a shorter duration than the referents. Later, this was also shown in a Swedish case-referent study based on prevalent cases (72 cases/288 referents)[34] Whether breast-feeding had a direct causal effect, or if the protective effect was indirect through postponed introduction of infant formula[32] or reduction in the amount of dietary gluten consumed[34] was also discussed.

Our incident case-referent study (491 cases/781 referents) was population-based. Furthermore, an attempt was made to develop an analytical model for causality based on immunological and epidemiological knowledge.[43] Multivariate analyses were used to adjust risk estimates for confounding and suggest causal relationships. Our study clearly demonstrated that breast-feeding had an independent protective effect against celiac disease if ongoing at the time when gluten-containing foods were introduced. This effect was even more pronounced if the infant also continued to be breast-fed beyond the time at which gluten was introduced. Confounding by other known dietary factors was unlikely because the risk estimates were influenced only slightly by adjustments for co-variates, e.g. amount of gluten consumed. Our study was performed in a population where the majority of infants were breast-fed for six months or longer. Thus, the cessation of breast-feeding was most commonly not accompanied by the introduction of infant formula.

The protective effect of breast-feeding was further supported by our ecological study of the Swedish epidemic[42]. Both the rise and later fall in incidence had a temporal relationship to a change in the proportion of infants introduced to gluten while still being breast-fed.

It was recently suggested that the intestinal lesion in celiac disease might be caused by an uncontrolled production of IFN-γ by intraepithelial T-cells that is not sufficiently counteracted by production of TGF-β1 (Forsberg G, personal communication). It is thus tempting to speculate that the TGF-β1 of breast-milk compensates for this deficiency, although this is just one of several possible mechanisms for a protective effect.

Thus, a protective effect of being breast-fed when dietary gluten is introduced is supported by several different epidemiological studies, which also have different designs. Moreover, the protective effect is biologically plausible in consideration of our present knowledge of breast-milk composition and celiac disease etiology.

Acknowledgements

We acknowledge the contributions by Susanne Walther, administrative assistant, Epidemiology, Department of Public Health and Clinical Medicine, Umeå University. The project was financially supported by grants from the Swedish Council for Forestry and Agricultural Research, the Swedish Foundation for Health Care Sciences and allergy Research, the Swedish Foundation for Research on Asthma and Allergy, and the Västerbotten County Council.

References

1. Mylotte M, Egan-Mitchell B, McCarthy CF, McNicholl B. Inci-dence of coeliac disease in the west of Ireland, BMJ 703-705 (1973).
2. Gumaa SN, McNicholl B, Egan-Mitchell B, Connolly K, Loftus BG. Coeliac disease in Galway, Ireland 1971-1990, Ir. Med. J. 90:60-61 (1997).
3. Hed J, Lieden G, Ottosson E, Ström M, Walan A, Groth O, et al. IgA anti-gliadin antibodies and jejunal mucosal lesions in healthy blood donors [letter], Lancet 2:215 (1986)
4. Grodzinsky E, Franzen L, Hed J, Ström M. High prevalence of celiac disease in healthy adults revealed by antigliadin antibodies, Ann. Allergy 69:66-70 (1992).
5. Hallert C, Gotthard R, Jansson G, Norrby K, Walan A. Similar prevalence of coeliac disease in children and middle-aged adults in a district of Sweden, Gut 24:389-391 (1983).
6. Midhagen G, Järnerot G, Kraaz W. Adult coeliac disease within a defined geographic area in Sweden. A study of prevalence and associated diseases, Scand. J. Gastroenterol. 23:1000-1004 (1988).
7. Uibo O, Uibo R, Kleimola V, Jogi T, Mäki M. Serum IgA anti-gliadin antibodies in an adult population sample. High prevalence without celiac disease, Dig. Dis. Sci. 38:2034-2037 (1993).

8. Meloni G, Dore A, Fanciulli G, Tanda F, Bottazzo GF. Sub-clinical coeliac disease in schoolchildren from northern Sardinia [letter], Lancet 353:37 (1999).
9. Catassi C, Rätsch IM, Gandolfi L, Pratesi R, Fabiani E, El Asmar R, et al. Why is coeliac disease endemic in the people of the Sahara? [letter], Lancet 354:647-648 (1999).
10. Gandolfi L, Pratesi R, Cordoba JC, Tauil PL, Gasparin M, Catassi C. Prevalence of celiac disease among blood donors in Brazil, Am. J. Gastroenterol. 95:689-692 (2000).
11. Walia BNS, Sidhu JK, Tandon BN, Ghai OP, Bhargava S. Coeliac disease in North Indian children, Br. Med. J. 2:1233-1234 (1966).
12. Rabassa EB, Sagaró E, Fragoso T, Castañeda C, Gra B. Coeliac disease in Cuban children, Arch. Dis. Child. 56:128-131 (1981).
13. Sher KS, Fraser RC, Wicks AC, Mayberry JF. High risk of coeliac disease in Punjabis. Epidemiological study in the south Asian and European populations of Leicestershire, Digestion 54:178-182 (1993).
14. Rawashdeh MO, Khalil B, Raweily E. Celiac disease in Arabs, J. Pediatr. Gastroenterol. Nutr. 123:415-418 (1996).
15. Demir H, Yuce A, Kocak N, Ozen H, Gurakan F. Celiac disease in Turkish children: presentation of 104 cases, Pediatr. Int. 42:483-487 (2000).
16. Ascher H, Krantz I, Rydberg L, Nordin P, Kristiansson B. In-fluence of infant feeding and gluten intake on coeliac disease, Arch. Dis. Child. 76:113-7 (1997).
17. Ascher H, Kristiansson B. The highest incidence of celiac disease in Europe: The Swedish experience, J. Pediatr. Gastroenterol. Nutr. 24:3-6 (1997).
18. Ascher H. The role of quantity and quality of gluten-containing cereals in the epidemiology of coeliac disease. In: Mäki M, Collin P, Visakorpi JK, ed. *Coeliac disease: Proceedings of the seventh international symposium on coeliac disease, 1996, Tampere, Finland.* Coeliac Disease Study Group, Tampere: 1997; 15-22.
19. Greco L. Epidemiology of coeliac disease. In: Mäki M, Collin P, Visakorpi JK, ed. *Coeliac disease: Proceedings of the seventh interna-tional symposium on coeliac disease, 1996, Tampere, Finland.* Coeliac Disease Study Group, Tampere: 1997; 9-14.
20. Troncone R, Greco L, Auricchio S. The controversial epidemi-ology of coeliac disease [comment], Acta Paediatr. 89:140-141 (2000).
21. Morris MA, Ciclitira PJ. Coeliac disease, J. R. Coll. Physicians Lond. 31:614-618 (1997).
22. Present-day practice in infant feeding. *Report of a working party of the panel on child nutrition.* Committee on Medical Aspects of Food Policy. Her Majesty's Stationery Office;London, 9:24-26 (1974).
23. Dossetor JFB, Gibson AAM, McNeish AS. Childhood coeliac disease is disappearing [letter], Lancet i:322-323 (1981).
24. Logan RFA, Rifkind EA, Busuttil A, Gilmour HM, Ferguson A. Prevalence and "incidence" of celiac disease in Edinburgh and the Lothian region of Scotland, Gastroenterology 90:334-342 (1996).
25. Challacombe DN, Mecrow IK, Elliott K, Clarke FJ, Wheeler EE. Changing infant feeding practices and declining incidence of coeliac disease in West Somerset, Arch. Dis. Child. 77:206-209 (1997).
26. ESPGAN committee on nutrition. Guidelines on infant nutrition III. Recommendations for infant feeding, Acta Paediatr. Scand. Suppl. 302:16-20 (1982).
27. Mäki M, Holm K. Incidence and prevalence of coeliac disease in Tampere: Coeliac disease is not disappearing, Acta Paediatr. Scand. 79:980-982 (1990).
28. Greco L, Mäki M, Di Donato F, Visakorpi JK. Epidemiology of coeliac disease in Europe and the Mediterranean area. A summary report on the multicentre study by the European Society of Paediatric Gastroenterology and Nutrition. In: Auricchio S, Visakorpi JK, ed. *Common food intolerances 1: Epidemiology of coeliac disease*, Karger, Basel: 1992; 25-44.
29. Lindberg T. Coeliac disease and infant feeding practices [letter], Lancet i:449 (1981).
30. Stenhammar L, Ansved P, Jansson G, Jansson U. The incidence of childhood celiac disease in Sweden, J. Pediatr. Gastroenterol. Nutr. 6:707-709 (1987).
31. Andersen DH, Di Sant'Agnese PA. Idiopathic celiac disease: I. Mode of onset and diagnosis, Pediatrics 11:207-222 (1953).
32. Auricchio S, Follo D, De Ritis G, Giunta A, Marzorati D, Prampolini L, et al. Does breast feeding protect against the development of clinical symptoms of celiac disease in children? J. Pediatr. Gastroenterol. Nutr. 2:428-433 (1983).
33. Greco L, Auricchio S, Mayer M, Grimaldi M. Case control study on nutritional risk factors in celiac disease, J. Pediatr. Gastroenterol. Nutr. 7:395-399 (1988).

34. Fälth-Magnusson K, Franzén L, Jansson G, Laurin P, Stenhammar L. Infant feeding history shows distinct differences between Swedish celiac and reference children, Pediatr. Allergy Immunol. 7:1-5 (1996).
35. Mäki M, Holm K, Ascher H, Greco L. Factors affecting clinical presentation of coeliac disease: Role of type and amount of gluten-containing cereals in the diet. In: Auricchio S, Visakorpi JK, ed. *Common food intolerances 1: Epidemiology of coeliac disease*. Karger, Basel: 1992; 76-82.
36. Ascher H, Holm K, Kristiansson B, Mäki M. Different features of coeliac disease in two neighbouring countries, Arch. Dis. Child. 69:375-380 (1993).
37. Weile B, Cavell B, Nivenius K, Krasilnikoff PA. Striking diff-erences in the incidence of childhood celiac disease between Denmark and Sweden: A plausible explanation, J. Pediatr. Gastroenterol. Nutr. 21:64-68 (1995).
38. Mitt K, Uibo O. Low cereal intake in Estonian infants: the possible explanation for the low frequency of coeliac disease in Estonia, Eur. J. Clin. Nutr. 52:85-88 (1998).
39. Marsh MN. Gluten, major histocompatibility complex, and the small intestine. A molecular and immunobiologic approach to the spectrum of gluten sensitivity ('celiac sprue'), Gastroenterology 102:330-354 (1992).
40. Ascher H, Krantz I, Kristiansson B. Increasing incidence of coeliac disease in Sweden, Arch. Dis. Child. 66:608-611 (1991).
41. Cavell B, Stenhammar L, Ascher H, Danielsson L, Dannaeus A, Lindberg T, et al. Increasing incidence of childhood coeliac disease in Sweden. Results of a national study, Acta Paediatr. 81:589-592 (1992).
42. Ivarsson A, Persson LÅ, Nyström L, Ascher H, Cavell B, Danielsson L, Dannaeus A, Lindberg T, Lindquist B, Stenhammar L, Hernell O. Epidemic of coeliac disease in Swedish children, Acta Paediatr. 89:165-171 (2000).
43. Ivarsson A, Hernell O, Stenlund H, Persson LÅ. Breast-feeding protects against celiac disease, Am. J. Clin. Nutr. In press, (2001).
44. Hanson LA. Breastfeeding provides passive and likely long-lasting active immunity, Ann. Allergy Asthma Immunol. 81:523-537 (1998).
45. Goldman AS. Modulation of the gastrointestinal tract of infants by human milk. Interfaces and interactions. An evolutionary perspective, J. Nutr. 426S-431S (2000).
46. Greco L, Mayer M, Grimaldi M, Follo D, De Ritis G, Auricchio S. The effect of early feeding on the onset of symptoms in celiac disease, J. Pediatr. Gastroenterol. Nutr. 4:52-55 (1985).
47. Mäki M, Kallonen K, Lähdeaho ML, Visakorpi JK. Changing pattern of childhood coeliac disease in Finland, Acta Paediatr. Scand. 77:408-412 (1988).
48. Kelly DA, Phillips AD, Elliott EJ, Dias JA, Walker-Smith JA. Rise and fall of coeliac disease 1960-85, Arch. Dis. Child. 64:1157-1160 (1989).
49. Stevens FM, Egan-Mitchell B, Cryan E, McCarthy CF, McNicholl B. Decreasing incidence of coeliac disease, Arch. Dis. Child. 62:465-468 (1987).

HAMLET - A COMPLEX FROM HUMAN MILK THAT INDUCES APOPTOSIS IN TUMOR CELLS BUT SPARES HEALTHY CELLS

Malin Svensson, Caroline Düringer, Oskar Hallgren, Ann-Kristine Mossberg, Anders Håkansson, Sara Linse and Catharina Svanborg*

1. INTRODUCTION

Breast-feeding has been proposed to protect both mother and child against cancer. The overall incidence of childhood cancer is reduced in breast-fed compared to bottle-fed individuals and decreases with the length of breast-feeding. Davis *et al.* compared 201 children diagnosed with cancer at an age of 1.5 - 15 years to 181 matched healthy controls, and found a crude odds ratio of 1.75 favoring infants breast-fed longer that 6 months.[1] The epidemiological association was especially strong for lymphomas (crude odds ratio 5.62). This association has been verified in several other studies (reviewed in).[2]

While these associations are likely to have many and complex explanations, they suggest that molecules in milk directly influence tissue development and homeostasis in the nursing child. We have discovered a protein complex in human milk that induces apoptosis in tumor cells but spares healthy cells. This complex was purified from casein and the main constituent was found to be a-lactalbumin (ALA). This is the most abundant protein in human milk with a concentration of about 2 mg/ml.[3] The structure of ALA has been elucidated using material from whey, and the crystal structure has been solved (Fig 1).[4] In the mammary gland ALA functions as a substrate specifier for galactosyl transferase, enabling the formation of lactose.[5]

2. Identification of the Active Component

To purify the apoptosis-inducing complex, the casein fraction from human milk was subjected to ion exchange chromatography. The active component was retained on the ion exchange matrix, and eluted only after the gradient was increased to 1 M NaCl. The eluate, which retained all of the apoptosis-inducing activity, contained multiple

* Department of Microbiology, Immunology and Glycobiology (MIG), Institute of Laboratory Medicine, Lund University, Sölvegatan 23, S-223 62 Lund, Sweden. Sara Linse, Department of Physical Chemistry 2, Lund University, P. O. box 124, S-221 00 Lund, Sweden

Integrating Population Outcomes, Biological Mechanisms and Research Methods in the Study of Human Milk and Lactation
Edited by Davis *et al.*, Kluwer Academic/Plenum Publishers, 2002

proteins with molecular masses ranging from 14 to 100 kDa. All proteins showed N-terminal amino acid sequence homology with human ALA, but native ALA from human milk whey was inactive in the apoptosis assays, suggesting that the active form of the molecule had undergone structural modifications.[6, 7] ALA is known to vary its fold, and has been used as a model in structural studies on protein folding intermediates.[8] A characteristic intermediary state of ALA is the molten globule state.[8-12] This is characterized by a native like secondary structure but lacks a well-defined tertiary structure.

Figure 1. Human ALA with its strongly bound Ca^{2+} ion (black circle).

We analyzed the folding state of the active form of ALA by circular dichroidism (CD) and by fluorescence spectroscopy with the hydrophobic dye ANS. The active fraction showed distinct spectral differences compared to the native ALA[6] control. The near UV CD spectrum showed a loss of signal, indicating less restrained tyrosines and tryptophanes. The large intensity increase and wavelength decrease in the ANS spectrum suggested that hydrophobic surfaces had become accessible. These results showed that the apoptosis-inducing ALA in casein was conserved in a molten globule-like state, and that activity was associated with this state.

2.1 Conversion of ALA to the Apoptosis Inducing form (HAMLET)

These findings suggested that the relative folding instability allowed ALA to undergo structural transitions and attain new essential functions. To prove this hypothesis we set out to deliberately convert native ALA to the apoptosis inducing form.[13] The starting material was ALA purified from human milk whey and recombinant ALA. EDTA treatment of native ALA resulted in a partially folded molten globule protein. Protein in the native or molten globule configurations were added to ion-exchange matrices, and material eluting after 1 M NaCl was characterized.

Using a clean matrix the ALA was not converted to the active form. Using a matrix previously exposed to human milk casein, however, molten globule ALA was converted to the apoptosis inducing form. A sharp peak eluted after high salt and this material induced apoptosis in L1210 cells. Spectroscopic analyses showed that the peak retained a molten globule configuration. Native ALA was not converted on this column, suggesting that partial unfolding of the protein was necessary.

As the clean column matrix could not convert molten globule ALA to the active complex, there must be a cofactor present in casein that associated with ALA to form the active complex. In order to identify the cofactor, casein conditioned matrix was extracted with organic solvents. This released various lipid classes that were identified by TLC. The free fatty acids were consumed during conversion suggesting that they formed a part of the active complex.

The individual fatty acids were identified using gas chromatography and clean matrixes were subsequently conditioned with each one, separately.[13] Matrix conditioned with oleic acid (C 18:1) was shown to convert molten globule ALA to the apoptosis inducing form, subsequently named HAMLET (Human α-lactalbumin Made Lethal to tumor cells). Both recombinant and whey derived molten globule ALA eluted as a sharp peak after high salt and these peaks induced DNA fragmentation, and was localized to the nucleus in L1210 cells. Spectroscopic analyses showed that HAMLET strongly resembled the molten globule control.

These results demonstrate that HAMET consists of ALA and the stabilizing cofactor C 18:1. ALA is an example of a protein where an altered tertiary structure results in a different biological function.

3. Biologic Activity

3.1 Hamlet-induced Apoptosis

Apoptotic cell death is characterized by loss of cytoplasmic material, nuclear changes with marginalization of chromatin and the formation of apoptotic bodies.[14-16] The reduction in cell viability is accompanied by stepwise chromatin fragmentation with initial formation of high molecular weight (HMW) DNA fragments (50-300 kbp) followed by oligonucleosome-length DNA fragments consisting of oligomers of approximately 200 bp.[17-19] Tumor cells treated with HAMLET displayed changes characteristic of apoptosis. The cells decreased in volume and the nuclear material became pyknotic. Associated with the morphologic change apoptotic DNA fragmentation was observed (Fig. 2).

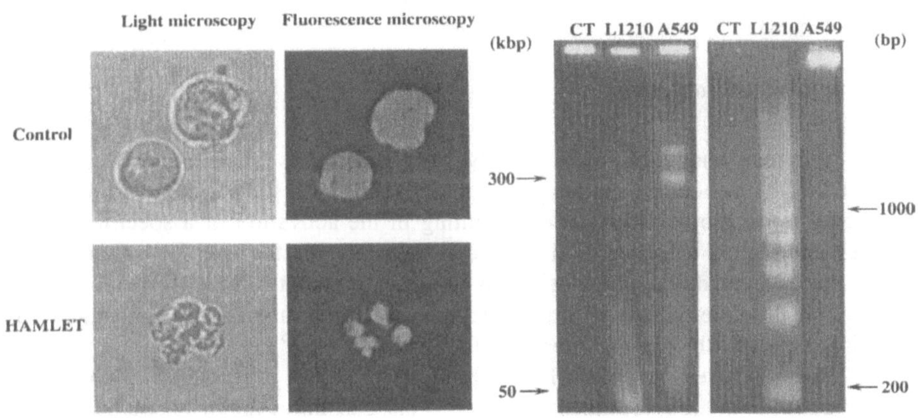

Figure 2. HAMLET induces apoptosis in tumor cells

Tumor cells and mature non-malignant cells differ in sensitivity to HAMLET. To date, over 50 cell lines of both human and animal origin have been tested. Cells of lymphoid origin are found to be very sensitive with an LD_{50} of 0.15 mg/ml. Carcinomas are somewhat more resistant having an LD_{50} 0.7 mg/ml and primary healthy cells are completely resistant to the effect.

3.2 Subcellular Localization of HAMLET

To understand the mechanisms underlying the difference in sensitivity, we compared the subcellular distribution of HAMLET in sensitive and resistant cells. By confocal microscopy and subcellular fractionation, HAMLET bound to the surface of sensitive cells and eventually accumulated in the nucleus. Resistant cells displayed only surface binding; no intracellular localization of the active complex was detected at any time point.

3.2.1 Nuclear Targeting

The results demonstrate that the active complex can target the cell nucleus and that the nuclear targeting mechanisms are more readily available in cells that are sensitive to apoptosis than in resistant cells.

In order to identify molecular targets for HAMLET in the nuclei of sensitive cells, extracts of purified nuclei were investigated for reactivity with HAMLET by overlay techniques. HAMLET overlays identified several protein bands in the nuclear extract of sensitive A549 and Jurkat cells (Düringer C *et al.*, manuscript in preparation). One of these bands was shown by N-terminal amino acid sequence analysis and immuno blotting with monoclonal antibodies to consist of histone H3. Histone H3 is one of the nucleosome core histones and is important in the regulation of chromatin structure. Acetylation and phosphorylation of H3 leads to changes in the level of chromatin condensation. This is important for cell cycle control and transcriptional regulation.

These studies demonstrated that HAMLET interacts specifically with distinct molecular species in the nucleus. By interfering with molecules involved in maintaining DNA integrity HAMLET may cause the observed effects on host cell DNA, and trigger apoptosis (Fig 3).

3.2.2 Mitochondrial Targeting

Apoptosis induction by a variety of agonists has been shown to involve mitochondria. Mitochondria contain pro-apoptotic molecules, such as cytochrome C, that are released upon apoptosis-activation, resulting in the activation of a specific class of proteases now called caspases.[20] They are responsible for the specific proteolysis of nuclear proteins, involved in cytoskeletal integrity, apoptotic body formation, DNA integrity and structure or having other functions. The cleavage of these substrates will favor disassembly of the cell and prepare it for elimination.[21-25]

Cells exposed to HAMLET showed granular intracellular fluorescence suggesting that HAMLET localized to a distinct class of intracellular organelles. Using Mitotracker and other mitochondria-specific reagents it was possible to show that

HAMLET targets mitochondria in intact cells.[26] Furthermore, there was a rapid release of cytochrome C following HAMLET exposure of the cells, suggesting mitochondrial activation. A direct effect of HAMLET on mitochondria was confirmed when HAMLET was shown to bind to the surface of isolated mitochondria and to trigger the release of cytochrome c (Fig. 3).

In the presence of dATP (ATP) cytochrome C forms a complex with the Apaf-1 and pro-caspase-9, leading to pro-caspase-9 activation.[27-30] Activated caspase-9, in turn, cleaves and activates pro-caspase-3, the main caspase in the execution-phase of apoptosis.[31] We have studied caspase activation in cells exposed to the apoptosis-inducing fraction. The results show that the caspase-3-like enzymes and, to a less extent, the caspase-6-like enzymes are activated in the sensitive but not in the resistant cells.[26] Caspase activation was completely inhibited by the specific substrate analogue zVAD-fmk and DEVD-cmk. Interestingly, pre-treatment of cells with zVAD-fmk only partially inhibited DNA fragmentation.

These results demonstrated that HAMLET activates caspases in sensitive cells, but that this activation pathway does not explain cell death.

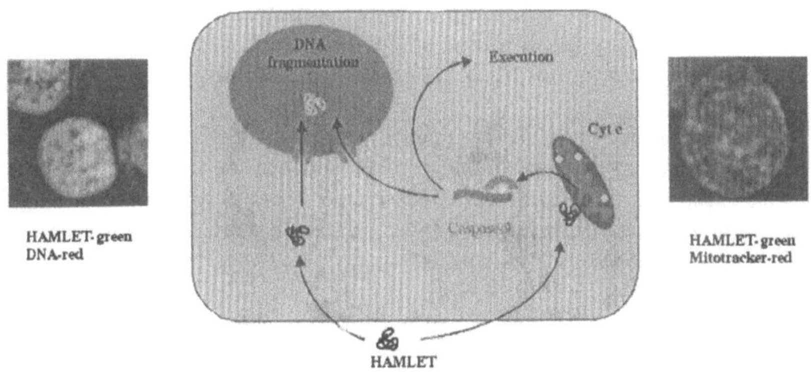

Figure 3. Intracellular targets for HAMLET.

3.3 Bcl-2 Independent Apoptosis

Bcl-2 was initially observed at the breakpoint of translocation in a follicular lymphoma,[32, 34] and was shown to influence the longevity of cells, as over-expressing cells survived longer than normal cells. Today there are at least 20 members of the Bcl-2 family known in mammalian cells, they have both anti-apoptotic and pro-apoptotic functions.[35, 36]

The anti-apoptotic members of the Bcl-2 family include Bcl-2 and Bcl-X_L. They protect the mitochondria, by maintaining the mitochondrial membrane potential and inhibit the release of apoptogenic factors.[36-38] To examine the role of Bcl-2 in HAMLET induced apoptosis we exposed a myeloid leukemia cell line, K562, to the protein. In parallel clones overexpressing Bcl-2 were tested for sensitivity to HAMLET (Håkansson *et a.l*, submitted). There was no difference in sensitivity between the bcl-2 over expressing clones and the vector control. All cells had an LD50 of 0.2 mg/ml. Cytochrome c was released followed by caspase activation. The caspase activation was functional shown by the cleavage of specific intracellular targets such as PARP, and there was DNA fragmentation. These results suggest that HAMLET differ from the majority of apoptosis-inducing stimuli in that apoptosis can be induced irrespective of Bcl-2 expression.

3.3.1 Analysis of Caspase or Bcl-2 mRNAs in Sensitive and Resistant cells

The difference in sensitivity to HAMLET between sensitive and resistant cells is intriguing. Uncovering of the underlying mechanisms might teach us about fundamental cellular processes that distinguish cancer cells from normal cells. We attempted to study such differences by analysis of mRNAs of genes involved in apoptosis. Using RNA protection methods, the caspase and Bcl-2 family mRNAs were quantified in sensitive or resistant cells, and before or after exposure of the cells to HAMLET. The results showed a difference in Bcl-2 family mRNA profile between two cell types with different sensitivity to HAMLET, but no regulation of Bcl-2 family in HAMLET exposed cells or difference in the responses to HAMLET between the cells (Hallgren O *et al.*, manuscript in preparation).

A similar approach was used to quantify mRNAs specific for different caspases. There was a difference in caspase mRNA profiles between the cells, but not in caspase mRNA levels except for a 2-fold increase in caspase 9 in one cell type and in caspase 2(l) in another cell type. These experiments showed no clear pattern linking the sensitivity to HAMLET to specific caspase mRNA species.

Conclusions

We here describe a naturally occurring molecular complex in human milk, that targets immature cells and tumor cells and that activates programmed cell death in those cells, sparing healthy cells. It may be speculated that molecules like HAMLET can have a protective function in the breast fed child, and that HAMLET is one of several naturally occurring surveillance molecules that purge unwanted cells from the local tissues and drive the intestinal mucosa towards maturity By inducing apoptosis, HAMLET may reduce the pool of potentially malignant cells that could serve as nuclei for future tumor development and explain the reduced frequency of cancer in breast-fed individuals. Further studies are required to understand the *in vivo* relevance of these findings.

References

1. M.K. Davis, D.A. Savitz, and B.I. Graubard, Infant feeding and childhood cancer, Lancet 2 (8607):365-368 (1988).
2. M.K. Davis, Review of the evidence for an association between infant feeding and childhood cancer, Int. J. Cancer Suppl. 11:29-33 (1998).

3. H.A. McKenzie, and F.H. White, Jr., Lysozyme and alpha-lactalbumin: structure, function, and interrelationships, Adv. Protein Chem. 41:173-315 (1991).
4. K.R. Acharya, J.S. Ren, D.I. Stuart, D.C. Phillips, and R.E. Fenna, Crystal structure of human alpha-lactalbumin at 1.7 A resolution, J. Mol. Biol. 221:(2), 571-581 (1991).
5. B.P. Ram, and D.D. Munjal, Galactosyltransferases: physical, chemical, and biological aspects, CRC Crit. Rev. Biochem. 17:(3), 257-311 (1985).
6. M. Svensson, H. Sabharwal, A. Hakansson, A.K. Mossberg, P. Lipniunas, H. Leffler, C. Svanborg, and S. Linse, Molecular characterization of alpha-lactalbumin folding variants that induce apoptosis in tumor cells, J. Biol. Chem. 274 (10):6388-6396 (1999).
7. A. Håkansson, B. Zhivotovsky, S. Orrenius, H. Sabharwal, and C. Svanborg, Apoptosis induced by a human milk protein, Proc. Natl. Acad. Sci. USA 92 (17):8064-8068 (1995).
8. K. Kuwajima, The molten globule state of alpha-lactalbumin, FASEB J 10 (1):102-9 (1996).
9. E.A. Permyakov, and L.J. Berliner, Alpha-Lactalbumin: structure and function, FEBS Lett 473 (3):269-274 (2000).
10. O.B. Ptitsyn, Molten globule and protein folding, Adv. Protein Chem. 47:83-229 (1995).
11. T.E. Creighton, How important is the molten globule for correct protein folding?, Trends Biochem. Sci. 22 (1):6-10 (1997).
12. A.T. Alexandrescu, P.A. Evans, M. Pitkeathly, J. Baum, and C.M. Dobson, Structure and dynamics of the acid-denatured molten globule state of α-lactalbumin: A two-dimentional NMR study, Biochemistry 32:1707-1718 (1993).
13. M. Svensson, A. Håkansson, A. Mossberg, S. Linse, and C. Svanborg, Conversion of alpha -lactalbumin to a protein inducing apoptosis [In Process Citation], Proc. Natl. Acad. Sci. USA 97 (8):4221-4226 (2000).
14. A.H. Wyllie, J.F. Kerr, and A.R. Currie, Cell death: the significance of apoptosis, Int. Rev. Cytol. 68 251-306 (1980).
15. J.F.R. Kerr, A.H. Wyllie, and A.R. Currie, Apoptosis: A basic biological phenomenon with wide-ranging implications in tissue kinetics, Br J Cancer 26:239-257 (1972).
16. M. Raff, Cell suicide for beginners, Nature 396 (6707):119-122 (1998).
17. A.H. Wyllie, Glucocorticoid-induced thymocyte apoptosis is associated with endogenous endonuclease activation, Nature 284:555-556 (1980).
18. B. Zhivotovsky, B. Cedervall, S. Jiang, P. Nicotera, and S. Orrenius, Involvement of Ca2+ in the formation of high molecular weight DNA fragments in thymocyte apoptosis, Biochem Biophys Res Commun 202 (1):120-127 (1994).
19. M.A. Lagarkova, O.V. Iarovaia, and RS.V. azin, Large-scale fragmentation of mammalian DNA in the course of apoptosis proceeds via excision of chromosomal DNA loops and their oligomers, J. Biol. Chem. 270:20239-20241 (1995).
20. E.S. Alnemri, D.J. Livingston, D.W. Nicholson, G. Salvesen, N.A. Thornberry, W.W. Wong, and J. Yuan, Human ICE/CED-3 protease nomenclature, Cell 87 (2):171 (1996).
21. N.A. Thornberry and Y. Lazebnik, Caspases: enemies within, Science 281:(5381), 1312-1316 (1998).
22. A. Rosen, and L. Casciola-Rosen, Macromolecular substrates for the ICE-like proteases during apoptosis, J. Cell Biochem. 64 (1):50-54 (1997).
23. S.J. Martin, G.A. O'Brien, W.K. Nishioka, A.J. McGahon, A. Mahboubi, T.C. Saido, and D.R. Green, Proteolysis of fodrin (non-erythroid spectrin) during apoptosis, J. Biol. Chem. 270 (12):6425-6428 (1995).
24. S.B. Brown, K. Bailey, and J. Savill, Actin is cleaved during constitutive apoptosis, Biochem. J. 323 (Pt 1):233-237 (1997).
25. C. Gueth-Hallonet, K. Weber, and M. Osborn, Cleavage of the nuclear matrix protein NuMA during apoptosis, Ex.p Cell Res. 233 (1):21-24 (1997).
26. C. Kohler, A. Håkansson, C. Svanborg, S. Orrenius, and B. Zhivotovsky, Protease activation in apoptosis induced by MAL, Exp. Cell Res. 249:260-268 (1999).
27. C. Adrain, E.A. Slee, M.T. Harte, and S.J. Martin, Regulation of Apoptotic Protease Activating Factor-1 Oligomerization and Apoptosis by the WD-40 Repeat Region, J. Biol. Chem. 274 (30):20855-20860 (1999).
28. P. Li, D. Nijhawan, I. Budihardjo, S.M. Srinivasula, M. Ahmad, E.S. Alnemri, and X. Wang, Cytochrome c and dATP-dependent formation of Apaf-1/caspase-9 complex initiates an apoptotic protease cascade, Cell 91 (4):479-489 (1997).
29. G. Pan, K. O'Rourke, and V.M. Dixit, Caspase-9, Bcl-XL, and Apaf-1 form a ternary complex, J. Biol. Chem. 273 (10):5841-5845 (1998).

30. A. Saleh, S.M. Srinivasula, S. Acharya, R. Fishel, and E.S. Alnemri, Cytochrome c and dATP-mediated oligomerization of Apaf-1 is a prerequisite for procaspase-9 activation, J. Biol. Chem. 274 (25):17941-17945 (1999).
31. E.A. Slee, M.T. Harte, R.M. Kluck, B.B. Wolf, C.A. Casiano, D.D. Newmeyer, H.G. Wang, J.C. Reed, D.W. Nicholson, E.S. Alnemri, D.R. Green, and S.J. Martin, Ordering the cytochrome c-initiated caspase cascade: hierarchical activation of caspases-2, -3, -6, -7, -8, and -10 in a caspase-9- dependent manner, J. Cell Biol. 144 (2):281-292 (1999).
32. Y. Tsujimoto, J. Cossman, E. Jaffe, and C.M. Croce, Involvement of the bcl-2 gene in human follicular lymphoma, Science 228 (4706):1440-1443 (1985).
33. M.L. Cleary, and J. Sklar, Nucleotide sequence of a t(14;18) chromosomal breakpoint in follicular lymphoma and demonstration of a breakpoint-cluster region near a transcriptionally active locus on chromosome 18, Proc. Natl. Acad. Sci. USA 82 (21):7439-7443 (1985).
34. A. Bakhshi, J.P. Jensen, P. Goldman, J.J. Wright, O.W. McBride, A.L. Epstein, and S.J. Korsmeyer, Cloning the chromosomal breakpoint of t(14;18) human lymphomas: clustering around JH on chromosome 14 and near a transcriptional unit on 18, Cell 41 (3):899-906 (1985).
35. J.C. Reed, J.M. Jurgensmeier, and S. Matsuyama, Bcl-2 family proteins and mitochondria, *Biochim Biophys* Acta 1366 (1-2):127-137 (1998).
36. G. Kroemer, and J.C. Reed, Mitochondrial control of cell death, Nat. Med. 6 (5):513-519 (2000).
37. A.Gross, J.M. McDonnell, and S.J .Korsmeyer, BCL-2 family members and the mitochondria in apoptosis, Genes Dev. 13 (15):1899-1911 (1999).
38. D.R. Green, and J.C. Reed, Mitochondria and apoptosis, Science 281 (5381), 1309-1312 (1998).

INFANT FEEDING AND AUTOIMMUNE DIABETES

H-Michael Dosch* and D.J. Becker#

1. INTRODUCTION

Type 1 (insulin-dependent) Diabetes Mellitus is a chronic autoimmune disorder of genetically susceptible hosts, caused by the interaction of uncertain environmental factors with the products of 20+ predisposing genes.[1] Self-reactive T lymphocytes, targeting components of the pancreatic islets of Langerhans, accumulate in that locale early in life and begin a slow process, called "prediabetes". Progressive prediabetes leads to the expansion of autoreactive T cells,[2] with increasingly efficient destruction of insulin producing beta cells and their precursors.

After usually years, even decades of prediabetes, the remaining beta cell mass is insufficient. Overt insulin deficiency declares itself with often life-threatening ketoacidosis. Insulin deficiency is permanent and requires daily insulin replacement therapy. While life saving, insulin treatment cannot prevent major long-term side effects of improper insulin supply-demand balances, and Type 1 Diabetes Mellitus (T1DM) remains the leading cause of blindness, kidney failure, stroke, non-traumatic loss of limbs etc, all caused by wide-spread micro- and late macrovascular disease. The *annual* cost of T1DM exceeds 14 Billion dollars in Canada, about 10 times that figure in the U.S., consuming at least one in seven health care dollars.[3]

There is an alarming, global rise in diabetes incidence, whose pace cannot be explained by genetics. Together with the low twin concordance rates of ~25%, it is clear that the driving force behind the growth of diabetes is environmental in origin.[4-6] Definitive signs of islet autoimmunity are present in 1-8% of the general population, but only a fraction (1-5%) of these subjects will proceed to overt T1DM, depending on ethnic background and geography.[7-8] The solid observation that prediabetes is common, but overt disease is not, has inspired major, continuing efforts:

*H.-Michael Dosch, The Hospital For Sick Children, Research Institute IIIR Program, University of Toronto, 555 University Ave., Toronto, ON, Canada M5G 1X8, hmd@sickkids.on.ca, 416-813-6260 (6255 fax)

Integrating Population Outcomes, Biological Mechanisms and Research Methods in the Study of Human Milk and Lactation
Edited by Davis *et al.*, Kluwer Academic/Plenum Publishers, 2002

1) To define and understand the elements that either promote or halt the progression of prediabetes to overt disease, and

2) To identify those individuals with a high risk of progression towards T1DM, the latter now a clinical reality.[8-16]

Two views are presently shared by many: 1) T1DM develops as a result of 'clean' living in modern, highly hygienic environments; 2) the progression of prediabetes is a promising and realistic target for immune intervention therapies, in particular when applied early. A panoply of intervention trials has been designed and implemented, several large efforts are currently in progress.[17] While successes are still limited, there is an air of cautious optimism in the field that the rise of diabetes incidence will be curbed and reversed in the present decade.

2. Animal Models of t1dm

Progress in diabetes research relies heavily on two excellent rodent models of T1DM, the spontaneously diabetic BB rat, and the non-obese diabetic (NOD) mouse.[8, 18, 19] Together, these models cover much of the spectrum of human T1DM, and they have allowed in depth analyses of the genetic and environmental disease causes, as well as its prevention by a spectrum of therapeutic strategies. Highly hygienic, protected housing is a prerequisite for high incidence rodent diabetes.[20] In the NOD mouse, disease development proceeds along a set program, with specific checkpoints that delay progression for some time.[21, 22] It is probably these checkpoints that are targeted by the over 100 regimen that prevent diabetes if applied early in life.[19]

Prediabetes begins with sparse accumulations of lymphoid cells surrounding the islet of 5-week-old mice. The process proceeds to massive peri-insulitis, which persists even if disease progression is prevented. By eight weeks of age, the islet mantel is breached in nearly all females. This gender bias is untypical for the common Caucasian T1DM, but it is typical for the relatively late onset autoimmune diabetes in Japanese. The islet invasive phase of prediabetes is associated with marked tissue destruction, as heralded by the appearance of autoantibodies to islet components.[23] The increasingly accurate disease prediction in humans is based on analysis of the spectrum of such autoantibodies.[10] Even after frank islet invasion, prediabetes progression still requires a prolonged period of time to generate massive beta cell death and overt T1DM. This final grace period is poorly understood, but very few of the presently known intervention regimen in rodents have promise after the islet is breached (reviewed in (24)).

3. Early Diet and Diabetes Risk

Beginning in the 1970's, ecological surveys of Multiple Sclerosis, the cousin autoimmune disease of T1DM, suggested that early nutrition patterns could enhance the risk of developing autoimmunity, with the single most important factor identified as high liquid cow milk consumption.[25, 26] This was followed by early epidemiological studies in type 1 diabetes families, which suggested that early weaning from exclusive breast feeding was associated with elevated risk to develop T1DM.[27] A plethora of studies have since addressed this issue, which at times polarized the field (reviewed in (28)). The controlled conditions of prospective animal studies reproducibly found from its beginnings,[29] that in order to develop diabetes, rodents had to be weaned to complex, multi-antigenic diets, which generated abnormal immunity to cow milk proteins in typical rodent chow.[30] However, the underlying mechanisms of this effect have resisted analysis and the disease prerequisite roles of weaning diets remain obscure.[31] Nevertheless, there is solid consensus that weaning of properly breast-fed, diabetes-prone rodents to amino acid or extensively hydrolyzed protein diets, is a potent strategy to prevent T1DM life long. In practical terms, the diet most often employed in these animal studies is the established human infant formula Nutramigen™.[31-36]

Human data on the link between weaning and childhood diets and the risk to develop T1DM are predictably more difficult to ascertain. In most of the high incidence Western Countries, weaning from exclusive breast feeding almost invariably involves cow milk-based formula. This explains the focus of research on cow milk protein, although other weaning diets have shown diabetogenic potential in animal studies.[37, 38] The large scale use of xenogeneic lactation products in essentially raw and non-denatured state represents a recent addition to human diets.

The *per capita* milk consumption and diabetes risk, analyzed globally or within a single Country, show strong, positive correlations.[39-41] However, retrospective case-control studies of feeding histories in diabetic children did not uniformly link early weaning and/or cow milk exposure with diabetes risk (reviewed in (1, 42-45)).

Exposure rates of susceptible hosts to a ubiquitous dietary element is not necessarily different from the exposure of non-susceptible individuals – the underlying premise of these case-control studies. In addition, these retrospective studies requires excellent memory of mothers for events often a decade or more ago. Maternal recall bias has been a consistent criticism.[38, 44] In 1994, an expert panel of the American Academy of Pediatrics concluded that the evidence was sufficiently strong to recommend emphasis of breast feeding and weaning to non-antigenic diets in families with high diabetes risk. The panel called for formal trials to prospectively analyze the cow milk-diabetes link.[46]

A small study in 1993,[47] followed by a larger, independent analysis in 1996 changed the scene.[48] These investigators argued that cases and controls should be matched for genetic risk, i.e. diabetes-associated risk alleles in the HLA locus. Both studies found that early weaning of infants with high risk HLA genotypes to cow milk-based formula resulted in a major, more than 10-fold rise in diabetes risk.[48] The diabetes risk was highest in children with high risk HLA genotype, less (RR 2.5) in those with intermediate risk genotypes and absent in children with low risk HLA alleles. These studies thus ruled out maternal recall bias, as such bias, if it were present, would map to HLA genes – an unlikely possibility.

4. Abnormal Immunity to Cow Milk Protein

Beginning in 1988,[49] a rapidly growing number of publications from independent groups reported that children with recent onset diabetes had abnormal humoral and cellular immunity to cow milk proteins, although the specificities and extent of abnormalities varied and a few laboratories failed to detect such abnormalities (reviewed in (28)). On balance, there is little doubt that diabetic autoimmunity commonly includes abnormal sensitization of B and T lymphocytes to constituents of cow milk. This observation seemed to link epidemiological observations with the diabetes-prone immune system, which after all mediates the disease. It proved to be difficult to establish this link in a mechanistically solid fashion.

As an example, we will describe observations made in our laboratories over the past decade. We initially identified abnormal immunity, both antibodies and T cell-mediated, to one rather minor cow milk protein, bovine serum albumin (BSA).[50-52] Interestingly, anti-BSA antibodies from diabetic BB rats showed cross-reactivity to an unidentified protein expressed in islets, which we cloned and identified as ICA69,[53] a then newly discovered autoantigen in human diabetes.[54] It turned out that new onset diabetic children as well as NOD mice routinely generate T cells that target the same peptide epitope in BSA (called 'ABBOS') and a corresponding, structurally related epitope in ICA69 (called 'Tep69').[55, 56] Indeed, the appearance of these crossreactive T cells was associated with a high risk to develop overt diabetes in first degree relatives of diabetes patients.[2]

A model offered itself, where cow milk exposure of a susceptible host would generate T cells that, by accident of nature, also targeted the beta cell protein ICA69, possibly contributing to beta cell damage and progression of diabetic autoimmunity.[57] Indeed, small doses of ABBOS peptide were found to precipitate diabetes,[24] and animals weaned to a diabetes-protective, cow milk-free diet, Nutramigen, failed to develop ICA69-specific T cells.[36]

Antigenic crossreactivity, also called 'mimicry', is a common theme in the immunobiology of autoimmune diseases, proposed as a mechanism to overcome the innate tolerance of normal immune systems to self proteins. To determine if ABBOS-specific T cells sensitized by dietary BSA could be a trigger of autoimmunity in diabetes, we decided to develop NOD mice with the critical target self protein, ICA69, removed through gene knock-out. These animals failed to develop Tep69-reactive T cells, even after immunization with ABBOS, demonstrating that it is the endogenous self-protein which generates the mimicry T cell pool, and not the exogenous mimicry antigen.[24] Thus, it became unlikely that dietary BSA/ABBOS would trigger autoimmunity in diabetes-prone hosts. The possibility remains, that diet-derived ABBOS may maintain ICA69-specific autoreactive T cells spontaneously generated during prediabetes. The mechanisms that underlie the link between multi-antigenic weaning diets such as cow milk, and diabetes risk remain equally unexplained as the diabetes protective effect of extensively hydrolyzed weaning diets such as Nutramigen.

5. Primary Diabetes Prevention: the 'trigr' Trial

By the beginning of the last decade, a small group of diabetes researchers from Helsinki, Toronto and Pittsburgh began to consider the practical implications of a possible link between cow milk exposure and diabetes, considerable parts of this evidence then still developing (reviewed in (58)). Should early weaning from exclusive breast-feeding and attendant exposure to cow milk be considered 'innocent until proven guilty', or should we proceed with suspicion 'until proven innocent'? As it was already clear that weaning to elemental diets protected diabetes-prone rodents from the disease,[29, 32, 33] the implication of a practical, medically harmless diabetes prevention appeared feasible and attractive: emphasize breast feeding and avoid early exposure to intact cow milk protein. While it was unclear if exclusive breast feeding provided protection, or if cow milk exposure itself was the culprit, animal studies then and since, employed natural (~3 week) breast feeding duration. The large Finnish DiMe study eventually answered this question for human diabetes: early (<3 months) and sustained cow milk exposure are risk factors for subsequent diabetes development, independently of overall breast feeding duration.[59-62]

The decision was made in the early 90's to begin the planning for a clinical trial that would test first the feasibility and then the promise of translating the animal experiments into a clinical trial, termed TRIGR (Trial to Reduce IDDM in the Genetically at Risk, IDDM or insulin-dependent diabetes mellitus, was then still the commonly used term for T1DM). The principal question was simple: would Nutramigen protect diabetes-prone humans from the disease, as it does in rodents.

The trial design would identify newborns with a first degree relative that has T1DM, and the trial would recruit newborns with high and intermediate risk HLA genotypes.[17] The trial would be randomized, prospective and placebo-controlled, assigning each proband to one of two weaning diets: Nutramigen or a cow milk-based infant formula. Mothers would be encouraged to breast feed, and the formulas used upon cessation of exclusive breast-feeding. The intervention would last till 6-8 months of age, depending on weaning time. There would be two trial outcome measures: the presence of diabetes-associated and –predictive autoantibodies by age 5 years, and the presence of overt diabetes by age 10 years.

Power calculations were based on prospective diabetes risks, dropout estimates and expected differences between the two trial cohorts. To detect a 40% difference in diabetes between the groups, 2800 newborns would need to be enrolled, 6000 would need to be screened by genetic tissue typing. These figures mandated a large, international multi-center trial, which now includes over 40 centers on 4 Continents, with nation-wide efforts in Finland, Holland and Canada.

Several pilot studies were conducted to establish trial feasibility, infrastructure, rule out an adverse effect of Nutramigen on surrogate markers of autoimmunity and reconfirm the animal data.[27, 36] The last pilot study enrolled 243 eligible Finnish newborns.[63] Blinding was lifted when the last proband reached age 2 years. The study outcome was diabetes-associated autoantibodies as surrogate markers of diabetic autoimmunity. Despite the comparatively small study size, there were significantly more seroconversions in the control vs. the Nutramigen group ($p=0.013$). The international trial proper is expected to begin in 2001, pending the acquisition of appropriate funding.

Selected References

1. W. Karges, J. Ilonen, B. H. Robinson, and H.-M. Dosch, Self and Non-Self Antigen in Diabetic Autoimmunity: Molecules and Mechanisms, Molec. Aspects Med. 16:79-213 (1995).
2. H.-M. Dosch, R. K. Cheung, W. Karges, M. Pietropaolo, and D. J. Becker, Persistent T cell anergy in human type 1 diabetes, J. Immunol., 163:6933-6940 (1999).
3. American Diabetes Association. Cost of Diabetes on the rise, ADA Professional Section Quarterly, Winter:5-7 (1998).
4. R. B. Lipton, J. Atchison, J. S. Dorman, R. J. Duquesnoy, K. Eckenrode, T. J. Orchard, R. E. LaPorte, W. J. Riley, L. H. Kuller, A. L. Drash, D. J. Becker, and M. Trucco, Genetic, immunological, and metabolic determinants of risk for type 1 diabetes mellitus in families, Diabet. Med. 9:224-232 (1992).
5. R. B. Lipton, M. Kocova, R. E. LaPorte, J. S. Dorman, T. J. Orchard, W. J. Riley, A. L. Drash, D. J. Becker, and M. Trucco, Autoimmunity and genetics contribute to the risk of insulin-dependent diabetes mellitus in families: islet cell antibodies and HLA DQ heterodimers, Amer. J. Epidemiol. 136:503-512 (1992).

6. D. Kumar, N. S. Gemayel, D. Deapen, D. Kapadia, P. H. Yamashita, M. Lee, J. H. Dwyer, P. Roy-Burman, G. A. Bray, and T. M. Mack, North-American twins with IDDM, genetic, etiological and clinical significance of disease concordance according to age, zygosity and the intervall after diagnosis of the first twin, Diabetes 42:1351-1363 (1993).
7. J. S. Dorman, R. E. LaPorte, R. A. Stone, and M. Trucco, Worldwide differences in the incidence of type I diabetes are associated with amino acid variation at position 57 of the HLA-DQ beta chain, Proc. Natl. Acad. Sci. U.S.A 87:7370-7374 (1990).
8. D. B. Schranz, and A. Lernmark, Immunology in diabetes: an update [see comments]. [Review] [248 refs], Diabetes Metab.Rev. 14:3-29 (1998).
9. A. Lernmark. Type 1 diabetes, Clin. Chem. 45:1331-1338 (1999).
10. P. J. Bingley, E. Bonifacio, A. J. Williams, S. Genovese, G. F. Bottazzo, and E. A. Gale, Prediction of IDDM in the general population: strategies based on combinations of autoantibody markers, Diabetes 46:1701-1710 (1997).
11. J. Hahl, T. Simell, J. Ilonen, M. Knip, and O. Simell, Costs of predicting IDDM, Diabetologia 41:79-85 (1998).
12. H. K. Akerblom, M. Knip, and O. Simell, From pathomechanisms to prediction, prevention and improved care of insulin-dependent diabetes mellitus in children, Ann. Med. 29:383-385 (1997).
13. G. S. Eisenbarth, R. Gianani, L. Yu, M. Pietropaolo, C. F. Verge, H. P. Chase, M. J. Redondo, P. Colman, L. Harrison, and R. Jackson, Dual-parameter model for prediction of type I diabetes mellitus, Proc. Assoc. Am. Physicians 110:126-135 (1998).
14. C. Levy-Marchal, F. Dubois, M. Noel, J. Tichet, and P. Czernichow, Immunogenetic determinants and prediction of IDDM in French schoolchildren, Diabetes 44:1029-1032 (1995).
15. C. F. Verge, R. Gianani, E. Kawasaki, L. Yu, M. Pietropaolo, R. A. Jackson, H. P. Chase, and G. S. Eisenbarth, Prediction of type I diabetes in first-degree relatives using a combination of insulin, GAD, and ICA512bdc/IA-2 autoantibodies, Diabetes 45:926-933 (1996).
16. P. Kulmala, K. Savola, H. Reijonen, R. Veijola, P. Vahasalo, J. Karjalainen, E. Tuomilehto-Wolf, J. Ilonen, J. Tuomilehto, H. K. Akerblom, and M. Knip, Genetic markers, humoral autoimmunity, and prediction of type 1 diabetes in siblings of affected children. Childhood Diabetes in Finland Study Group, Diabetes 49:48-58 (2000).
17. M. Knip, and H. K. Åkerblom, IDDM prevention trials in progress - a critical assessment, J. Ped. Endocrinol. & Metab. 11:371-377 (1998).
18. A. A. Rossini, E. S. Handler, J. P. Mordes, and D. L. Greiner, Human autoimmune diabetes mellitus: lessons from BB rats and NOD mice--Caveat emptor, Clin. Immunol. Immunopathol. 74:2-9 (1995).
19. M. A. Atkinson, and E. H. Leiter, The NOD mouse model of type 1 diabetes: as good as it gets?, Nat. Med. 5:601-604 (1999).
20. P. Pozzilli, A. Signore, A. J. Williams, and P. E. Beales, NOD mouse colonies around the world: recent facts and figures, Immunol. Today 14:193-196 (1993).
21. I. Andre, A. Gonzalez, B. Wang, J. Katz, C. Benoist, and D. Mathis, Checkpoints in the progression of autoimmune disease: lessons from diabetes models. [Review], Proc. Natl. Acad. Sci. USA 93:2260-2263 (1996).
22. P. Hoglund, J. Mintern, C. Waltzinger, W. Heath, C. Benoist, and D. Mathis, Initiation of autoimmune diabetes by developmentally regulated presentation of islet cell antigens in the pancreatic lymph nodes, J. Exp. Med. 189:331-339 (1999).
23. S. Reddy, N. Bibby, and R. B. Elliott, Longitudinal study of islet cell antibodies and insulin autoantibodies and development of diabetes in non-obese diabetic (NOD) mice, Clin. Exp. Immunol. 81:400-405 (1990).
24. S. Winer, L. Gunaratnam, I. Astsatourov, R. K. Cheung, V. Kubiak, W. Karges, D. Hammond-McKibben, R. Gaedigk, D. Graziano, M. Trucco, D. J. Becker, and H.-M. Dosch, Peptide Dose, MHC-Affinity and Target Self-Antigen Expression are Critical for Effective Immunotherapy of NOD Mouse Prediabetes, J. Immunol. (in press) (2000).
25. J. Butcher, The distribution of multiple sclerosis in relation to the dairy industry and milk consumption, N. Z. Med. J. 83:427-430 (1976).
26. P. J. Butcher, Milk consumption and multiple sclerosis--an etiological hypothesis, Med. Hypotheses 19:169-178 (1986).
27. K. Borch-Johnsen, T. Mandrup-Poulsen, B. Zachau-Christiansen, G. Joner, M. Christy, K. Kastrup, and J. Nerup, Relation between breast-feeding and incidence rates of insulin-dependent diabetes mellitus, Lancet 2:1083-1086 (1984).
28. D. Hammond-McKibben, and H.-M. Dosch, Cow milk, BSA and IDDM: can we settle the controversies?, Diab. Care 20:897-901 (1997).

29. R. B. Elliott, and J. M. Martin, Dietary protein: a trigger of insulin-dependent diabetes in the BB rat? Diabetologia 26:297-299 (1984).
30. H. Beppu, W. E. Winter, M. A. Atkinson, N. K. Maclaren, K. Fujita, and H. Takahashi, Bovine albumin antibodies in NOD mice, Diabetes Res. 6:67-79 (1987).
31. S. Malkani, D. Nompleggi, J. W. Hansen, D. L. Greiner, J. P. Mordes, and A. A. Rossini, Dietary cow's milk protein does not alter the frequency of diabetes in the BB rat, Diabetes 46:1133-1140 (1997).
32. D. Daneman, L. Fishman, C. Clarson, and J. M. Martin, Dietary triggers of insulin-dependent diabetes in the BB rat, Diabetes Res. 5:93-97 (1987).
33. H.-M. Dosch, J. Karjalainen, J. Morkowski, J. M. Martin, and B. H. Robinson, Nutritional triggers of IDDM, in *Epidemiology and etiology of Insulin-dependent diabetes in the young*, edited by C. Lévy-Marchal, and P. Czernichow (Karger, Basel, 1992), pp. 202-17.
34. X. B. Li, F. W. Scott, Y. H. Park, and J. W. Yoon, Low incidence of autoimmune type I diabetes in BB rats fed a hydrolysed casein-based diet associated with early inhibition of non-macrophage-dependent hyperexpression of MHC class I molecules on beta cells, Diabetologia 38:1138-1147 (1995).
35. F. W. Scott, D. Daneman, and J. M. Martin, Evidence for a critical role of diet in the development of insulin-dependent diabetes mellitus, Diabetes Res. 7:153-157 (1988).
36. W. Karges, D. Hammond-McKibben, R. K. Cheung, M. Visconti, N. Shibuya, D. Kemp, and H. M. Dosch, Immunological Aspects of Nutritional Diabetes Prevention in NOD Mice. A Pilot Study for the Cow's Milk-Based IDDM Prevention Trial, Diabetes 46:557-564 (1997).
37. J. Hoorfar, F. Scott, and H. E. Cloutier, Dietary plant materials and development of diabetes in the BB rat, J. Nutr. 121:908-916 (1991).
38. F. W. Scott, J. M. Norris, and H. Kolb, Milk and type I diabetes: examining the evidence and broadening the focus, Diab. Care 19:379-383 (1996).
39. F. W. Scott, Cow milk and insulin-dependent diabetes mellitus: is there a relationship, Am. J. Nutrition 51:489-491 (1990).
40. K. Dahl-Jørgensen, G. Joner, and K. F. Hanssen, Relationship between cow milk consumption and incidence of IDDM in childhood, Diabetes Care 14:1081-1083 (1991).
41. D. Di Fava, R. D. G. Leslie, and P. Pozzilli, Relationship between dairy product consumption an incidence of IDDM in childhood in Italy, Diabetes Care 17:1488-1490 (1994).
42. H. Gerstein, Cow's milk exposure and type 1 diabetes mellitus, Diabetes Care 17:13-19 (1994).
43. W. Karges, and H.-M. Dosch,Environmental Factors: Cow milk and others, in *Diabetes Prediction, Prevention and Genetic Counselling in IDDM*, edited by J. P. Palmer (John Wiley & Sons LTD, Chichester, UK, 1996), pp. 167-210.
44. J. M. Norris, and F. W. Scott, A metaanalysis of infant diet and insulin-dependent diabetes mellitus: do biases play a role?, Epidemiol. 7:87-92 (1996).
45. H. K. Åkerblom, and M. Knip, Putative environmental factors in Type 1 diabetes [see comments]. [Review] [273 refs], Diabetes Metab. Rev. 14:31-67 (1998).
46. American Academy of Pediatrics-Work Group and Cow's Milk Protein and Diabetes, Infant feeding practices and their possible relationship to the etiology of diabetes mellitus, Pediatrics 94:752-754 (1994).
47. J. N. Kostraba, K. J. Cruickshanks, J. Lawler-Heavner, L. F. Jobim, M. J. Rewers, E. C. Gay, H. P. Chase, G. Klingensmith, and R. F. Hamman, Early exposure to cow's milk, and solid foods in infancy, genetic predisposition and risk of IDDM, Diabetes 42:288-294 (1993).
48. F. Perez-Bravo, E. Carrasco, M. D. Gutierrez-Lopez, M. T. Martinez, G. Lopez, and M. Garcia de los Rios, Genetic predisposition and environmental factors leading to the development of insulin-dependent diabetes mellitus in Chilean children, J. Mol. Med. 74:105-109 (1996).
49. E. Savilahti, H. K. Åkerblom, V.-M. Tainio, and S. Koskimies, Children with newly diagnosed insulin dependent diabetes mellitus have increased levels of cow's milk antibodies, Diabetes Res. 7:137-140 (1988).
50. J. Karjalainen, J. M. Martin, M. Knip, J. Ilonen, B. H. Robinson, E. Savilahti, H. K. Åkerblom, and H.-M. Dosch, A bovine albumin peptide as a possible trigger of insulin dependent diabetes mellitus, N. Engl. J. Med. 327:302-307 (1992).
51. J. Karjalainen, T. Saukkonen, E. Savilahti, and H.-M. Dosch, Disease-associated anti-BSA antibodies in Type I (insulin-dependent) diabetes mellitus are detected by particle concentration fluoroimmunoassay but not by enzyme linked immunoassay, Diabetologia 35:985-990 (1992).
52. R. K. Cheung, J. Karjalainen, J. VanderMeulen, D. Singal, and H.-M. Dosch, T cells of children with insulin dependent diabetes are sensitized to bovine serum albumin, Scand. J. Immunol. 40:623-628 (1994).

53. I. Miyazaki, R. Gaedigk, M. F. Hui, R. K. Cheung, J. Morkowski, R. V. Rajotte, and H.-M. Dosch, Cloning of human and rat p69, a candidate autoimmune target in Type I diabetes, Biochim. Biophys. Acta 1227:101-104 (1994).
54. M. Pietropaolo, L. Castano, S. Babu, R. Buelow, Y.-L. Kuo, S., S. Martin, A. Martin, A. C. Powers, M. Prochazka, J. Naggert, E. H. Leiter, and G. S. Eisenbarth, Islet cell autoantigen 69 kD (ICA69). Molecular cloning and characterization of a novel diabetes-associated autoantigen, J. Clin. Invest 92:359-371 (1993).
55. I. Miyazaki, R. K. Cheung, R. Gaedigk, M. F. Hui, J. Van der Meulen, R. V. Rajotte, and H.-M. Dosch, T cell activation and anergy to islet cell antigen in type 1 diabetes, J. Immunol. 154:1461-1469 (1995).
56. W. Karges, R. Gaedigk, M. F. Hui, R. K. Cheung, and H.-M, Dosch. Molecular Cloning of Murine ICA69: Diabetes-prone Mice Recognize the Human Autoimmune-Epitope, Tep69, Conserved in Splice Variants from both Species, Biochim. Biophys. Acta 1360:97-101 (1997).
57. W. Karges, D. Hammond-McKibben, R. Gaedigk, N. Shibuya, R. Cheung, and H.-M. Dosch, Loss of self-tolerance to ICA69 in non-obese diabetic mice, Diabetes 46:1548-1556 (1997).
58. J. M. Martin, B. Trink, D. Daneman, H.-M. Dosch, and B. H. Robinson, Milk proteins in the etiology of insulin-dependent diabetes mellitus (IDDM), Ann. Med. 23:447-452 (1992).
59. S. M. Virtanen, L. Räsänen, A. Aro, J. Lindström, H. Sippola, R. Lounamaa, L. Toivanen, J. Tuomilehto, H. K. Åkerblom, and theChildhood Diabetes in Finland Study Group, Infant feeding in Finnish children less than 7 yr of age with newly diagnosed IDDM, Diabetes Care 14:415-417 (1991).
60. S. M. Virtanen, L. Räsänen, L. Aro, K. Ylönen, R. Lounamaa, J. Tuomilehto, H. K. Åkerblom, and the Childhood Diabetes in Finland Study Group, Feeding in infancy and the risk of Type 1 diabetes mellitus in Finland, Diabetic Med. 9:815-819 (1992).
61. S. M. Virtanen, L. Räsänen, K. Ylönen, A. Aro, D. Clayton, B. Langholz, J. Pitkäniemi, E. Savilahti, R. Lounamaa, J. Tuomilehto, H. K. Åkerblom, and the Childhood Diabetes in Finland Study Group, Early introduction of dairy products associated with increased risk of IDDM in Finnish children, Diabetes 42:1786-1790 (1993).
62. S. M. Virtanen, T. Saukkonen, E. Savilahti, K. Ylonen, L. Rasanen, A. Aro, M. Knip, J. Tuomilehto, H. K. Akerblom, and the Childhood Diabetes in Finland Study Group, Diet, cows milk protein antibodies and the risk of IDDM in Finnish children, Diabetologia 37:381-387 (1994).
63. H. K. Åkerblom, S. M. Virtanen, A. Hämäläinen, J. Ilonen, E. Savilahti, O. Vaarala, A. Reunanen, K. Teramo, and M. Knip, Emergence of diabetes-associated autoantibodies in the nutritional prevention of IDDM (TRIGR) project, Diabetes (1999).

PROTECTIVE EFFECT OF BREASTFEEDING AGAINST INFECTION IN THE FIRST AND SECOND SIX MONTHS OF LIFE

Peter W. Howie

INTRODUCTION

The American Academy of Pediatrics (1997) has recommended that babies be exclusively breastfed for the first six months of life, partially breastfed for the second six months and thereafter for as long as is mutually desired. This corresponds closely to the recommendations from the Department of Health in the United Kingdom (1994) of full breastfeeding for 4 months, the introduction of weaning foods between 4 and 6 months and the continuing use of breast milk as an important part of the diet for up to a year or more. Several factors, especially nutritional considerations, contribute to these recommendations but the beneficial health effects for the baby also play an important part.

Although there is widespread consensus about the protective effect of breastfeeding in non-industrialised countries (Victora et al., 1992), there has been less agreement about the benefits in industrialised countries which enjoy clean water supplies. This paper discusses the nature and strength of the protective effects of breastfeeding against infection in infancy and childhood.

Breastfeeding and Infection – Methodological Considerations

The most vigorous way of studying a cause and effect relationship between breastfeeding and infection would be through randomised controlled studies. Because this is not possible, careful observational studies are required giving due consideration to the important methodological issues necessary to reduce potential bias.

In 1986, Bauchner and colleagues critically reviewed all the studies published after 1970 in the English language, which had studied the association between breastfeeding and infant infection in industrialised countries. They identified 14 cohort

*Peter W. Howie, Department of Obstetrics and Gynaecology, University of Dundee, Scotland DD1 9SY

and 6 case control studies and applied four key methodological criteria namely, the avoidance of detection bias, the careful definition of the outcome variables of illness, the exact definition of modes of infant feeding and correction for confounding variables. While 8 of the 14 cohort studies and 4 of the 6 case control studies reported a protective effect of breastfeeding against infection with none showing an adverse effect, Bauchner et al reported that only two of the reports fulfilled all 4 of the methodology criteria. Even these two were unsatisfactory on the basis of small sample size leading Bauchner et al (1986) to conclude that "in industrialised countries, breastfeeding has, at most, a minimal protective effect". As bias can both exaggerate or conceal true differences, the clear need to emerge from Bauchner's work was to conduct methodologically sound studies of sufficient size to examine fully the relationships between breastfeeding and infection in infancy.

Dundee Infant Feeding and Health Study, Phase I

The Dundee Infant Feeding and Health Study (Howie et al., 1990) had two initial aims. The first was to conduct a study of sufficient size to determine the protective effect of breastfeeding against infection in the first year of life, giving due consideration to the required methodological criteria. The second was to determine if a short exposure to breastmilk conferred any protective effect.

On the basis of previous reports, the required sample size was estimated to be 540 mother/infant pairs. Consequently, 750 were initially recruited and then reduced to a cohort of 674 after excluding babies that were preterm, small for dates or ill at the time of birth. This cohort has allowed very good follow up rates and have been studied in three phases as follows:

Phase I

Monitored for the first 24 months of life to determine exact infant feeding patterns and disease in infancy.

Phase II

Studied at aged 7 years to determine patterns of childhood illness, body growth and blood pressure.

Phase III

Currently being conducted at aged 15 years to examine adolescent health, growth, glucose tolerance and markers of cardiovascular health.

Phase I of the study was designed to meet the methodological criteria set out by Bauchner et al (1986) in the following ways:

Avoidance of Detection Bias

All mother infant pairs were visited on a regular basis to minimise recall failure with data being collected by trained staff according to standardised protocols. These data were reinforced by checks against their health records held by their general practitioners.

Definition of Outcome Events

Illnesses in infancy, with emphasis on gastrointestinal and respiratory disease, were defined clearly and included events which lasted for 48 hours or more. Doubts about the status of any episode were resolved by a single paediatrician who was blind to the infant feeding status of the baby.

Definition of Infant Feeding

Data on infant feeding was collected contemporaneously allowing exact categorisation into bottle feeding, early weaning (breast feeding discontinued by 13 weeks), breast feeding for more than 13 weeks with supplements and breastfeeding for more than 13 weeks without supplements.

Allowance for Confounding Variables

Data on a total of 40 potential confounding variables were gathered and analysed using regression analysis. Three key confounders were social class, age of mother and maternal smoking. After allowing for these, no other confounders modified the relationship between infant feeding and infection in infancy.

Results

The major effect of breastfeeding in the first year was a reduction in gastrointestinal illness. Bottle fed babies had an incidence of 19.5% per cent of gastrointestinal illness during the first three months of life compared with 2.2 per cent in exclusively breastfed infants, an eight-fold difference. After allowing for confounders, this difference fell to five-fold but was still highly significant. Babies who stopped breastfeeding before 13 weeks appeared to gain no benefit but significant protection persisted in breastfed babies given supplements before 13 weeks.

The protective effect of breastfeeding against gastrointestinal disease remained strong up to 6 months and persisted, at a reduced level, up to 1 year. A similar, but much smaller, protective effect of breastfeeding against respiratory illness was also observed. The conclusion of the study was that breastfeeding in an industrialised society offered substantial protection against infection in infancy (Howie et al., 1990).

Breastfeeding and Infection in Infancy – a Systematic Review

To explore further the relationship between breastfeeding and infection in infancy, a systematic review has been performed of all relevant studies from industrialised countries identified between 1966 and 1998 (Chien and Howie, 2000). These studies were identified through searches of Medline and the Science Citation Index. The review included studies which gave specific data on infectious morbidity during the first year of life and gave clear data on infant feeding status.

For the purpose of analysis, babies were categorised as either bottle fed from birth or breastfed, either partially or exclusively. By combining partial and exclusive breastfeeding together, this would tend to present the most conservative estimate of the effect of breastfeeding on infection and may counter any selection bias in the reverse direction. The review confirmed the conclusion of Bauchner et al (1986) that several of the studies suffered from methodological shortcomings although some were methodologically sound.

Statistical Methods

Odd ratios and their confidence intervals were calculated for each study according to their outcome measures. Odd ratios were pooled and combined using the fixed effects model (Mantel and Haensil, 1952). Heterogeneity of treatment effects in the meta-analysis was formally assessed by comparing the odd ratios for each study using the Breslow-Day test (1980).

Results

Sixteen studies were analysed in respect of breastfeeding and gastrointestinal infection, of which 9/16 (56%) showed statistically significant protective effects, with the rest showing non-significant protection. For the cohort studies, the pooled odds ratio for the protective effect of breastfeeding against gastrointestinal infection was 0.36 (95% CI 0.32 – 0.41, $p < 0.0001$; heterogeneity $p < 0.001$).

In respect of respiratory disease including otitis media, there were also sixteen studies, of which 8/16 (50%) showed significant protective effects from breastfeeding. The pooled odds ratio of the cohort studies for protection against respiratory disease against infection was 0.56 (95% CI 0.49 – 0.64, $p < 0.0001$, heterogeneity $p < 0.0001$).

In respect of both gastrointestinal and respiratory infection, the studies of highest methodological quality all showed distinct trends towards protective effects from breastfeeding.

Breastfeeding, Meningitis and Other Illnesses

A total of five reports were identified which examined the potential effect of breastfeeding against meningitis, most commonly with Haemophilus influenzae virus, type b. Four of these reports (Lum et al., 1982; Cochi et al., 1986; Arnold et al., 1993; Gessner et al., 1995) showed protective effects of breastfeeding in the first 6 months of life and two showed that this extended to the second six months.

Silfverdal et al (1997) have looked at the relationship between early feeding and meningitis up to 10 years of age, and reported a long lasting protective effect increasing with each additional month of breastfeeding.

There are a small number of reports supporting the view that breastfeeding offers protection against urinary tract infection.

Summary of Systemic Review

The data from the systematic review confirmed the methodological shortcomings in some of the studies identified by previous reviewers (Kovar et al., 1984; Bauchner et al., 1986). It is possible that bias could be in either direction but the evidence of benefit is most apparent in the highest quality studies. The combined data points to a distinct protective benefit of breastfeeding against gastrointestinal disease in the first year of life with an important but smaller protection against respiratory illness in industrialised countries. The benefit is most clearly seen in the first six months and diminishes in the second six months. The review is supportive of currently recommended policies on the duration of breastfeeding.

Breastfeeding and Childhood Illness

There has been increasing interest in the possibility that events in fetal life and early childhood may influence subsequent health. This raises the possibility that infant feeding may be an important independent variable in determining childhood health.

Dundee Infant Feeding and Health Study – Phase II

The second phase of this study had the aim of determining the relationship between infant feeding and subsequent childhood health as shown by incidence of disease, body growth and blood pressure. Accordingly, at the age of 7 years, 544/674(81%) of the original cohort were contacted and 410/674(61%) were available for physical examination (Wilson et al., 1998).

In this study, available children completed a validated disease questionnaire and underwent height, weight, body composition and blood pressure measurements. In the analysis, the timing of first introduction of solid food was found to be an important confounding factor.

Results

The results of the seven-year examination showed that children who had been breastfed for 4 months had suffered less childhood respiratory infection and had lower blood pressures. These associations persisted after correcting for potential confounding variables and were independent of the timing of the introduction of solid feeds.

By contrast, babies who had been given solids at an early age had suffered more wheeze and had greater body fat. These associations were found for both breast and bottle-fed infants.

These associations do not establish a cause and effect relationship but raise the possibility that both breastfeeding and the early introduction of solids infant feeding practices may influence childhood health, body composition and markers of subsequent risk of cardiovascular disease. These potentially important associations are being further pursued currently in the Phase III of the Dundee Infant Feeding and Health Study at aged 15 years.

Conclusions

The accumulated evidence clearly supports the view that, in industrialised countries, breastfeeding offers protection against gastrointestinal and respiratory illness in infancy. The protection is strongest against gastrointestinal illness and most obvious in the first six months of life. Breastfeeding must continue for at least 3 months to obtain the observed benefits.

There are also reports which suggest that breastfeeding offers protection against meningitis, otitis media and urinary tract infection.

A potentially important new feature is that early infant feeding practices, including both breastfeeding and the timing of first solid food, may influence childhood illness and development and could have consequences for subsequent cardiovascular health.

The combined evidence is supportive of current recommendations of desired durations of exclusive and partial breastfeeding.

Future Research

There remains a strong need for further high quality studies investigating the protective effect of breastfeeding in different cultural settings. Studies which could show a benefit from effective interventions on infant feeding practices would be most welcome.

The relationship between infant feeding and long term health benefits are of great importance and merit further research as does improved understanding of the biological mechanisms involved.

References

American Academy of Pediatrics Work Group on Breastfeeding, Breastfeeding and the use of human milk, Pediatrics 100:1035-1039 (1997).

Arnold, C., Makintube, S., and Istre, G.R., Day care attendance and other risk factors for invasive haemophilus influenzae type b disease, Amer. J. Epidemiol. 138:333-340 (1993).

Bauchner, H., Leventhal, J.M., and Shapiro, E.D., Studies of breastfeeding and infections. How good is the evidence? JAMA 256:887-892 (1986).

Breslow, N.E., and Day, N.E., Statistical Methods in Cancer Research Volume 1, The Analysis of Case Control Studies, LARC Scientific Publications, Lyon:France (1980).

Chien, P.F.W., and Howie, P.W., Breast milk and the risk of opportunistic infection in infancy in industrialised and non-industrialised settings, Adv. Nutr. Res. (in press) (2001).

Cochi, S.L., Fleming, D.W., Hightower, A.W., Limpakarnjanarat, K., Faeklam, R.R., Smith, J.D., Sikes, R.K., and Broome, C.V., Primary invasive haemophilus influenzae type b disease; a population-based assessment of risk factors, J. Pediatr. 108:887-896 (1986).

Department of Health and Social Security, Weaning and the weaning diet. Report of the working group on the weaning diet of the Committee on Medical Aspects of Food Policy, London HMSO (1994).

Gessner, B.D., Ussery, X.T., Parkinson, A.J., and Breiman, R.F., Risk factors for invasive disease caused by streptococcus pneumoniae among Alaska native children younger than two years of age, Paediatr. Infect. Dis. J. 123-128 (1995).

Howie, P.W., Forsyth, J.S., Ogston, S.A. and Florey, C.D., Protective effect of breastfeeding against infection, BMJ 300:11-16 (1990).

Kovar, M.G., Serdula, M.K., Marks, J.S., and Fraser, D.W., Review of the epidemiologic evidence for an association between infant feeding and infant health, Pediatrics 74:615-638 (1984).

Lum, M.K., Ward, J.I., and Bender, T.R., Protective influence of breastfeeding on the risk of developing invasive H. influenzae type b disease (Abstract), Pediatr. Res. 16:151A (1982).

Mantel, N., and Haenszel, W., Statistical aspects of the analysis of data from retrospective studies of disease, J. Natl. Cancer Inst. 22:719-748 (1959).

Silfverdal, S.A., Bodin, I., Hugosson, S., Garpenholt, O., Werner, B., and Estjorner, E. et al., Protective effect of breastfeeding on invasive haemophilus influenzae infection: a case control study in Swedish preschool cihldren, Int. J. Epidemiol. 26:443-450 (1997).
Victora, C.G., Huttley, S.R., Fuchs, S.C. Nobre, L., and Barras, F.C., Death due to dysentry, acute and persistent diarrhoea among Brazilian infants, Acta Paediatr. Supplement, 381:7-11 (1992).
Wilson, A.C., Forsyth, J.S., Greene, S.A., Irvine, L., Hau, C., and Howie, P.W., Relation of infant diet to childhood health; seven years of follow-up of cohort of children in Dundee infant feeding study, BMJ 316:21-25 (1988).

BREASTFEEDING - AN EVOLUTIONARY AND NEUROENDOCRINE PERSPECTIVE

Jan Winberg*

INTRODUCTION

Breastfeeding (BF) is part of a complex spectrum of intimately related mother-infant interactions unfolding after delivery, serving energy economy, feeding, protection and attachment. It is hypothesized that these behaviors are partly genetically based, selected by evolution because of their survival value and that physical proximity promotes their expression and perception. With this hypothesis as a backbone it is possible to make a number of testable predictions about behaviours and reactions of mothers and newborns when cared for together and when separated. The results suggest that exposure to contexts contingencies that are evolutionary unexpected - such as separation, Caesarean section, stress in delivery rooms, neonatal and maternal wards - can create pathology, for example failure of BF. To promote BF, care in delivery units/maternal and neonatal wards should aim at a fit with the biology of mother/newborn and a strengthening of maternal selfconfidence, which seems to be fragile in many mothers. If we believe in the health benefits of breastmilk, it would be a first priority to improve the prerequisites for successful BF, by improving the conditions for mothers both at the hospital and the society level.

Lactation Crises (Insufficient Milk Syndrome)

Out of 119 mothers intending to breastfeed their babies, 45% had dried up two weeks after birth[1]. This is the "milk insufficiency syndrome" or better "lactation crisis", since it is not the milk synthesis but the let down reflex which fails, leading to a rapid dry up[2]. It was prevalent thirty years ago and still is so. When we face a problem which persists in spite of years or decades of research, it may be worth while to take a step back and reconsider the hypotheses on which research and therapy have been based.

*Dpt of Paediatrics, Karolinska Hospital Q208, 17176 Stockholm, Sweden

Immediate Postnatal Care

We have undertaken a series of studies guided by the following hypothesis:

Breastfeeding is an aspect of mother/infant interaction

1. Depends on exchange of signals between mother and baby
2. These signals are adaptive, i.e. selected by evolution
3. Physical closeness promotes their expression and perception
4. The signals are sensitive to interference and to stress

With this in mind and considering that human babies belong to the so called altricial mammals, ie they are born helpless and depend on a caregiver for their basic needs of warmth, food, protection and attachment, it is possible to define a number of testable predictions. Since our needs and behaviours were shaped in our prehistoric ancestors tens of thousands of years ago, when the maternal body was the most proximate source of these needs, one would predict that newborns placed skin-to-skin with the mother would be comfortable, warm, protected and easily latch on to the breast. This position would presumably support lactation as well as maternal attachment and attentiveness to the baby´s needs. In obstetric/paediatric care the mother and her baby have often been separated. What is the evidence that this might be against an "agenda" selected for us by evolution?

Separation Distress Cries

A baby´s cry tells "I am uncomfortable, help me". If natural selection has programmed the newborn to "expect" the sensory stimuli associated with close contact with the mother's body, one would predict that human babies, like other mammals with

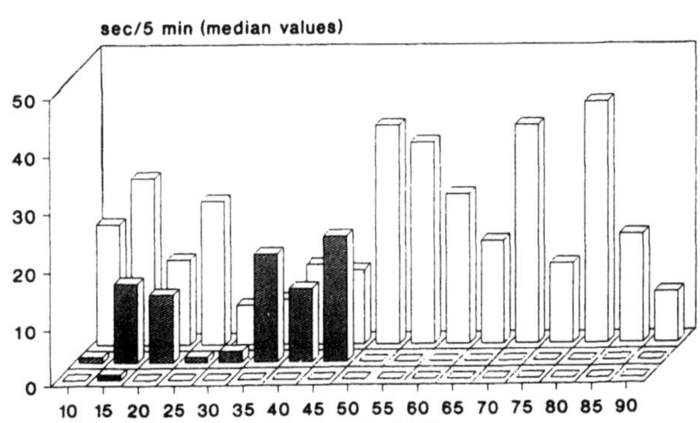

Figure 1. Medians for crying time - seconds/5 min periods - during the first 90 minutes after birth. The babies were kept either skin-to-skin (1st row),or in a cot (3rd row). A third group was placed in a cot the first 45 minutes and then skin-to-skin (2nd row).

an immature thermoregulatory system,[3] would signal when separated from the mother and become silent at reunion.

Newborns were randomly selected to be placed either in a warm cot or skin-to skin on the mother's chest during the first 90 minutes post partum. The latter babies were quiet all the time, while the separated babies cried in bouts for the 90 minutes.[4] At reunion with the mother crying stopped immediately, Fig. 1. The cry signal probably serves to achieve and maintain proximity to the mother.

Babies in body contact with their mothers cried very little compared to babies placed in a warm cot beneath the mother's bed. When babies were transferred from the cot to the mother at 45 minutes after birth their crying stopped immediately (Reproduced from 4).

Temperature and Metabolic Adjustment

Is there a risk for hypothermia when babies are cared for naked with the mother? In a controlled study the opposite was shown to be true: babies placed skin-to-skin with the mother for 90 minutes immediately post partum were significantly warmer than those placed in a warm cot, their acid-base balance and respiration stabilized more rapidly and most significantly their blood glucose level was higher at 90 minutes ($p<0.001$).[5] Thus the close contact seemed to conserve rather than spend energy. Considering the conditions under which our prehistoric ancestors reproduced it seems probable that proximity beween mother and baby has been selected by evolution to increase survival and wellbeing.

Promotion of bf and Post-delivery Skin-to-skin Contact

Positioning the newborn on the mother's abdomen and chest within minutes after birth will elicit a prefeeding behaviour: they root, achieve hand-mouth contact, show mouthing, lift the head, move without help to the breast, attach to the nipple and eventually begin to suck after about 60 minutes.[6] These behaviours and their timing are highly consistent from baby to baby suggesting that they are based on an innate central program, which activates locomotion generating centers after appropriate sensory stimulation. The experience that her newborn baby can find the breast and suck already within an hour after birth, does increase the mother's confidence in her ability to breastfeed and may stabilize the let down reflex and reduce the risk of "lactation crises". In fact body- and suckling contact immediately after delivery does increase the duration of breastfeeding.[1, 7]

Role of Odour

A number of maternal sensory stimuli (tactile, thermal, visual, auditory, odourus) influence the newborn's behaviour. The role of odour has often been overlooked. Breast odours are attractive to the newborn, and when babies had to choose between a washed and a non-washed breast the first hour after birth, they went for the unwashed one.[8] They are able to distinguish the own mother's odours from those of other women and to form odour memories (for review see[9]). Maternal breast odours activate the typical prefeeding behaviour mentioned above. When newborns were exposed to the smell of colostrum a few hours after birth this caused an increase in the blood flow to certain regions of the brain.[10] Since odours seem to play a role for infantile nursing behaviour a practical consequence would be not to wash the breasts before feeding, and to avoid perfumes or deodorants.

Effects on Breastfeeding of Interference with Suggested Innate Behaviors

Nipple-areola Attachment

An important aspect of BF is that the baby latches on to the breast with mouth wide open, tongue low in the mouth, with both nipple and a large part of the areola in mouth when sucking starts. This behaviour is favoured when the baby is given time to find the breast itself.[11] When babies attach with only the nipple in mouth, possibly due to external interference with the rooting reflex, this shortens the duration of BF and increases the frequency of sore nipples.[11, 12] An undisturbed first hour after birth - so often violated in busy delivery units - promotes an adquate breast attachment.

In a study, one subgroup of infants were removed from the mother about 20 minutes post partum, weighed, bathed, dressed and returned to her after another 20 minutes.[11] In another subgroup the babies had continuous skin-to-skin contact with their mothers for at least one hour. In the former group fewer babies attached to the breast with nipple and areola in mouth than in the latter group ($p < 0.001$). In a subsequent study it was shown that improper attachment to the breast at discharge from the maternity ward (only nipple, not areola, in the mouth), was associated with shorter breastfeeding period.[13] Thus a subtle intervention can disturb an innate behaviour, and give breastfeeding a suboptimal start.

Pain Relief

Newborn babies, whose mothers have received pethidine (meperidine) and certain other analgetics are either slow to attach to the breast or do not attach at all. Sucking, when it occurs, is of low intensity.[11, 14] This is not to discourage pain relief, but babies under the influence of sedative drugs should be allowed ample time to find the breast at their own pace.

Pushing the Baby's Head

When the baby is slow, the nursing staff or the mother may be tempted to "help" the baby by pushing the head towards the breast or manipulate the nipple into the "unwilling" mouth. This can elicit a reflex by which the tongue bends upwards against the palate and makes it impossible to suck.[15] A crying baby, an unhappy mother and a stressed nurse may result.

Pacifier

A recent prospective, but not randomized study indicated that the use of pacifiers can interfere with breastfeeding,[16] but this is still controversial. In addition an extensive use of a pacifier might deprive the baby of the enriched sensory stimulation provided by physical contact with a caregiver.[17]

Maternal Responses to Proximity

Klaus & Kennell et al found that a total of 18 hours of extended contact during the first three days after delivery increased the mothers affectionate attention to their babies[18, 19] de Chateau et al.[1, 20, 21] found that 15-20 minutes of skin-to-skin contact immediately after birth had similar effects. Although some of these findings still are controversial, it is well documented that early sucking facilitates BF.[1, 7]

In a later study all babies were placed skin-to-skin with the mother for about 45 min post partum, one subgroup suckled or at least licked/touched the areola region, while the other subgroup had the first sucking contact at a median age of 8.8 hours. Maternal behaviour in these two groups differed significantly on three items. The mothers in the early areola-licking/touching group left their babies alone in the nursery significantly less, talked more to their babies during breastfeeding, and had hormonal changes suggesting less stress during feeding.[22]

The sensitive Areola and Oxytocin

The interpretation of these findings remain speculative. Just before delivery there is a pronounced increase in tactile sensitivity of the areola and breast skin,[23] and self-manipulation of the areola at term causes a surge of the plasma oxytocin level.[24] This "love hormone" is released in three phases of reproduction: intercourse, delivery and breastfeeding. It therefore makes sense that the baby´s touch of the highly sensitive areola should increase the mother´s proximity seeking behaviour and perhaps her drive to talk to the baby.[22] The latter behaviour corresponds well with the newborn's sophisticated ability to perceive the mother's voice and the fact that the human voice seems to have an organizing influence on the development of the growing brain.[25] Thus any care routine that stimulates maternal talking to her newborn may be of some benefit.

Maternal Responses to Stress

Rooming in During Illness

The duration of BF was compared in two groups of newborn infants with minor illnesses. are in the mother´s room was associated with a significantly longer duration of BF as compared to care in a neonatal ward.[26]

Test Weighing and Supplementary Feeds

It has been a common practice to weigh the baby before and after each breastfeed and add milk substitutes when the baby had sucked "too little". hen weighing and supplementary formula were cancelled, there was an immediate and dramatic increase in breast feeding success, while weight gain during the first as well as the following five weeks was not influenced.[1] Resonable interpretation is that these routines stressed the mothers, inhibited oxytocin release and the let down reflex and paved the way for "lactation crises".

Breastfeeding After Caesarean Section (CS)

It is a clinical experience that after CS, mothers have difficulties to sense the let down reflex during the first days. The duration of BF may also be shorter in CS mothers (for ref see[27]). One possible explanation is that the pulsatile release pattern of oxytocin during suckling is less pronounced after CS than after vaginal delivery.[27] A practical consequence would be to give CS mothers more time to let the let down reflex mature and to give them adequate support and reassurance meanwhile.

Comments

In Sweden there has been a dramatic increase in breastfeeding frequency and duration since the mid 1970s. The reasons for this are probably multiple but the increase has at least coincided with the introduction of new hospital routines keeping mother and baby together.[28] Similar findings have been reported from Norway.[29]

It is shown above that skin-to-skin care immediately after birth helps the newborn to conserve energy, restore acid-base balance, adjust respiration, become comfortable (not crying) and begin sucking. This indicates that the newborn's brain is programmed for an environment enriched by the various sensory stimuli, that the mother provides. It is reasonable to assume that skin-to-skin care has been selected during evolution because of its survival value. Some of the functions of the mother can be replaced by modern technology, but not the sensory stimulation that she provides. This is true whether the newborn is healthy or sick, fullterm or prematurely born.

The potential for early sensory stimuli to shape and organize complex brain networks is unbelievably great. Thus, one hour of visual stimulation of the newborn human infant is sufficient to turn on the development of the visual capacity (for ref. see [30]). Other sensory modalities may also be activated by early stimulation. Gentle touch/stroking and other discrete stimuli[31, 32] and "Kangaroo" care seem to support neurodevelopmental processes, and Fifer has shown the organizing effect of the mother's voice on brain function in the newborn.[25] Hofer has shown the perturbation of regulatory factors when pups are separated from their mothers and other animal data show the potential impact of neonatal experience on future behavior.[33] Whether we like it or not, there are some striking similarities between rodents and man.

How is the impact of early sucking on maternal behaviour and breastfeeding mediated? "In principle evolution is conservative on mechanisms".[34] With this in mind it might be of interest to mention that in nonhuman primates, beta-endorphin and oxytocin are central as regulators of mother-infant interaction as well as for social interaction with group members.[34, 35]

In human mothers, plasma levels of endorphins double during suckling,[36] which might contribute to the wellbeing that mothers often report while breastfeeding. Furthermore frequent and prolonged BF might contribute to the attachment process due to the repeated activation of the opioid system. It is possible that beta-endorphin at least in part acts through the release of peptide transmitters like oxytocin.[37, 38, 39] The latter peptide is an "affiliative" hormone released i.a. during successive phases of the human reproductive cycle - intercourse, parturition, suckling - but also in response to non-noxious tactile stimuli. In some mammals it is involved in activation of maternal behaviour and bonding.[37, 40] Oxytocin is produced in hypothalamic nuclei (for ref see[38, 39]) and it is interesting that immediately after delivery these nuclei undergo structural rebuilding (animal studies) which facilitatates pulsatile oxytocin release seen during feeding.

In human mothers plasma oxytocin levels are high with several distinct peaks during the first 15-60 minutes post partum.[42] These peaks appear during the period when significant effects on maternal behaviour and on breastfeeding performance were induced by the newborn. If this early increase in plasma levels of oxytocin is matched by a release of oxytocin in oxytocinergic nerves in the brain, for which there is some experimental evidence,[35] it is tempting to speculate that the effects on maternal behaviour of early skin-to-skin and suckling contact were related to an activation of the oxytocin system.

If close body contact is important for the developing mother-infant relations as suggested above, how can adoptive parents attach warmly to their babies without support of pregnancy and delivery, and how can parents of premature and sick babies compensate for a long separation from their babies? Innate, genetically determined programs can be "closed" or "open", i.e. open programs can incorporate experience and reason.[42] This presupposes a large brain and is probably of selective advantage in mammals with a long period of dependence upon parental care. Thus, in humans cognitive factors and experience can take over neuroendocrine ones in reinforcing attachment between parents and the baby, although the neuro-endocrine support might help mothers to feel selfconfidence and to read the baby's signals. Vulnerable families or parents with certain personality traits may be especially helped by the neuro-endocrine support in establishing a reliable tie to and in coping with the new family member.

Concluding Remarks

If breastfeeding is as important for early and future health as suggested by present knowledge, it would be a first priority to find ways to support initiation and maintenance of BF where rates of these are low. Dealing with social/cultural factors which undermine BF is a political issue, where lay organizations like La Leche League and the medical profession can act as pressure groups. To organize obstetric and neonatal care so that it optimizes initiation and maintenance of BF is the responsibility of doctors, midwifes and nurses. It matters![43]

BF is not a cognitive skill, but depends on psycho-neuro-endocrine factors which may be strengthened by some of the measures discussed in this paper. Reliable information on BF epidemiology is a necessary basis for evaluating attempts to reduce BF failures - here much remains to be done.

Acknowledgement

The author is grateful for stimulating collaboration during many years with present and former students M Bartocci, P de Chateau, K Christensson, G Marchini, E Nissen, V Wahlberg, H Varendi, A-M Widström and present and past coworkers R Porter and K Uvnäs-Moberg.

References

1. P. de Chateau, H. Holmberg, K. Jakobson, J. Winberg, A study of factors promoting and inhibiting lactation, *Dev. Med. Child. Neurol.* 19,575-584 (1977).
2. C. Hillervik-Lindquist, Studies on perceived breast milk insufficiency, *Acta Paediatr. Scand.* Suppl. 376 (1991).
3. M.A.Hofer, H.N. Shair, Ultrasonic vocalization, laryngeal braking, and thermogenesis in rat pups: a reappraisal, *Behavioral Neuroscience* 107,354-362 (1993).
4. H. Varendi, R.H. Porter, J. Winberg, Does the newborn baby find the nipple by smell?, *Lancet* 344,989-990 (1994).
5. K. Christensson, C. Siles, L. Moreno, A. Belastequi, P. de la Fuente, H. Lagercrantz, P. Puyol, J. Winberg, Temperature, metabolic adaptation and crying in healthy, full-term newborns cared for skin to skin or in a cot, *Acta Paediatr.* 81,488-493 (1992).
6. A.M. Widström, A.B. Ransjö-Arvidson, K. Christensson, A.S.Mathiesen, J. Winberg, K. Uvnäs-Moberg, Gastric suction in healthy newborn infants. Effects on circulation and developing feeding behaviour, *Acta Paediatr. Scand.* 76,566-72 (1987).
7. E.M. Salariya, P.M. Easton, J.I. Cater, Duration of breast-feeding after early initiation and frequent feeding, *Lancet* II,1141-1143 (1978).
8. K. Christensson, T. Cabrera, E. Christensson, K. Uvnäs-Moberg, J. Winberg, Separation distress call in the human neonate in the abscence of maternal body contact, *Acta Paediatr.* 84,468-473 (1995).
9. J. Winberg, R. Porter, Olfaction and human neonatal behaviour:clinical implications, *Acta Paediatr,* 87,6-10 (1998).
10. M. Bartocci, J. Winberg, C. Ruggiero, L.L. Bergqvist, G. Serra, H. Lagercrantz, Activation of olfactory cortex in newborn infants after odor stimulation: a functional near-infrared spectroscopy study, *Pediatr. Res.* 48,18-23 (2000).
11. L. Righard, M.O. Alade, Effect of delivery room routines on success of first breast-feed, *Lancet* 336,1105-1106 (1990).
12. M.W. Woolridge, Aetiology of sore nipples, *Midwifery* 2,172-76 (1986).
13. L. Righard, M.O. Alade, Sucking technique and its effect on success of breastfeeding, *Birth* 19,185-189 (1992).
14. E. Nissen, G. Lilja, A-S.Mathiessen, A.B. Ransjö-Arvidsson, K. Uvnäs-Moberg, A.M. Widström, Effects of maternal pethidine on infants' developing breastfeeding behaviour, *Acta Paediatr.* 84,140-145 (1995).
15. A.M. Widström, J. Thingström-Paulsson, The position of the tongue during rooting reflexes elicited in newborn infants before the first suckle. *Acta Paediatr.* 82,281-283 (1993).
16. C.R. Howard, F.M. Howard, B. Lanphear, E.A. deBliek, S. Eberly, R.A. Lawrence, The effects of early pacifier use on breastfeeding duration, *Pediatrics* 103,659(e33) (1999).
17. J. Winberg, R.H. Porter, Olfaction and human neonatal behaviour:clinical implications, *Acta Paediatr.* 87,6-10 (1998).
18. M.H. Klaus, R. Jerauld, N. Kreger, W. McAlpine, M. Steffa, J.H. Kennell, Maternal attachment: importance of the first post partum days, *New Engl. J. Med.* 286,460-463 (1972).
19. M.H. Klaus, J.H. Kennell, *Maternal infant bonding* (C.V. Mosby Co, St Louis, 1976).
20. P. de Chateau, B. Wiberg, Long term effect on mother-infant behaviour of extra contact during the first hour post partum. I. First observations at 36 hours, *Acta Paediatr. Scand.* 66,137-143 (1977).
21. P. de Chateau, B. Wiberg, Long-term effect on mother-infant behaviour of extra contact during the first hour post partum. II. A follow-up at three months, *Acta Paediatr. Scand.* 66,145-151 (1977).
22. A.M. Widström, V. Wahlberg, A-S. Mathiesen, P. Eneroth, K. Uvnäs-Moberg, S. Werner, J. Winberg, Short-term effects of early suckling and touch of the nipple on maternal behaviour, *Early Human Development* 21, 153-163 (1990).
23. J.E. Robinson, R.V. Short, Changes in breast sensitivity at puberty, during menstrual cycle and paturition, *Brit. Med. J.* 1,1188-1191 (1977).
24. K. Christensson, B.A. Nilsson, S. Stock, A-S. Matthiesen, K. Uvnäs-Moberg, Effect of nipple stimulation on uterine activity and on plasma levels of oxytocuin in full term, healthy, pregnant women, *Acta Obstet. Gynecol. Scand.* 68,205-210 (1989).
25. W.P. Fifer, C.M. Moon, The role of mother's voice in the organization of brain function in the newborn. *Acta Paediatr.* 83 (Suppl. 397),86-93 (1994).
26. G. Elander, T. Lindberg, Short mother-infant separation during first week of life influences the duration of breastfeeding, *Acta Paediatr. Scand.* 73,237-240 (1984).

27. E. Nissen, K. Uvnäs-Moberg, K. Svensson, S. Stock, A.M. Widström, J. Winberg, Different patterns of oxytocin, prolactin but not cortisol release during breastfeeding in women delivered by Caesarean section or by the vaginal route, *Early Human Dev.* 45,103-118 (1996).
28. J. Winberg, V. Wahlberg, P. de Chateau, Breast feeding - early initiation and duration. An epidemiologic study, in: *Human Milk. Its biological and social value,* edited by S. Freier and A.I. Eidelman Amsterdam, (Excerpta Medica, Int. Congress Series No 518, Amsterdam 1980), pp. 283-286.
29. E. Heiberg-Endresen, E. Helsing, Changes in breastfeeding practices in Norwegian maternity wards - national surveys 1973, 1982 and 1991, *Acta Paediatr.* 84,719-724 (1995).
30. R. Sireteanu, Switching on the infant brain, *Science* 286:59-61 (1999).
31. T.M. Field, S.M. Schanberg, F. Scafidi et al, Tactile/kinesthetic stimulation effects on preterm neonates, *Pediatrics* 77,654-658 (1986).
32. H. Als, F.H. Duffy, G.B. McAnulty, Effectiveness of individualized neurodevelopmental care in the newborn intensive care unit (NICU), *Acta Paediatr.* 85 (Suppl. 416),21-30 (1996).
33. M.A. Hofer, Early relationships as regulators of infant physiology and behavior, *Acta Paediatr.* 83 (Suppl 397),9-18 (1994).
34. E.B. Keverne, C.M. Nevison, F.L. Martel, Early learning and the social bond, in: *The Integrative Neurobiology Of Affiliation,* edited by C.S. Carter, I.I Lederhendler and B. Kirkpatrick, Ann. N.Y. Acad. Sci. 807,329-339 (1997).
35. E.B. Keverne, K.M. Kendrick, Maternal behaviour in sheep and its neuroendocrine regulation, *Acta Paediatr.* 83 (Suppl 397),47-56 (1994).
36. R. Franceschini, P.L. Venturini, A. Cataldi, T. Barrecs, N. Ragni, E. Rolandi, Plasma beta-endorphin concentration during suckling in lactating women, *Br. J. Obst. Gynecol.* 96,711-713 (1989).
37. C.A. Pedersen, A.J. Prange Jr, Induction of maternal behavior in virgin rats after intracerebroventricular administration of oxytocin, *Proc. Nat. Acad. Sci.* 76,:6661-6665 (1979).
38. K. Uvnäs-Moberg, Neuroendocrinology of the mother-child interaction, *Trends Endocrinol. Metab.* 7,126-131 (1996).
39. K. Uvnäs-Moberg, Physiological and endocrine effects of social contact, in: *The Integrative Neurobiology Of Affiliation,* edited by C.S. Carter, I.I Lederhendler and B. Kirkpatrick, Ann. N.Y. Acad. Sci. 807,146-163 (1997).
40. K.M. Kendrick, E.B. Keverne, B.A. Baldwin, Intracerebroventricular oxytocin stimulates maternal behaviour in he sheep, *Neuroendocrinology* 46,56-61 (1987).
41. B.K. Modney, G.I. Hatton, Maternal behaviors: evidence that they feed back to alter brain morphology and function, *Acta Paediatr* 83, (Suppl 397), 29-32 (1994).
42. E. Nissen, G. Lilja, A.M. Widström, K. Uvnäs-Moberg, Elevation of oxytocin levels early post partum in women. *Acta Obstet. Gynecol. Scand.* 74,530-33 (1995).
43. A. Wright, S. Rice, S. Wells, Changing hospital practices to increase the duration of breastfeeding, *Pediatrics* 97,669-675 (1996).

LACTOGENESIS AND INFANT WEIGHT CHANGE IN THE FIRST WEEKS OF LIFE

Kathryn G. Dewey, Laurie A. Nommsen-Rivers, M. Jane Heinig, and Roberta J. Cohen

INTRODUCTION

Breastfeeding success depends greatly on what occurs during the first week or two postpartum. This is the time when lactogenesis stage II takes place and both the mother and the infant are learning how to breastfeed. It is also the peak period for breastfeeding problems. Early supplementation with non breast milk fluids during this time often leads to premature weaning. Therefore, it is essential to understand the factors that influence lactogenesis and the adequacy of milk transfer to the infant.

Inadequate milk transfer can lead to poor infant weight gain and other, more serious consequences if the situation is not handled appropriately. Mothers do not always realize that something is wrong because their infants often seem content; this is due to the fact that the underfed infant is often very sleepy. As a result, cases of hypernatremia, dehydration and malnutrition among breastfed infants have been reported in the literature for nearly 30 years.[1-4] Recent reports suggest that the incidence of "breastfeeding malnutrition" has increased as shorter hospital stays have become more common.[5,6] Without adequate follow-up, hypernatremic infants may not be identified until significant medical complications occur.

This paper will review the literature regarding the incidence of and potential risk factors for insufficient milk intake by breastfed infants, and will report preliminary results from a community-based prospective study designed to address these issues. It is important to keep in mind that breast milk production is a function of both maternal *capacity* for milk synthesis and infant *demand* for milk. Thus, "insufficient milk" may be directly related to characteristics of the mother, or may be secondarily induced by the feeding behavior and sucking ability of the infant.

*Kathryn G. Dewey, University of California, Department of Nutrition, Davis, CA 95616-8669

Incidence of Insufficient Breast Milk Intake by Newborns

"Insufficient milk" is one of the most common reasons mothers give for early termination of breastfeeding or supplementation with infant formula, but there has been considerable debate over whether this is more often perceived than real.[7-11] In a cross-cultural study of Anglo-American, Mexican-American, and Jamaican mothers, the percentage of breastfeeding mothers complaining of insufficient milk during the first 6 wk postpartum was about 40% in all groups, despite major differences in socioeconomic status and maternal anthropometric indices.[9] It is generally believed that 90-95% of mothers have the capacity to breastfeed successfully, so it is unclear why such a large proportion of women complain of this phenomenon. Some investigators have suggested that the mother's perception of insufficient milk does not reflect a true inadequacy but is due primarily to maternal ambivalence about breastfeeding or misinterpretation of cues such as infant fussiness or lack of fullness in the breasts.[9, 11] Data from Swedish mothers provide some evidence for this interpretation: breast milk volume (measured by test-weighing) during self-reported transient lactation "crises" was no lower than volume measured one week later, after the "symptoms" of insufficient milk had disappeared.[11] However, it should be noted that these data were collected after the mothers had already successfully established lactation; the situation could be very different during the first few weeks postpartum. In many cases, the response to perceived insufficient milk, such as supplementary bottle-feeding, may lead to a truly inadequate milk supply (due to reduced infant demand) even if the perception was incorrect to begin with. Thus, it is likely that maternal lack of confidence in the ability to breastfeed coupled with lactation mismanagement is responsible for many cases of "insufficient milk".

On the other hand, there are clearly cases in which infant breast milk intake is inadequate even when the mother is highly motivated to breastfeed and adequate lactation guidance is provided.[12] At present, there is no good estimate of the incidence of this phenomenon. Neifert el al.[12] conducted a prospective study of 319 healthy, motivated, primiparous women in the Denver area and evaluated adequacy of lactation on the basis of infant weight gain during the first three weeks postpartum. They found that 15% of the subjects had "persistent milk insufficiency" despite receiving breastfeeding guidance. However, there are several limitations to the study that cast doubt on this estimate. First, the subjects were not a random sample of breastfeeding mothers, but were recruited through announcements and flyers. Thus, it is possible that women who anticipated having problems with breastfeeding, and wanted the lactation guidance provided by the study, were more likely to enroll. There were a surprisingly high percentage of women who had prior breast surgery (6.9%), which may reflect self-selection bias. Second, the subjects were all primiparous, with no previous breastfeeding experience. Finally, the investigators classified a woman as having insufficient milk if her infant gained less than 28.5 g/d between a visit at 9-14 days and a subsequent visit at or before 21 days of age. Reference data for growth velocity of breastfed infants published by Nelson et al.[13] indicate that this cut-off is too high: for male infants this value is above the 10th percentile and for females it is above the 25th percentile during this time interval. Thus, one would normally expect at least 10-25% of breastfed infants to gain less than 28.5 g/d. Neifert et al. reported that the percentage of infants with a weight gain less than 20 g/d between the two visits was 10.7%. This is a more reasonable cut-off but may still overestimate the percentage with insufficient milk among mothers of female infants.

Potential Risk Factors for Impaired Lactogenesis or Insufficient Milk Intake

Factors potentially associated with lactogenesis can be classified as biological or behavioral, and attributed primarily to either the mother or the infant (although sometimes it is a mismatch between the mother's and infant's characteristics that creates a problem, such as a nipple that is too large for the infant's mouth). This paper will deal primarily with biological factors. Because relatively few studies have included direct measurement of lactogenesis (e.g. timing of onset of breast fullness or biochemical markers of milk synthesis), studies that assessed other outcomes such as duration of breastfeeding or infant suckling ability are also included.

The mother's and infant's experiences during labor and delivery may influence lactation in several ways. Although mothers who have Cesarean sections, prolonged labor, or other complications during labor and delivery can breastfeed successfully, they may be at higher risk for problems in the first weeks postpartum.[14, 15] A report based on case studies has suggested that maternal postpartum hemorrhage may be a risk factor for insufficient milk production.[16] Studies on the impact of Cesarean delivery on breastfeeding duration have shown mixed results: some have reported a shorter duration among mothers with C-sections,[17-19] while others have shown no difference.[20-22] The timing of the first breastfeeding is often delayed after a C-section,[14, 22] but given a supportive hospital and home environment this need not adversely affect breastfeeding success.

There is inadequate information on whether anesthesia or analgesia used during labor and delivery have any influence on lactation. Danish women with C-sections given epidural anesthesia were more likely to continue breastfeeding to 3 or 6 months postpartum than women given general anesthesia, but differences in maternal motivation could not be ruled out.[23] Delayed "effective feeding" (by maternal report) has been linked to use of the narcotic analgesic alphaprodine 1-3 hr prior to delivery.[24] Use of epidural anesthesia has been shown to be associated with lower scores for the orientation and motor clusters of the Neonatal Behavioral Assessment Scale during the first month of life, after controlling for potentially confounding variables.[25] Mothers given epidural anesthesia also spent less time with their infants in the hospital. Although this study did not assess infant feeding behavior, breastfeeding success may be compromised by decreased infant alertness and ability to orient, and reduced maternal-infant interaction. In a study of the effect of labor analgesia on infant suckling ability in the early postpartum period, infants of unmedicated mothers had higher IBFAT suckling scores than those of medicated mothers.[26] While IBFAT suckling scores were similar for infants whose mothers received either intravenous or epidural analgesia, scores of infants of mothers receiving both types of medications were lower. Breastfeeding duration, assessed at 6 weeks postpartum, did not differ among groups.

The presence of labor support persons has been shown to affect later breastfeeding success: in South Africa, women who were randomly assigned to receive supportive companionship during labor from community volunteers were almost twice as likely to be exclusively breastfeeding at six weeks postpartum than women assigned to the control group.[27]

One of the most vexing problems faced by breastfeeding mothers is delayed onset of milk production. Normally, the amount of milk produced is minimal for the first 1-2 days, but increases dramatically by about 2-3 days postpartum.[28] However, there is a relatively wide range in the timing of this phenomenon, with some mothers not experiencing an increase in milk production until 4-7 days postpartum. This can be a very stressful situation for the mother who wishes to exclusively breastfeed but whose infant is clearly hungry. Other than abnormal conditions such as placental retention,[29] the causes of delayed onset of milk production have only recently been investigated. Chen et al.[30] reported that primiparity, long duration of labor and maternal exhaustion during labor and delivery were associated with a later onset of milk production. In addition, infant stress during labor and delivery (as reflected by cord blood cortisol and glucose concentrations) was independently associated with later onset of lactogenesis. Chapman and Perez-Escamilla[15] observed that delayed onset of lactogenesis was associated with prolonged duration of stage II labor for vaginal deliveries and with unscheduled (but not elective) Cesarian section delivery. Unscheduled C-sections are often the result of fetal stress or serious maternal complications. Thus, these two studies are consistent in showing that maternal and infant stress during labor and delivery can adversely affect the onset of milk production. There are several possible mechanisms for this relationship. Maternal stress is known to affect the release of oxytocin, and thus may inhibit the milk-ejection reflex.[31] Infant stress may affect lactogenesis via weak or inadequate suckling ability or reduced infant demand. Alternatively, there are several confounding variables, such as greater use of labor medications during long or difficult deliveries, that may play a role.

Infant characteristics can influence the timing of onset of milk production, the milk-ejection reflex, and ultimately the volume of milk produced. Mothers of low birth weight infants report problems of insufficient milk more often.[9] Even within the normal range of birth weight, smaller size at birth has been associated with delayed lactogenesis.[15] This may be due to reduced sucking strength, frequency or duration of feeds among smaller infants.[32] Gestational age, i.e., immaturity, may also play a role via the coordination and strength of the infant's suck. Infants who are excessively sleepy or passive during the first week or two may not stimulate an optimal milk supply: it is well known that nursing frequency during the early postpartum period is associated with milk production.[32]

It has been hypothesized that use of artificial nipples or pacifiers will lead to "nipple confusion" by newborns,[33] and/or to reduced breastfeeding frequency. The physical movements of the mouth and tongue required to suckle effectively from the breast are quite different from those used to suck from a bottle or pacifier.[34] Infants must learn how to breastfeed effectively when first born, and there is anecdotal evidence that some newborns have difficulty latching on to the breast and extracting milk if there is exposure to artificial nipples before breastfeeding has been successfully established.[35] To date there has not been a randomized trial of the effects of pacifier use, but two studies have demonstrated that pacifier use is associated with shorter breastfeeding duration.[36, 37]

In relatively affluent populations, maternal anthropometric indices such as weight gain during pregnancy, body mass index, and weight loss postpartum have generally not been associated with milk volume once lactation is established,[32] but their relationship to milk production during the initiation of lactation is less clear. Some postpartum weight loss is expected, particularly in the first few weeks as the fluid retained during pregnancy is lost. However, it is possible that women who lose at a very rapid rate, or who are relatively lean initially and lose fat postpartum, may have more difficulty in establishing an optimal level of milk production. On the other hand, there is evidence from two studies that maternal obesity is a risk factor for delayed onset of lactogenesis[15] and short duration of breastfeeding.[38] It has been speculated that this could be due to differences in steroid hormone levels or difficulty in latching on to the breast by infants of overweight women.

Maternal age does not appear to be related to milk production[32] or lactogenesis[12], but parity is an important factor. Compared to multiparous mothers, Hildebrandt[39] reported that the onset of lactogenesis occurred 14 hours later in primiparous mothers, and Zuppa et al.[40] found that their milk volume on days 2-4 postpartum was significantly lower. On the other hand, Chapman and Perez-Escamilla[15] reported that parity was not a significant predictor of delayed onset of milk production after controlling for duration of labor, which tends to be longer in primiparous women.

Other characteristics of the mother, such as previous breast surgery, prenatal breast enlargement, and nipple protuberance, may be related to the risk of insufficient milk. Neifert et al.[12] found that women who had had periareolar breast incisions were 4.6 times more likely to have lactation insufficiency than women with no surgery, whereas for those with nonperiareolar incisions the relative risk (1.9) was not statistically significant. The type of breast surgery performed is thus of importance, with women having had reduction mammoplasty at greatest risk. Even in such cases, however, the ability to breastfeed can be preserved if appropriate surgical methods are used. In the study by Neifert et al.[12], breast size did not predict lactation success but prenatal and postpartum breast enlargement, which reflect mammary development, were significant factors. They had previously reported that breast asymmetry was linked to insufficient milk production[41], but did not find this association in the prospective study[12]. The relative risk of insufficient milk for women with inverted nipples was 2.9, which was marginally significant ($p=0.07$).

Maternal smoking and alcohol consumption may also be risk factors for insufficient milk, as both have been shown to inhibit prolactin and oxytocin release in animal studies.[32, 42] In humans, smoking has been associated with lower milk volume among mothers of both preterm[43] and term[44] infants. Inhibition of the milk-ejection reflex by maternal alcohol intake appears to be dose dependent.[42] Thus, the influence of alcohol on the establishment of lactation probably depends on the frequency and amount consumed, although even small amounts apparently can affect the infant's feeding behavior.[45]

There is little information on the adequacy of lactation among mothers with chronic health conditions. With good metabolic and dietary management, diabetic women are able to breastfeed successfully,[46, 47] although the onset of milk production occurs later than in non-diabetic women.[48] Unless medications used to treat women with chronic medical conditions are contraindicated during lactation, breastfeeding is generally feasible.[46] However, the risk of insufficient milk production may be higher in such women if their infants are more likely to be born prematurely, small-for-gestational age, or with health problems that interfere with feeding.[46] Most non-prescription and prescription medications are not thought to interfere with milk production (except for those that are specifically contraindicated).[46] However, combined estrogen-progestin oral contraceptives are known to reduce milk volume and lead to earlier cessation of breastfeeding.[49] There is no evidence that progestin-only methods (including implants) adversely affect milk volume.[50, 51]

A Prospective, Community-Based Study of Lactogenesis and Infant Weight Gain

Because of the lack of information on the incidence o f insufficient milk and risk factors for this phenomenon, we undertook a prospective, community-based study of 280 mother-infant pairs in Davis, CA. Mothers were recruited at 5 area hospitals, generally within 24 h postpartum, and given lactation guidance by trained professionals in the hospital and in the home on days 3, 7, 14 and additionally as needed. All mothers planned to breastfeed exclusively for at least one month, and all infants were healthy, single and born at term (> 37 wk). About 56% of mothers were primiparous, and the average educational level was very high (16.6 years). Social support for breastfeeding was very high in this community. The key outcome variables were infant weight change, infant suckling ability (assessed using the Infant Breastfeeding Assessment Tool[52]), and the timing of onset of milk production (defined by when the mother first reported that her breasts were noticeably fuller (3 on a scale of 5), which was previously shown to correlate highly with biochemical markers of lactogenesis and milk volume on day 5).[30]

The preliminary results can be summarized as follows: 1) Delayed onset of milk production (> 72 h postpartum) was common (24%), especially in primiparous women (34%); 2) Weight loss ≥ 10% of birth weight occurred in 12% of infants by day 3, and was 6 times more likely if there was delayed onset of milk production (35 vs. 5.6%); 3) Delayed onset of milk production was a key risk factor for subsequent formula use; 4) About half of infants scored ≤ 10 on the Infant Breastfeeding Assessment Tool on day 1,but this decreased to 22% on day 3 and 14% on day 7; 5) The vast majority (81%) of mothers reported breastfeeding problems on day 3. The most common problems were sore nipples (severe/moderate: 49%) and breast pain (38%); 6) Delayed onset of milk production was more common if there was a long duration of labor (> 14 hr), the mother had a C-section, or the infant had a low IBFAT score on day 3; 7) Infant weight loss ≥ 10% of birth weight was more common if there was a long duration of labor (in primiparous women only) or if the mother had an urgent C-section.

The results demonstrate the need for follow-up of all mothers at about 3 days postpartum, with particular attention given to primiparous women, those with a long duration of labor or C-section, and those whose infants do not suckle well by day 3.

References

1. Cooper WO, Atherton HD, Kahana M, and Kotagal UR, Increased incidence of severe breastfeeding malnutrition and hypernatremia in a metropolitan area, Pediatrics 96:957-960 (1995).
2. Kaplan JA, Siegler RW, and Schmunk GA, Fatal hypernatremic dehydration in exclusively breastfed newborns due to maternal lactation failure, Am. J. Forensic Med. Pathol. 19:19-22 (1998).
3. Ng PC, Chan HB, Fok TF et al., Early onset of hypernatraemic dehydration and fever in exclusively breast-fed infants, J. Pediatr. Child. Health. 35:585-587 (1997).
4. Livingstone VH, Willis CE, Abdel-Wareth LO, et al., Neonatal hypernatremic dehydration associated with breastfeeding malnutrition: a retrospective survey, CMAJ 162:647-652 (2000).
5. Pascale JA, Brittian L, Lenfestey CC, and Jarrett-Pullman C, Breastfeeding, dehydration, and shorter maternity stays, Neonatal Network 15:37-43 (1996).
6. Heimler R, Shekhawat P, Hoffman RG, et al., Hospital readmission and morbidity following early newborn discharge, Clin. Pediatr. 37:609-615 (1998).
7. Gussler JD, and Briesemeister LH, The insufficient milk syndrome: a biocultural explanation, Med. Anthro. 4:1-24 (1980).
8. Greiner T, Van Esterik P, and Latham M, The insufficient milk syndrome: An alternative explanation, Med. Anthro. 5:233-247 (1981).
9. Tully J, and Dewey KG, Private fears, global loss: a cross-cultural study of insufficient milk syndrome, Med. Anthro. 9:225-243 (1985).
10. Hill PD, The enigma of insufficient milk supply, Mat. Child. Nurs. 16:312-316 (1991).
11. Hillervik-Lindquist C, Studies on perceived breast milk insufficiency, Acta Paed. Scand. (Suppl) 376:1-27 (1991).
12. Neifert M, DeMarzo S, Seacat J, Young D, Leff M, and Orleans M, The influence of breast surgery, breast appearance, pregnancy-induced breast changes on lactation sufficiency as measured by infant weight gain, Birth 17:31-38 (199).
13. Nelson SE, Rogers RR, Ziegler EE, and Fomon SJ, Gain in weight and length during early infancy, Early Hum. Dev. 19:223-239 (1989).

14. Kearney MH, Cronenwett LR, and Reinhardt R, Cesarean delivery and breastfeeding outcomes, Birth 17:98-103 (199).
15. Chapman DJ, and Perez-Escamilla R, Identification of risk factors for delayed onset of lactation, J. Am. Diet Assoc. 99:450-454 (1999).
16. Willis CE, and Livingstone V, Infant insufficient milk syndrome associated with maternal postpartum hemorrhage, J. Human Lact. 11:123-126 (1995).
17. Whichelow M, Factors associated with the duration of breastfeeding in a privileged society, Early Hum. Dev. 7:273-280 (1982).
18. Procianoy R, Fernandes-Filho P, Lazaro L, and Sartori N, Factors affecting breastfeeding: The influence of cesarean section, J. Trop. Pediatr. 30:39-42 (1984).
19. Samuels S, Margen S, and Schoen E, Incidence and duration of breastfeeding in a health maintenance organization population, Am. J. Clin. Nutr. 42:504-510 (1985).
20. Loughlin H, Clapp-Channing N, Gehlbach S, Pllard J, and McCutchen T, Early termination of breastfeeding: identifying those at risk, Pediatr. 75:508-513 (1985).
21. Tamminen T, Verronen P, Saarikoski S, Goransson A, and Tuomiranta H, The influence of perinatal factors on breastfeeding, Acta Paed. Scand. 72:9-12 (1983).
22. Janke J, Breastfeeding duration following cesarean and vaginal births, J. Nurs. Midwifery 33:159-164 (1988).
23. Lie B, and Juul J, Effect of epidural vs. general anesthesia on breastfeeding, Acta Obstet. Gynecol. Scand. 67:207-209 (1988).
24. Crowell MK, Hill PD, and Humenick SS, Relationship between obstetric analgesia and time of effective breast feeding, J. Nurse-Midwifery 39:150-156 (1994).
25. Sepkoski CM, Lester BM, Ostheimer GW, and Brazelton TB, The effects of maternal epidural anesthesia on neonatal behavior during the first month, Develop. Med. Child. Neurol. 34:1072-1080 (1992).
26. Riordan J, Gross A, Angeron J, Krumwiede B, and Melin J, The effect of labor pain relief medication on neonatal suckling and breastfeeding duration, J. Hum. Lact. 16:7-12 (2000).
27. Hofmeyr GJ, Nikodem VC, Wolman WL, Chalmers BE, and Kramer T, Companionship to modify the clinical birth environment: effects on progress and perceptions of labor, and breastfeeding, Br. J. Obstet. Gynecol. 98:756-764 (1991).
28. Neville MC, Keller R, Seacat J, Lutes V, Neifert M, Casey C, Allen J, and Archer PA, Studies in human lactation: milk volumes in lactating women during the onset of lactation and full lactation, Am. J. Clin. Nutr. 48:1375-1386 (1988).
29 Neifert MR, McDonough SL, and Neville MC, Failure of lactogenesis associated with placental retention, Am. J. Obstet. Gynecol.140:477-478 (1981).
30. Chen DC, Nommsen-Rivers L, Dewey KG, and Lonnerdal B, Stress during labor and delivery and early lactation performance, Am. J. Clin. Nutr. Aug;68(2):335-344 (1998).
31. Newton M, and Newton NR, The let-down reflex in human lactation, J. Pediatr. 33:698-704 (1948).
32. Institute of Medicine. Nutrition During Lactation. National Academy Press, Washington DC, 1991.
33. Neifert M, Lawrence R, and Seacat J, Nipple confusion: toward a formal definition, J. Pediatr. 126:S125-S129 (1995).
34. Woolridge MW, The >anatomy= of infant sucking, Midwifery 2:164-171 (1986).
35. Newman J, Breast-feeding problems associated with the early introduction of bottles and pacifiers, J. Hum. Lact. 6:59-63 (1990).
36. Barros FC, Victora CG, Semer TC, Tonioli Filho S, Tomasi E, and Weiderpass E, Use of pacifiers is associated with decreased breast-feeding duration, Pediatrics 95:497-499 (1995).
37. Howard CR, Howard FM, Lanphear B, deBlieck EA, Eberly S, and Lawrence RA, The effects of early pacifier use on breastfeeding duration, Pediatrics 103:e33 (1999).
38. Hilson JA, Rasmussen KM, and Kjolhede CL, Maternal obesity and breastfeeding success in a rural population of white women, Am. J. Clin. Nutr. 66:1371-1378 (1997).
39. Hildebrandt HM, Maternal perception of lactogenesis time: A clinical report, J. Hum. Lact. 15:317-323 (1999).
40. Zuppa AA, Tornesello A, Papacci P, Tortorolo G, Segni G, Lafuenti G, Moneta E, Diodato A, Sorcini M, and Carta S, Relationship between maternal parity, basal prolactin levels and neonatal breast milk intake, Biol. Neonate 53:144-147 (1998).
41. Neifert MR, Seacat JM, and Jobe WE, Lactation failure due to insufficient glandular development of the breast, Pediatr. 76:823-828 (1985).
42. Subramanian MG, Chen XG, Bergeski BA, and Savoy-Moore RT, Alcohol inhibition of suckling-induced prolactin release in lactating rats: threshold evaluation, Alcohol. 8:203-206 (1991).
43. Hopkinson JM, Schanler RJ, Fraley JK, and Garza C, Milk production by mothers of premature infants: influence of cigarette smoking, Pediatr. 90:934-938 (1992).

44. Vio, F, Salazar G, and Infante C, Smoking during pregnancy and lactation and its effects on breast milk volume, Am. J. Clin. Nutr. 54:1011-1016 (1991).
45. Mennella JA, and Beauchamp GK, The transfer of alcohol to human milk, New Eng. J. Med. 325:981-985 (1991).
46. Lawrence RA, Breast Feeding: A Guide for the Medical Profession, Fifth Edition. Mosby-Year Book Inc., St. Louis, MO 1999.
47. Neubauer SH, Lactation in insulin-dependent diabetes, Prog. Food Nutr. Sei 14:333-370 (199).
48. Neubauer SH, Ferris AM, Chase CG, Fanelli J, Thompson, CA, Lammi-Keefe CJ, Clark RM, Jensen RG, Bendel RB, and Green KW, Delayed lactogenesis in women with insulin-dependent diabetes mellitus, Am. J. Clin. Nutr. 58:54-60 (1999).
49. Saarikoski S, Contraception during lactation, Annals Med. 25:181-184 (1993).
50. Dunson TR, McLaurin VL, Grubb GS, and Rosman AW, A multicenter clinical trial of a progestin-only oral contraceptive in lactating women, Contraception 47:23-35 (1993).
51. WHO Task Force for Epidemiological Research on Reproductive Health. Progestogen-only contraceptives during lactation: I. Infant growth, Contraception 50:35-53 (1994).
52. Matthews MK, Developing an instrument to assess infant breastfeeding behavior in the early neonatal period, Midwifery 4:154-165 (1988).

THE PRACTICE OF BREASTFEEDING AND INFANT/CHILD MORTALITY IN DEVELOPING WORLD IN THE CONTEXT OF HIV PANDEMIC

Elsa R. J Giugliani*

1. INTRODUCTION

The protective effect of breastfeeding against infant death has been suspected for a long time. There are some indications that in Europe, in the eighteenth and nineteenth centuries, the number of deaths among non-breastfed infants was 2 to 3 times higher compared to breastfed infants. (Wickes, 1953; Knodel and Van der Valle, 1967). When Victora (1996) reanalyzed data collected in Brighton between 1903 and 1905, he found that the risk of dying from epidemic diarrhea was 91 times higher among infants receiving condensed milk and 40 times higher for those on cow's milk when compared with breastfed infants.

In the last 20 years several studies on breastfeeding and infant/child mortality were published. The main conclusions that may be drawn from these studies are: 1) there is a significant protective effects of breastfeeding, at least for specific age groups; 2) the main reason for the reduction of infant mortality among breastfed infants is the ability of breast milk to protect against common infections of children such as diarrheal diseases and respiratory infections, 3) the protective effect of breastfeeding against infant mortality is modified by a number of demographic, socioeconomic, environmental, and dietary factors - greater protection is expected among the very young exclusively breastfed infants who live in places where there is poverty, crowding, unsafe drinking water, contaminated or low energy density complementary foods.

A simulation exercise based on the literature requested by the World Health Organization estimated that full or partial breastfeeding would reduce by 50% the death from respiratory diseases and by 66% the death from diarrhea. (Victora *et al.*, 1999). An increase of 40% in the rates of breastfeeding would prevent up to 15% of diarrhea death and 7% of pneumonia deaths occurring in regions with short breastfeeding duration, such as urban Latin America.

*Elsa R.J. Giuliani, Department of Pediatrics of Federal University of Rio Grande do Sul, Porto Alegre, Brazil

Until recently there was no question about the advantages and the safety of breastfeeding for all women and children in the world, except in very rare circumstances. But this scenario is changing with the HIV pandemic. The knowledge that HIV can be transmitted through breast milk brought a major public health dilemma in areas where HIV is highly prevalent and the main causes of infant mortality are infectious diseases and malnutrition. This is the case for many places in sub-Saharan Africa, where more than 90% of HIV-infected children worldwide live. Replacement feeding to reduce the risk of HIV transmission is only acceptable if the risk of illness and death from artificial feeding is less than the risk of HIV transmission. Benefits and risks of breastfeeding withdrawal must be balanced. Some simulation models were developed to weigh the advantages and disadvantages of breastfeeding for HIV positive women in order to provide a basis for policy decisions. But one of the major limitations of these simulations is the poor quantification of the relative risks for mortality associated with lack of breastfeeding (Nduati, 1998).

2. QUANTIFICATION OF RELATIVE RISKS FOR MORTALITY ASSOCIATED WITH LACK OF BREASTFEEDING

One example of how important it is to have an accurate quantification of the relative risks for mortality associated with infant feeding can be shown using the mathematical modeling proposed by Smith and Kuhn (1999). This model compares the risks of HIV transmission through breastfeeding against the increased risks of non-HIV mortality associated with lack of breastfeeding. According to this model the threshold infant mortality rate (i.e. breastfeeding has fewer adverse outcomes - HIV infections and infant death - above the threshold while no breastfeeding has fewer adverse outcomes below the threshold) when the HIV transmission rate is 14% would be 149 if the relative risk of not breastfeeding is 2.0, and 80 if the risk is 3.0. This means that breastfeeding can be a better alternative than replacement feeding in more settings if the risks of death associated with lack of breastfeeding is higher.

The lack of accurate quantification of the relative risks for mortality associated with lack of breastfeeding is mainly the result of common methodological problems of observational studies comparing morbidity and mortality of breastfed and artificially fed infants. These problems include self-selection (if mothers stop breastfeeding when infants are not growing well on breast milk), reverse causality (if a disease is the cause of the weaning and not the opposite), and confounding (factors associated with breastfeeding as well as morbidity (Victora, 1966). In addition, many studies fail in defining breastfeeding properly.

Randomized controlled trials may avoid most of the methodological problems, but they are not used for being difficult, expensive and not ethical. Compliance with a recommended feeding practice is another obstacle.

A recent meta-analysis was published in an attempt to provide adequate quantification of the risk of artificial feeding compared to breastfeeding. (WHO, 2000). This study attempted to control for bias such as reverse causality (through records of breastfeeding before the disease that caused the death) and confounding (through stratification by maternal education).

Six data sets with information on infant feeding practices and causes of death were reanalyzed. The studies (two case-control and four cohort studies) were carried out between 1983 and 1991 in six different settings - Brazil, Gambia, Ghana, Pakistan, Philippines and Senegal - and provided data for 1,223 death of children under two years of age. All death in the first week of life was excluded as well as death not attributed to infectious diseases or of unknown causes.

The data from the three African sites could not be used in the infant mortality analyses because almost all children were breastfed during the first year of life. As a result, the pooled odds ratios for infant mortality associated with lack of breastfeeding were based on the other three studies. In the first year of life, breast milk protection declined with age: odds ranged from 5.8 for infants under two months to 1.4 for infants between nine and eleven months. This finding is of particular interest, as simulations of the impact of replacement feedings for infants from HIV positive mothers usually do not take in account different levels of protection according to age.

In the first six months of life, protection against diarrhea was much higher (OR=6.1) than protection against acute respiratory infection deaths (OR=2.4). But during the second half of the first year, similar levels of protection were observed for diarrhea and respiratory disease death. It is worthy to note that the level of protection against respiratory infection death did not change with age.

During the second year of life, the pooled odd ratios based in five studies ranged from 1.6 and 2.1. In Pakistan, Gambia and Philippines there was no significant protective effect of breastfeeding during the second year of life. Nevertheless the authors concluded that the results for the second year are still compatible with a protective effect. This conclusion is important as most simulations assume that there is no effect of breastfeeding on mortality in the second year of life.

Even though the odd ratios provided by this study seem to be more accurate than existing data from single studies, still some caution is necessary when using this information in simulations of the impact of replacement feedings for infants from HIV positive mothers. This study did not take in account breastfeeding pattern. It is known that the protection provided by human milk against death due to infections is much more evident when the infant is being exclusively breastfed. In Brazil, a case-control study showed that, in the first year of life, infants who were exclusively breast fed had a much lower risk of dying from diarrhea or respiratory diseases than those partially or not breastfed (Victora *et al.*, 1987).

Besides providing more protection against common **infections** in children, exclusive breastfeeding, at least for short periods, may offer some protection against vertical HIV transmission. A recent study, in Durbin, South Africa, showed that exclusively breastfed infants of HIV-positive mothers had a lower probability of being infected at three months (14.6%) when compared with children who were partially breastfed (24.1%) or not breastfed (18.8%) (Coutsoudis *et al.*, 1999). If these findings are confirmed, infants of HIV-positive mothers would receive all the benefits of exclusive breastfeeding in the first months of life without facing the additional risk of HIV infection. Also, these results would have implications for simulation models that have compared the risks of HIV transmission through breastfeeding against the risks of non-HIV mortality associated with not breastfeeding. The rate of transmission through breastfeeding would be lower and consequently the threshold infant mortality rate would be lower, increasing the range of settings in which breastfeeding would cause less death compared with breastmilk substitutes (Smith and Kuhn, 1999).

Another possible limitation of the study is the exclusion of death in the first weeks of life, which may result in an underestimation of the protection of breast milk. Several reports from developing countries have described an important reduction in the incidence of neonatal infectious diseases after breastfeeding rates were increased (Clavano, 1982; Pichaipat, 1993). A randomized controlled trial in India showed that newborns at higher risk of infections were less vulnerable if fed only with raw human milk (Narayanan *et al.*, 1984). In Europe, a multicentre prospective study showed that mortality due to necrotising enterocolitis was 10.6 and 3.5 times higher among preterm newborns receiving artificial milk only and artificial plus human milk, respectively, compared with newborns fed exclusively human milk (Lucas and Cole, 1990).

Since this study considered only death due to infectious diseases, the relative weight of infections in overall infant and child mortality should be considered. In developing countries, it is estimated that infections are responsible for two-thirds of death of under-five children.

CONCLUSION

In this paper only the risks of infant mortality related to lack of breastfeeding were considered. Other factors must be considered too when recommending replacement feedings for children from HIV infected mothers such as infant morbidity, nutrition, mother fertility (more infants at risk of HIV and of becoming orphan), mother social stigmatization and "spill over" effect.

A revised statement by WHO, UNICEF and UNAIDS recommended that voluntary counseling and testing should be offered to all pregnant women, that HIV-infected women should be made aware of the transmission risks of breastfeeding to make informed decisions, and that they should be supported in their choices (Join United Nations Programme on HIV/AIDS, 1996). One major obstacle to this recommendation is that in places with high prevalence of HIV the choices are limited because of unsafe water or cultural norms. HIV-positive mothers who chose not to breastfeed must be informed about the risks of their infants to die from diarrhea, respiratory diseases, and other infections. This risk is likely to be different according to the characteristics of the population, making it difficult to generalize specific recommendations. Also, information on the risks is limited in many settings. In Africa, for example, there are no reliable data because almost all babies are breastfed during the first year of life. Therefore, in order to face the dilemma regarding the better infant feeding practices in areas where HIV is highly prevalent and the main cause of infant mortality are infectious diseases and malnutrition it is crucial to better quantify the risks of replacement feeding.

REFERENCES

1. Clavano, N.R., 1982, Mode of feeding and its effect on infant mortality and morbidity, *J. Trop. Pediatr.* **28**:287- 293.

2. Cooutsoudis, A., Pillay, K., Spooner, E., Kuhn, L., and Coovadia, H. M., 1999, Influence of infant-feeding patterns on early mother-to-child transmission of HIV-1 in Durban, South Africa: A prospective cohort study, Lancet 354:471-476.
3. Joint United Nations Programme on HIV/AIDS, 1998, HIV and infant feeding: An interim statement, *Wkly. Epidemiol. Rec.* **71**:289-291.
4. Knodel, J. and Van der Valle, E., 1967, Breast feeding, fertility and infant mortality. An analysis of early German data, *Pop. Stud.* **21**:109-31.
5. Lucas, A. and Cole, T. J., 1990, Breast milk and neonatal necrotising enterocolitis, *Lancet* **336:**1519-1523.
6. arayanan, I., Prakash, K., Murthy, N. S., and Gujral, V. V., 1984, Randomised controlled trial of effect of raw and holder pasteurised human milk and of formula supplements on incidence of neonatal infection, *Lancet* **ii:**1111-1113.
7. Nduati, R., 1998, HIV and infant feeding: A review of HIV transmission through breastfeeding. WHO/UNAIDS, Geneva.
8. Pichaipat, V., Thanomsingh, P., and Tongpenyai, Y., 1984, Reduction of postnatal morbidity, mortality and budget in Natchasima Hospital during breast-feeding program period, *Thai J. Epidemiol.* **1**:45-52.
9. Smith, M. M. and Kuhn, L., 1999, Infant feeding and HIV-1 transmission {letter}, *Lancet* **354:**103-1904.
10. Victora, C. G., 1996, Infection and disease: The impact of early weaning, *Food Nutr. Bull.* **17**:390-396.
11. Victora, C. G., Smith, P.G., Vaughan, J.P., Nobre, L. C., Lombardi, C., Teixeira, [a] M. B., Fuchs, S. M. C., Moreira, L. B., Gigante, L. P., and Barros, F. C., 1997, Evidence for a strong protective effect of breastfeeding against infant death due to infectious diseases in Brazil, *Lancet* **2:**319-322.
13. Victora, C. G., Kirkwood, B. R., Ashworth, A., Black, R. E., Rogers, S., Sazawal S., Campbell H., and Gove, S., 1999, Potential interventions for the prevention of childhood pneumonia in developing countries: improving nutrition, *Am. J. Clin. Nutr.* **70:** 309-20.
14. Wickes, I. G., 1953, A history of infant feeding, *Arch. Dis. Child.* **4**:233-240, 416-422.
15. World Health Organization Collaborative Team, 2000, How much does breastfeeding protect against infant and child mortality due to infectious diseases? A pooled analysis of six studies from less developed countries, *Lancet* **355:**451-455.

SPECIFIC AND NONSPECIFIC PROTECTIVE FACTORS IN MILK: WHY DON'T THEY PREVENT VIRAL TRANSMISSION DURING BREASTFEEDING?

Charles E. Isaacs*

1. INTRODUCTION

It is generally agreed that breastfeeding reduces the incidence and severity of gastrointestinal and respiratory infections in infants,[1] including those caused by respiratory syncytial virus, rotavirus and enteroviruses.[2] Protection against viral infection is provided by a multiplicity of protective factors including secretory antibodies, lipids, lactoferrin, secretory leukocyte protease inhibitor (SLPI), and oligosaccharides.[3, 4] Despite the presence of antiviral compounds in human milk, some viruses are transmitted from mother to infant via milk. Reports in the literature suggest that most milk-transmitted viruses are enveloped (Table 1).

Table 1. Some Enveloped Viruses Transmitted Through Milk[a]	Virus Type[b]
Human Immunodeficiency Virus Type 1 (HIV)[5]	R
Human T-Lymphotrophic Virus Type 1 (HTLV-1)[6]	R
Human T-Lymphotrophic Virus Type 2 (HTLV-2)[7]	R
Murine Leukemia Virus[8]	R
Feline Immunodeficiency Virus[9]	R
Maedi/Visna Virus[10]	R
Caprine Arthritis-Encephalitis Virus[11]	R
Mouse Mammary Tumor Virus[12]	R
Rubella Virus[2]	T
Cytomegalovirus (CMV)[2, 13, 14]	H

[a] There are conflicting reports as to whether a number of other enveloped viruses are transmitted by human milk including herpes simplex viruses 1 and 2, hepatitis C virus, Varicella-Zoster virus and Epstein-Barr virus.[3, 15-19]

[b] Retrovirus (R); Herpes virus (H); Togavirus (T)

*Charles E. Isaacs, Ph.D., New York State Institute for Basic Research in Developmental Disabilities, 1050 Forest Hill Road, Staten Island, NY 10314, USA

Integrating Population Outcomes, Biological Mechanisms and Research Methods in the Study of Human Milk and Lactation
Edited by Davis *et al.*, Kluwer Academic/Plenum Publishers, 2002

Of these the majority are retroviruses including HIV and HTLV-1. The AIDS epidemic has led to a reopening of the question as to whether it is always best for mothers to breast feed their children. In developed countries HIV positive mothers refrain from breast feeding while in developing countries, when alternate feeding regimens are too expensive or clean water is not available, HIV infected mothers continue to breast feed. In the developing world infants who are not breastfed have a six-fold higher risk of death from infectious disease during their first two months of life than breastfed infants.[20] However, a recent study in Kenya[21] showed that the use of breast milk substitutes prevented 44% of infant HIV infections. Whether the risk of HIV-1 transmission in breast milk exceeds the potential risk of mortality from other infectious diseases in developing countries remains to be determined. Given this uncertainty, an examination of the known factors involved in the prevention or facilitation of viral transmission through breastfeeding was undertaken.

2. Antiviral Factors in Milk

Secretory IgA (sIgA) comprises greater than 90% of human milk immunoglobulin while the remaining 10% consists primarily of IgG and secretory IgM.[3] Colostrum and mature breast milk from HIV-1 infected women contain antibodies against HIV antigens.[22] IgG has been shown to be the predominant antibody type against HIV in breast milk and the evidence suggests that milk IgG is synthesized locally within the mammary gland.[23] When breast milk was taken from HIV positive women who had transmitted and those who had not transmitted HIV to their children during breastfeeding the specific activities of sIgA and IgG antibodies against viral proteins were similar in all milks.[22] This study indicates that humoral mucosal immunity against HIV is not a primary mechanism for protecting against HIV transmission in breast milk.

Milk also contains nonspecific protective factors, which often function against more than one pathogen. Lactoferrin, SLPI, lactoadherin and oligosaccharides which are present in human milk have all been shown to prevent adsorption of viruses to target cells.

Lactoferrin is an iron binding protein that has long been known to have antibacterial activity.[24, 25] Recent studies *in vitro* have shown that lactoferrin is also antiviral and inhibits the enveloped viruses herpes simplex virus types 1 and 2, CMV, HIV and hepatitis C virus.[24, 25] Additionally, lactoferrin has activity against nonenveloped viruses such as rotavirus and poliovirus.[25] Lactoferrin inhibition of viral infection is at an early stage and is most likely correlated to a competition for cell receptors since the N-terminal of lactoferrin binds to surface glycosaminoglycans which viruses also use for initial cell binding. Despite the potent activity of milk lactoferrin against human CMV however, this virus is transmitted by breast feeding.[3]

SLPI is a serine protease inhibitor, which is present in human milk and saliva and has been found to have antiviral activity.[4] SLPI interacts with a host cell molecule and not the virus itself and is able to inhibit HIV, Sendai virus and influenza virus but not CMV, herpes simplex virus or murine leukemia virus.[4] However, SLPI in human milk and colostrum has not been found to be a major determinant in preventing HIV transmission via human milk.[26] Whether this is due to insufficient SLPI concentration in milk or other factors is not known.

Lactadherin is a human milk mucin associated glycoprotein, which binds to rotavirus and inhibits its replication.[27] This antiviral glycoprotein is dependent upon the carbohydrate portion of lactahedrin. Human milk oligosaccharides that are not protein associated also protect against viral infections.[28] Glycosaminoglycans in milk inhibit the binding of the HIV envelope glycoprotein gp120 to the host cell CD4 receptor.[29] Milk oligosaccharides can function as decoy molecules, which bind to the viruses' carbohydrate binding proteins. Pathogens are then cleared from the gastrointestinal or respiratory tracts by the physiological mechanism characteristic of the tissue being protected.[28] These protective molecules can also bind directly to virus receptors and prevent virus adsorption and penetration of the host cell. Despite the presence of dozens of oligosaccharides and glycoproteins in human milk, which cover the gastrointestinal and nasopharyngeal mucosal surfaces and have been shown *in vitro* to block adhesion of infant pathogens, HIV is still transmitted by breast feeding.

Lipids in human milk have a dual role. They are both nutrients and nonspecific protective factors, which inactivate bacteria, protozoa and enveloped viruses.[30] The lipid-dependent antiviral activity delivered in milk to the suckling neonate results from the release of antiviral fatty acids and monoglycerides from human milk triglycerides by lipolytic activity in the infant's gastrointestinal tract.[31] Milk lipids are not unique in possessing antiviral activity. Similar fatty acid dependent antimicrobial activity is found in human skin and lung surfactant lipids as well as in marine algae.[32 - 34] Antiviral lipids disrupt the envelope of enveloped viruses such as HIV, CMV and HTLV-1 thereby destroying their infectivity.[30] Studies have shown that the stomach contents from infants fed human milk by a nasogastric tube have very strong lipid dependent antiviral activity which can be diluted 1/20 and still reduce viral infectivity by at least 10,000 fold.[35] Once again however despite the presence of strong lipid dependent antiviral activity, which has been shown to inactivate HIV,[35] the virus is still transmitted by human milk.

3. Factors Which May Promote Viral Transmission in Human Milk

Protective factors in human milk likely reduce the viral load of HIV and other breast milk transmitted viruses but not enough to completely prevent viral transmission. The initial HIV concentration in milk as well as any mechanisms that the virus has developed to prevent inactivation by protective factors may be able to tip the balance in favor of transmission in some instances. This would explain why the rate of HIV transmission due to breast feeding is approximately 16% and not all suckling infants are infected.[21]

A number of recent studies have suggested that mastitis, an inflammatory process in the breast, is associated with a higher HIV load in breast milk and increased mother to child transmission of HIV.[36] Mastitis, including subclinical mastitis, is also associated with higher milk concentrations of immunological, e.g. lactoferrin and SLPI, and inflammatory factors in breast milk.[37] During mastitis, inflammatory cells and extracellular fluid enter the milk and the concentration of cell free HIV in milk is elevated. The presence in human milk of potentially HIV infected inflammatory cells and increased extracellular HIV may increase the risk of HIV transmission through breast feeding. The association between mastitis and HIV transmission is likely also relevant to other vertically transmitted retroviruses including HTLV-1, HTLV-2 and perhaps also to CMV.

HIV in human milk is present as both cell free and intracellular virus.[38, 39] An unresolved issue is which form of the virus is responsible for vertical transmission. Intracellular virus would be protected from antiviral compounds in milk, but the increased cell free virus found with mastitis[36] may overwhelm the endogenous human milk protective factors. HTLV-1 has been shown to be transmitted in human milk by infected lymphocytes and approximately 25% of infants breast fed by infected mothers acquire the infection.[6] HTLV-2 shows a similar pattern of transmission by human milk to HTLV-1.[7] CMV is present in human milk as both cell free and intracellular virus and it has not been established which form of the virus is responsible for transmission.[13, 14] Studies done with animal retroviruses have shown that adult macaques exposed orally to cell free simian immunodeficiency virus became infected and developed AIDS.[40] It has also been found that feline immunodeficiency virus (FIV) can be transmitted by oral administration of cell free FIV, FIV infected mononuclear cells or milk from a queen infected following delivery.[9, 38] It is possible that while many milk transmitted viruses are present both intracellularly and extracellularly that different viruses are transmitted preferentially by one or the other route or by both means. Making the situation even more complex are recent findings[41] that HIV-1 can bind B cells through interactions between CD21 expressed on the surface of B cells and complement (C3 fragments) bound to HIV particles. Virions bound to B cells through immune complexes are far more infectious for T cells than free virions.[42] T and B cells comprise, respectively, 83% and 4-6% of lymphocytes in early human milk[43] and C3 is present in human milk.[3] The possibility exists that a portion of extracellular HIV in human milk is bound to B cells and protected from inactivation.

Free virus particles may also be protected from inactivation by the skim milk membrane fraction of human milk, which is derived from the plasma membrane of the lactating cell, leukocytes and milk fat globule membrane.[44 - 47] The skim milk membrane fraction of bovine milk is greatly increased in udders with mastitis and almost completely disappears following treatment with antibiotics.[45] An association between cell free virus and milk membrane material may protect the virus from antiviral compounds thereby facilitating vertical transmission. Milk from mothers with mastitis therefore not only has an increased concentration of cell free HIV and infected leukocytes but also a larger skim milk membrane fraction containing membrane vesicles which could help to transport and protect cell free HIV and other viruses in human milk.[36]

Membranes may also help to protect cell free HIV from inactivation when milk and saliva are mixed during infant suckling. The natural hypotonicity of saliva disrupts leukocytes which carry HIV.[48] However, when human milk or colostrum are mixed with saliva in physiological volumes, hypotonic saliva cannot lyse HIV infected cells due to the presence of solutes provided by human milk, which is isotonic. *In vivo* this inhibition will not be perfect and some percentage of HIV infected milk macrophages and lymphocytes will be lysed. Studies done by Sarka Southern[49] (presented in this volume) show that after saliva disrupts human milk cells, cell membrane fragments with HIV can be endocytosed into salivary epithelial cells and the virus remains infectious. Cellular complexes produced during the interaction between milk cells and saliva could protect viruses from the protective components in human milk.

In vitro studies have shown that in the presence of human milk growth of HIV-1 in lymphocyte cultures increases and the entry of HIV-1 into a mammary epithelial cell line is enhanced.[50] Both of these effects were related to the presence of cathepsin D activity in milk and could be eliminated using the protease inhibitor pepstatin A or anti-cathepsin D antibody. It was suggested that these HIV promoting effects may be due to conformational modification of viral gp120 by the proteolytic activity of cathepsin D allowing direct interaction with a co-receptor. Whether there are other enzymes present in milk whose activity also promotes HIV replication is not known.

It is as yet to be determined how HIV and other milk borne viruses, whether intracellular or cell free, cross the mucosal barrier and infect the suckling infant. Possibly M cells whose function is to sample luminal antigens in the gastrointestinal tract are exploited by milk borne viruses and facilitate entry from the lumen into epithelial cells, lymphocytes and dendritic cells.[51] It has been previously shown that poliovirus adheres to M cells and is then taken up in clathrin coated vesicles. M cells endocytose and transport any small particles that adhere to their surface.[51] Therefore, potentially both cell free and intracellular milk transmitted viruses can be transported by M cells. The portal of entry for HIV in the gastrointestinal tract however, has not been definitively established, and it could also result from breaks in the mucosal barrier or direct infection of epithelial cells such as enterocytes.[52]

As discussed above, the presence of antibodies in milk against HIV is not protective.[22] The same is also true for milk antibodies against CMV.[3] A possible explanation for this lack of protective efficacy can be found in the failure so far to develop effective vaccines against herpes simplex viruses. Antibodies do not provide sterilizing immunity.[53] Initial replication of viral pathogens in host tissue is not prevented but the host can eradicate the pathogen following initial infection. However, retroviruses and herpes viruses enter through mucosal surfaces and establish latent infection in contrast to other viruses which do not persist in host cells and eventually are eliminated. As a result of viral latency, human milk and colostral antibodies may never be able to completely prevent the transmission of herpes viruses and retroviruses.

Retroviruses are of course very well adapted to their environment. The presence of particular HTLV-1 haplotypes, whose primary mode of transmission is breast feeding, in 1,300 – 1,700 year old mummies found in the Andes has been used to examine the origin of present day Chileans.[54] Ancient HTLV-1 sequences are almost identical to those found in modern day Chilean and Japanese HTLV-1 seropositive individuals. HTLV-1 and other milk borne viruses have been using the process of breast feeding for their transmission for quite a long time and it will not be easy to interrupt the process while maintaining high rates of breast feeding as well as the integrity of protective mechanisms in milk.

4. Prevention of Vertical HIV Transmission

In developing countries, particularly in subsaharan Africa, the choice for HIV positive mothers is often between breast feeding with the attendant risk of transmitting the virus to their infants or using alternatives to human milk which substantially increase the risk of infant morbidity and mortality from other infectious agents. The question at present is whether the means exist through the use of drugs, changes in behavior or possibly treatment of banked human milk to substantially reduce HIV transmission by breast feeding.

Breast feeding cannot be considered in isolation when trying to reduce vertical HIV transmission. The solution to the problem is multidisciplinary. The most effective means of preventing HIV transmission by breast feeding is to prevent the initial infection of the mother.[55] A recent article by Anthony Fauci contains the statement that, "Unlike microbial scourges, such as malaria and tuberculosis (among many others), for which there is very little that people can do to prevent infection, HIV infection in adults is entirely preventable by behavior modificaiton."[56] When the practical limits of behavior modification have been reached other interventions are possible. The development of a vaginal microbicide would reduce the sexual transmission of HIV and consequently the vertical transmission of the virus to the infant in breast milk.[30] Additionally, when a mother is already HIV positive a vaginal microbicide can be used to inactivate the virus in the birth canal thereby reducing the risk of HIV transmission during the perinatal period. Another factor that could reduce sexually transmitted HIV is male circumcision. Circumcised males are two to eight times less likely to become infected with HIV and consequently spread the virus to women and infants.[57]

Antiretroviral drugs, such as zidovudine and nevirapine, when given to both mother and infant, can substantially reduce vertical HIV transmission during labor and delivery but do not reduce HIV transmission due to breast feeding.[58, 55] Antibiotics are another class of drugs that could help to reduce HIV transmission in milk.[45] Since mastitis increases the viral load in human milk and thus the risk of viral transmission, treating this infection using antibiotics could be used in combination with other strategies to reduce HIV transmission by breast milk.

Studies in Japan have shown that refraining from breast feeding effectively reduces the transmission of HTLV-1 from mother to infant.[59] Data from Africa shows that from 1/3 to 2/3 of vertical HIV-transmission is due to breast feeding and that the use of breast milk substitutes, where clean water is available, can reduce vertical HIV-transmission to the same extent as nevirapine or zidovudine.[21, 52, 55] However, infants would lose the nutritional and immunological benefits provided by human milk.

Another approach to make breast feeding safer, in areas where vertical HIV transmission rates are high, would be to collect milk from HIV positive mothers and kill cell free virus and infected cells while maintaining the nutritional and immunological integrity of the milk. Freezing and thawing human milk can significantly reduce the concentration of HTLV-1 while keeping the immunological benefits.[60] However, the facilities to safely accomplish this may not be readily available in developing countries. Pasteurization (heat treatment) of either the mother's own milk or donor milk is potentially another way to reduce HIV and HTLV-1 in human milk and still provide protection from diarrheal diseases in the developing world.[61, 62] Heat treatment at 56°C does not destroy sIgA and raising the temperature to 62.5°C still maintains 70 – 100% of sIgA activity.[61] IgG is decreased by 33% and sIgM is lost following treatment at 62.5°C for 30 minutes.[61] Nonspecific protective factors such as lipids and oligosaccharides are stable during pasteurization at 62.5°C for 30 minutes, whereas milk cells are destroyed under these conditions.[61] Pasteurization can therefore inactivate both cell free and intracellular viruses.

It may also be possible to inactivate HIV in milk by either releasing the antiviral fatty acids in banked milk prior to ingestion or by adding antiviral fatty acids to the collected milk.[63, 64] Cell free enveloped viruses present in milk are inactivated by added lipids[64] and lipids can lyse cells that are not protected by a mucosal barrier.[30] If facilities for short-term milk storage are available, both antiviral fatty acids which are inexpensive and lipases which are currently produced for industrial applications[65] should be cost effective.

5. Conclusions

Viruses, including HIV, are transmitted by breastfeeding despite the presence of multiple antiviral compounds in human milk and colostrum. The efficiency of viral transmission increases with the viral load. Viruses that initiate latent infections, e.g. HIV and CMV, represent the majority of those spread by human milk. Refraining from breast feeding in developing countries reduces the spread of HIV from mother to infant but also exposes infants to increased risk of dying from infectious diseases other than AIDS. The solution to this quandry is multidisciplinary. It will require changes in the behavior and attitudes of populations at risk, the development of affordable drugs and other interventions, and the means to deliver them to those at risk.

Acknowledgements

This work was supported by a grant to CEI from the National Institutes of Health (AI 39061) and by the New York State Office of Mental Retardation and Developmental Disabilities. Anna Parese is thanked for help in preparation of the manuscript.

References

1. A.S. Cunningham, D.B. Jelliffee, and E.F.P. Jelliffe, Breast-feeding and health in the 1980s: A global epidemiologic review, J. Pediatr. 118:659-666 (1991).
2. L.A. Hanson, I. Lonnroth, S. Lange, J. Bjersing, and U.I. Dahlgren, Nutrition resistance to viral propagation, Nutr. Reviews 58:S31-S37 (2000).
3. A.S. Goldman and P.L. Ogra, Anti-infectious and infectious agents in human milk, in: *Mucosal Immunology,* edited by P.L. Ogra, M.E. Lamm, J. Bienenstock, J. Mestecky, W. Strober, and J.R. McGhee (Academic Press, 1999), pp. 1511-1512.
4. D.C. Shugars, Endogenous mucosal antiviral factors of the oral cavity, J. Infect. Dis. 179(Suppl 3):S431-435 (1999).
5. J. Kreiss, Breastfeeding and vertical transmission of HIV-1. Acta Paediatr. Suppl., 421:113-117 (1997).
6. S.O. Southern and P.J. Southern, Persistent HTLV-1 infection of breast luminal epithelial cells: A role in HTLV transmission? Virology 241:200-214 (1998).
7. R.B. Lal, R.A. Gongora-Biachi, D. Pardi, W.M. Switzer, I. Goldman, and A.A. Lal, Evidence for mother-to-child transmission of human T lymphotropic virus type II. J. Infect. Dis. 168:586-591 (1993).
8. A.H. Sharpe, J.J. Ruprecht, and R. Jaenisch, Maternal transmission of retroviral disease and strategies for preventing infection of the neonate, J. Virol. 63:1049-1053 (1989).
9. R.K. Sellon, H.L. Jordan, K. Kennedy-Stoskopf, M.B. Tompkins, and W.A.F. Tompkins, Feline immunodeficiency virus can be experimentally transmitted via milk during acute maternal infection, J. Virol. 68:3380-3385, (1994).
10. M. Dawson, Pathogenesis of maedi-visna, Vet. Rec. 120:451-454 (1987).

11. D.S. Adams, P. Klevjer-Anderson, J.L. Carlson, T.C. McGuire, and J.R. Gorham, Transmission and control of caprine arthritis-encephalitis virus, Am. J. Vet. Res. 44:1670-1675 (1983).
12. P. Hainaut, D. Vaira, C. Francois, C.-M. Calberg-Bacq, and P.M. Osterrieth, Natural infection of Swiss mice with mouse mammary tumor virus (MMTV): Viral expression in milk and transmission of infection, Arch. Virol. 83:195-206 (1985).
13. K. Numazaki, Human cytomegalovirus infection of breast milk, FEMS Immun. Med. Microbiol. 18:91-98 (1997).
14. K. Hamprecht, M. Vochem, A. Baumeister, M. Boniek, C.P. Speer, and G. Jahn, Detection of cytomegaloviral DNA in human milk cells and cell free milk whey by nested PCR, J. Virol. Methods 70:167-176 (1998).
15. D. Kotronias and N. Kapranos, Detection of herpes simplex virus DNA in maternal breast milk by *in situ* hybridization with tyramide signal amplification, In Vivo 13:463-466 (1999).
16. S. Polywka, H. Feucht, B. Zollner, and R. Laufs, Hepatitis C virus infection in pregnancy and the risk of mother-to-child transmission, Eur. J. Clin. Microbiol. Infect. Dis. 16:121-124 (1997).
17. R.M. Kumar and S. Shahul, Role of breast-feeding in transmission of hepatitis C virus to infants of HCV-infected mothers, J. Hepatology 29:191-197 (1998).
18. M. Yoshida, N. Yamagami, T. Tezuka and R. Hondo, Case report: Detection of Varicella-Zoster virus DNA in maternal breast milk, J. Med. Virol. 38:108-110 (1992).
19. A.K. Junker, E.E. Thomas, A. Radcliffe, R.B. Forsyth, A.G.F. Davidson, and L. Rymo, Epstein-Barr virus shedding in breast milk, Am. J. Med. Sci. 302:220-223 (1991).
20. WHO collaborative study team on the role of breastfeeding on the prevention of infant mortality, Effect of breastfeeding on infant and child mortality due to infectious diseases in less developed countries: a pooled analysis, Lancet 355:451-455 (2000).
21. R. Nduati, G. John, D.Mbori-Ngacha, B. Richardson, J. Overbaugh, A. Mwatha, J. Ndinya-Achola, J. Bwayo, F.E. Onyango, J. Hughes, and J. Kreiss, Effect of breastfeeding and formula feeding on transmission of HIV-1, JAMA 283:1167-1174 (2000).
22. P. Becquart, H. Hocini, M. Levy, A. Sepou, M.D. Kazatchkine, and L. Belec, Secretory anti-human immunodeficiency virus (HIV) antibodies in colostrum and breast milk are not a major determinant of the protection of early postnatal transmission of HIV, J. Infect. Dis. 181:532-539 (2000).
23. P. Becquart, H. Hocini, B. Garin, A. Sepou, M.D. Kazatchkine, and L. Belec, Compartmentalization of the IgG immune response to HIV-1 in breast milk, Aids 13:1323-1331 (1999).
24. M.C. Harmsen, P.J. Swart, M-P. de Bethune, R. Pauwels, E. De Clercq, T. Hauw The, and D.K.F. Meijer, Antiviral effects of plasma and milk proteins: lactoferrin shows potent activity against both human immunodeficiency virus and human cytomegalovirus replication *in vitro*, J. Infect. Dis. 172:380-388 (1995).
25. R. Siciliano, B. Rega, M. Marchetti, L. Seganti, G. Antonini, and P. Valenti, Bovine lactoferrin peptidic fragments involved in inhibition of herpes simplex virus type 1 infection, Biochem. Biophys. Res. Comm. 264:19-23 (1999).
26. P. Becquart, G. Gresenguet, H. Hocini, M.D. Kazatchkine, and L. Belec, Secretory leukocyte protease inhibitor in colostrum and breast milk is not a major determinant of the protection of early postnatal transmission of HIV, AIDS 13:2599-2601 (1999).
27. D.S. Newburg, J.A. Peterson, G.M. Ruiz-Palacios, D.O. Matson, A.L. Morrow, J. Shults, M. de Lourdes Guerrero, P. Chaturvedi, S.O. Newburg, C.D. Scallan, M.R. Taylor, R.L. Ceriani, and L.K. Pickering, Role of human-milk lactadherin in protection against symptomatic rotavirus infection, Lancet 351:1160-1164 (1998).
28. D. Zopf and S. Roth, Oligosaccharide anti-infective agents, Lancet 347:1017-1021 (1996).
29. D.S. Newburg, R.J. Linhardt, S.A. Ampofo, and R.H. Yolken, Human milk glycosaminoglycans inhibit HIV glycoprotein gp 120 binding to its host cell CD4 receptor, J. Nutr. 125:419-424 (1995).
30. C.E. Isaacs and M.F. Lampe, Lactolipids, in: *Natural food antimicrobial systems,* edited by A.S. Naidu (CRC Press, 2000), pp. 159-182.
31. C.E. Isaacs, S. Kashyap, W.C. Heird, and H. Thormar, Antiviral and antibacterial lipids in human milk and infant formula feeds, Arch. Dis. Child. 65:861-864 (1990).
32. S.J. Miller, R. Aly, H.R. Shinefeld, and P.M. Elias, *In vitro* and *in vivo* antistaphylococcal activity of human stratum corneum lipids, Arch. Dermatol. 124:209-215 (1988).
33. J.D. Coonrod, Role of surfactant free fatty acids in antimicrobial defenses, Eur. J. Respir. Dis. 71:209-214 (1987).
34. K.-G. Rosell and L.M. Srivastava, Fatty acids as antimicrobial substances in brown algae, Hydrobiologia 151/152:471-475 (1987).

35. C.E. Isaacs and H. Thormar, Human milk lipids inactivate enveloped viruses, in: ***Breastfeeding, nutrition, infection and infant growth in developed and emerging countries,*** edited by S.A Atkinson, L.A. Hanson, R.K. Chandra (Arts Biomedical Publishers and Distributors, St. John's, Newfoundland, Canada, 1990) pp. 161-174.
36. R.D. Semba and M.C. Neville, Breast-feeding, mastitis, and HIV transmission: Nutritional implications, Nutr. Reviews 57:146-153 (1999).
37. R.D. Semba, N. Kumwenda, T.E. Taha, D.R. Hoover, Y. Lan, W. Eisinger, L. Mtimavalye, R. Broadhead, P.G. Miotti, L. Van Der Hoeven, and J.D. Chiphangwi, Mastitis and immunological factors in breast milk of lactating women in Malawi, Clin. Diag. Lab. Immun. 6:671-674 (1999).
38. P. Lewis, R. Nduati, J.K. Kreiss, G.C. John, B.A. Richardson, D. Mbori-Ngacha, J. Ndinya-Achola, and J. Overbaugh, Cell-free human immunodeficiency virus type 1 in breast milk, J. Infect. Dis. 177:34-39 (1998).
39. R. W. Nduati, G.C. John, B.A. Richardson, J. Overbaugh, M. Welch, J. Ndinya-Achola, S. Moses, K. Holmes, F. Onyango, and J.K. Kreiss, Human immunodeficiency virus type 1 – infected cells in breast milk: Association with immunosuppression and Vitamin A deficiency, J. Infect. Dis. 172:1461-1468 (1995).
40. T.W. Baba, A.M. Trichel, L. An, V. Liska, L.N. Martin, M. Murphey-Corb, and R.M. Ruprecht, Infection and AIDS in adult macaques after nontraumatic oral exposure to cell-free SIV, Science 272:1486-1489 (1996).
41. S. Moir, A. Malaspina, Y. Li, T.-W. Chun, T. Lowe, J. Adelsberger, M. Baseler, L.A. Ehler, S. Liu, R.T. Davey, Jr., J.A.M. Mican, and A.S. Fauci, B cells of HIV-1 infected patients bind virions through CD21-complement interactions and transmit infectious virus to activated T cells, J. Exper. Med. 192: 637-645 (2000).
42. J.J. Jakubik, M. Saifuddin, D.M. Takefman, and G.T. Spear, Immune complexes containing human immunodeficiency virus type 1 primary isolates bind to lymphoid tissue B lymphocytes and are infectious for T lymphocytes, J. Virol. 74:552-555, (2000).
43. A.S. Goldman and R.M. Goldblum, Transfer of maternal leukocytes to the infant by human milk, Current Topics Microbiol. Immun. 222:205-213 (1997).
44. C.E. Isaacs and R.C. Moretz, Skim milk membranes in human milk, in: *Human Lactation 2,* edited by M. Hamosh and A.S. Goldman, (Plenum Publishing Corp, 1986) pp. 617-620.
45. M. Anderson, B.E. Brooker, A.T. Andrews, and E. Alichanidis, Membrane material isolated from milk of mastitic and normal cows, J. Dairy Sci. 57: 1448-1458 (1974).
46. S. Patton and G.E. Huston, Membrane distribution and carotenoid content in human colostrum, J. Ped. Gastroenterol. Nutr.5:774-779 (1986).
47. M. Hamosh, J.A. Peterson, T.R. Henderson, C.D. Scallan, R. Kiwan, R.L. Ceriani, M. Armand, N.R. Mehta, and P. Hamosh, Protective function of human milk: The milk fat globule, Seminars in Perinatology 23:242-249 (1999).
48. S. Baron, J. Poast, C.J. Richardson, D. Nguyen, and M. Cloyd, Oral transmission of human immunodeficiency virus by infected seminal fluid and milk: A novel mechanism, J. Infect. Dis. 181:498-504 (2000).
49. S.O. Southern, Milk-borne transmission of HIV, J. Human Virol. 1:328-337 (1998).
50. K.E. Messaoudi, L.F. Thiry, C. Liesnard, N.V. Tieghem, A. Bollen, and N. Moguilevsky, A human milk factor susceptible to cathepsin D inhibitors enhances human immunodeficiency virus type 1 infectivity and allows virus entry into a mammary epithelial cell line, J. Virol. 74:1004-1007 (2000).
51. M.R. Neutra, A. Frey, and J.-P. Kraehenbuhl, Epithelial M cells: Gateways for mucosal infection and immunization, Cell 86:345-348 (1996).
52. P. Van de Perre, Transmission of human immunodeficiency virus type 1 through breast-feeding: How can it be prevented?, J. Infect. Dis. 179:S405-S407 (1999).
53. J.R. Mascola, Herpes simplex virus vaccines – Why don't antibodies protect?, JAMA 282:379-380 (1999).
54. H.-C. Li, T. Fujiyoshi, H. Lou, S. Yashiki, S. Sonoda, L. Cartier, L. Nunez, I. Munoz, S. Horai, and K. Tajima, The presence of ancient human T-cell lymphotropic virus type I provirus DNA in an Andean mummy, Nature Medicine 12:1428-1432 (1999).
55. K.M. De Cock, M.G. Fowler, E. Mercier, I. De Vincenzi, J. Saba, E. Hoff, D.J. Alnwick, M. Rogers, and N. Shaffer, Prevention of mother-to-child HIV transmission in resource-poor countries, JAMA 283:1175-1182 (2000).
56. A.S. Fauci, The Aids Epidemic, N. Engl. J. Med. 341:1046-1050 (1999).

57. R. Szabo and R.V. Short, How does male circumcision protect against HIV infection?, BMJ 320:1592-1594 (2000).
58. J. Cohen, The mother of all HIV challenges, Science 288:2160-2163 (2000).
59. S. Hino, Milk-borne transmission of HTLV-1 as a major route in the endemic cycle, Acta Paediatr. Jpn. 31:428-435 (1989).
60. Y. Ando, K. Kakimoto, T. Tanigawa, K. Furuki, K. Saito, S. Nakano, H. Hashimoto, I. Moriyama, M. Ichijo, and T. Toyama, Effect of freeze-thawing breast milk on vertical HTLV-I transmission from seropositive mothers to children, Jpn. J. Cancer Res. 80:405-407 (1989).
61. J.T. May, Microbial contaminants and antimicrobial properties of human milk, Microbiol. Sciences 5:42-46 (1988).
62. M.R. Tully, Is pasteurized mother's own or donor milk an answer to the HIV crisis?, J. Hum. Lact. 15:345-346 (1999).
63. C.E. Isaacs, R.E. Litov, P. Marie, and H. Thormar, Addition of lipases to infant formulas produces antiviral and antibacterial activity, J. Nutr. Biochem. 3:304-308 (1992).
64. C.E. Isaacs, R.E. Litov, and H. Thormar, Antimicrobial activity of lipids added to human milk, infant formula, and bovine milk, J. Nutr. Biochem. 6:362-366 (1995).
65. V.M. Balcao, A.L. Paiva, and F.X. Malcata, Bioreactors with immobilized lipases: State of the art, Enzyme Microbiol. Technol. 18:392-416 (1996).

CELLULAR MECHANISM FOR MILK-BORNE TRANSMISSION OF HIV AND HTLV

Sarka Southern and Peter Southern*

1. INTRODUCTION

Transmission of human immunodeficiency virus (HIV) and human T-cell leukemia viruses (HTLV) by breastfeeding contributes significantly to the global spread of human retroviral diseases including acquired immunodeficiency virus (AIDS), adult T-cell leukemia (ATL) and HTLV-associated myelopathy-tropical spastic paraparesis (HAM/TSP). It is imperative to develop novel strategies that could be used to improve maternal health, lower the transmission rate and secure disease-free status in the infected child. We have used a combination of *in vitro* studies and analysis of normal and seropositive mothers to describe the cellular composition of milk during long-term breastfeeding, and define the cellular constituents that are involved in transmission of HIV and HTLV infectivity in mammary tissue and milk.[1-4] Our results provide evidence that the breast can serve as a reservoir of retroviral infectivity, and that several constituent cell types in milk can support productive HIV and HTLV infections. During ingestion, milk is mixed with saliva prior to swallowing. We found that mixing milk and saliva triggers fundamental changes in the composition of milk by cellular disruption and the formation of complexes between milk cells and salivary epithelial cells. This process could alter both the infective and the protective properties of milk and play an important role in oral transmission of cell associated retroviruses.

We discuss a hypothesis suggesting that milk could provide specific immunological help for anti-retroviral protection of the breastfed child by transplanting virus-specific memory T cells, and by vaccinating against viral antigens.[3] Better understanding of the mechanisms underlying HIV and HTLV persistence and transmission, and the interactive immunological relationship between the mother and her breastfed child

* Sarka Southern, SWF Science Center, San Diego, CA, USA. Peter Southern, University of Minnesota, Minneapolis, MN, USA.

will provide a basis for effective anti-retroviral vaccination and drug treatment optimally suited for mothers and children.

2. Results and Discussion

3. Characterization of Maternal Cells in Milk during Long-term Breastfeeding

Traditionally, children have been breastfed for 2 to 3 years. In some cultures, women practice communal breastfeeding. They continuously lactate for over 20 years in order to breastfeed any needy child in the community.[1, 5] Long-term breastfeeding is also common in some other large mammals such as primates and whales.[6, 7] While the nutritional role of breastfeeding and the immunological role of maternal antibodies have been well established, little is known about milk-borne cells and their functional role in breastfeeding. We have investigated the cellular composition of milk during 5 years of continued lactation to determine what cell types are present, and at what concentrations. The cellular composition of breast milk was determined in a normal mother who exclusively breastfed her child for the first 11 months, and then breastfed at least 2 times daily for 4.5 years.[3] Over 200 samples of milk were collected at various times during the long-term breastfeeding. Colostrum and early milk were also analyzed in several other normal mothers. The cell types were determined using immunohistochemical staining, and by fluorescence-activated cell sorter (FACS). Colostrum and early milk contained variable concentrations of granulocytes and monocytes/macrophages (Table 1). After about 2 month of lactation, the cellular composition of milk changed. The mature milk contained B and T lymphocytes, luminal epithelial cells, and monocytes/macrophages at different stages of differentiation (Table 1). In contrast with T cells in blood (in the same mother), the T lymphocytes in milk were predominantly CD45RO memory cells. The cell population in the mature milk appeared to be an active system, characterized by high levels of cell motility, adhesion and phagocytosis. The specific cellular content and cell concentration of the mature milk remained the same for over 5 years of lactation. The milk-borne cells were viable in all milk isolates. Primary culture conditions were established that allowed survival of milk leukocytes and epithelial cells *in vitro* for approximately 3 weeks. These cultures were employed in studying the mechanism of milk-borne transmission of HTLV[1] and HIV.[3]

We conclude that the cellular content of milk is different in the colostrum/early milk and in the mature milk. During long-term breastfeeding a large population of viable maternal leukocytes and epithelial cells is continuously transferred into the breastfed child.

4. Maternal Cells Transferred to Child during Breastfeeding: Helping the Child's Immune System?

Considering that the milk-borne cell population has a nearly constant composition and high viability, it appears to result from a well-controlled process rather than random cell sloughing. We would like to discuss a hypothesis suggesting that milk-borne cells help the child's immune system, by transferring maternal immunological memory, and by vaccinating against milk-borne pathogens. Further investigation of the immunological functions that could be provided by milk-borne cells is particularly urgent in HIV research. The first proposed mechanism involves maternal leukocytes in mature milk. Large concentrations of T cells displaying the CD45RO memory phenotype, and macrophages are present in the mature milk (Table 1). The mature milk also contains B cells, albeit in much lower concentrations. Daily, over 100 million macrophages, 10

Table 1. Cellular Composition of Early Milk and Mature Milk During Long-Term Breastfeeding

Milk Cells	Colostrum[a]	Early Milk[a]	Mature Milk[b]
Approximate Breastfeeding Period After Birth	1-3 d	1-2 mo	3 mo-6 y
Cell Concentration (cells/ml)	10^6-10^7	10^6-10^5	10^5
Cell types (% positive)			
Macrophages and giant cells	40-60	40-60	60-70
T cells	0.1	1-2	15-20
*95% CD45RO memory phenotype CD4:CD8=1.7			
Luminal epithelial cells	0.1	0.1	5-10
Granulocytes	40-60	40-60	0.1
B cells	0.1	0.1	0.1

[a]Colostrum and early milk were donated by several healthy mothers and 3 HIV-seropositive mothers (see Methods). The cellular content and composition were assessed by immunocytochemical staining and FACS. The cellular content of early milk samples obtained from the healthy and the HIV-infected mothers was generally similar.

[b]Samples of mature milk were collected from a nursing mother during several years of continued lactation and the cellular content was analyzed by immunocytochemical staining and FACS. A detailed description of cell types identified in milk during long-term breastfeeding will be presented elsewhere (S. Southern, manuscript in preparation).

million memory T cells, and 0.1 million B cells are transferred from mother to child, during exclusive breastfeeding. If transplanted into a child, these cells could provide specific cellular immunity. Children do not have fully developed specific

immunity and so they could significantly benefit from help by mature leukocytes, in particular the memory cells, and macrophages. The memory lymphocytes are poised to respond to health threats encountered by the mother in the past, and mother and her child during lactation. Maternal memory cells may also include cells specific to health threats encountered only by the child because breastfed children efficiently transfer immunological information about their own (communicable) diseases to the mother. It was suggested previously that milk-borne lymphocytes, and probably also the other constituent cell types, could survive in the child.[8-11] The maternal memory T cells could provide help directly, and/or they could imprint the child's specific immunity. The second mechanism involves cells and cell fragments that harbor pathogens. These cells are known to be in the colostrum and the early milk. About 10^5 - 10^7 cells carrying viral antigens are transferred by HIV and HTLV infected mothers in each breastfeeding, during early lactation.[3, 12] We suggest that fragments of the infected milk cells could be processed as antigens in the child's tonsils and gut, and serve as a vaccine, triggering specific immune response in the naive lymphocytes of the child. It was shown that successful vaccination of neonates against retroviruses required sustained low doses of an antigen.[13] This requirement would be fulfilled by the putative milk-borne vaccine. The milk vaccine could continuously supply the child with the "latest version" of maternal HIV antigenic variants, thus "updating" the vaccine. The milk-borne macrophages could play a role in presentation of the milk-borne antigens to the child's naive gut lymphocytes, and to maternal memory T cells in milk. Perhaps also the milk-borne luminal epithelial cells could act as antigen presenting cells. Additional "danger" signaling needed to activate the specific immune response could be provided by stress proteins.[14] We found recently that stress activated proteins are over-expressed in the breast and milk-borne cells of HIV seropositive mothers (S. Southern, in preparation).

The two hypothetical mechanisms discussed here could be viewed as remarkable cases of natural transplantation and vaccination, that could provide a unique way, specific to mammals, for vertical transfer of the immunological information. The cell-mediated immunological help, provided during long-term breastfeeding, would complement the help provided by the maternal antibodies and the innate immunity factors in colostrum and the early milk. It is important to appreciate the unique features inherent in the specific immunological help that a mother can provide to her child. Each child can receive immunological help uniquely tailored to her needs, through interactive immunological communication in the mother-child dyad. The most complete, and therefore the optimal immunological help can be achieved in exclusive breastfeeding as compared to mixed breastfeeding. This unique flexibility and specialization would be particularly important in defense against HIV and other rapidly mutating pathogens, where having the "latest virus alert" is essential.

5. HIV and HTLV Infected Cells and Cellular Stress in Breast

We have found that primary basal epithelial cells (HMEC) and luminal epithelial cells (MCF 10) derived from normal human mammary tissues supported and transmitted HTLV infection in culture.[2] Productive HIV infection in cultured mammary epithelial cells has been described previously.[15] Cells harboring viral antigens were visualized in biopsies of breast tissue from an HIV positive pregnant woman, and from an HTLV positive man with chronic adult T cell leukemia (ATL) using immunohistochemical staining. HIV infected lymphocytes, macrophages and groups of alveolar luminal epithelial cells were detected in the mammary tissue of the HIV positive woman (S. Southern, in preparation). HTLV-infected groups of alveolar luminal epithelial cells, and a large lymphocytic infiltrate comprised of typical HTLV transformed T cells, were detected in the breast of the ATL patient.[4] High levels of stress activated proteins were detected in a large number of infected and uninfected cells in the mammary tissues from both patients, suggesting that the retroviral infection is associated with widespread cellular stress in breast (S. Southern, in preparation). Molecular stress response in breast could contribute to the disease process by facilitating the spread of the infection into stressed cells, and by significantly perturbing the physiology of uninfected bystander cells. Significant physiological perturbation may interfere with cellular functions, facilitate the spread of opportunistic infections, and trigger apoptosis.

We conclude that the glandular epithelium in the breast can constitute a HIV and HTLV reservoir comprised of productively infected cells and uninfected stressed cells. The retroviral reservoir in breast could directly contribute to the viral load in milk through seeding milk with infected cells and cell-free virus. Conversely, a reservoir of retroviral infectivity in the breast could also contribute to induction and release of antiviral milk-borne constituents. The infectious and the protective constituents could also spread from the breast to other mucosal sites in the mother. If the breast commonly constitutes a HIV reservoir, then prevention of productive infection in breast should be monitored in ongoing anti-HIV drug and vaccine studies.

6. HIV and HTLV Infection in Milk-Borne Cells

HIV infection in colostrum and early milk was investigated in milk samples from HIV seropositive mothers. Productively infected lymphocytes and macrophages were detected using immunohistochemical staining (ICS) of viral antigens, and in situ hybridization (ISH) to detect viral RNA.[3] Infected macrophages represented over 97% of the infected milk cells. Approximately 0.1 - 1% of all milk cells were productively infected in the first week of breastfeeding. Mothers infected with HTLV-I have a similar frequency of productively infected cells in colostrum and early milk (0.1% - 1% cells positive for HTLV-I capsid protein expression.[12] We estimate that 10^5-10^7 HIV infected milk cells can be transferred to a breastfed child daily, during early lactation. HIV and HTLV infection in mature milk was investigated in cultured cells that were infected *in vitro*. Primary cultures of normal mature milk cells were infected by co-culture with HIV or HTLV infected cells. Productively infected macrophages, lymphocytes and luminal epithelial cells were detected by ICS and ISH.[3]

We conclude that macrophages are the most numerous carriers of infectivity in colostrum and early milk. During long-term breastfeeding, lymphocytes and epithelial cells could also become prominent carriers of HIV and HTLV infectivity as their concentrations rise in the mature milk. The mature milk from HIV-infected mothers needs to be analyzed to assess the presence and possible role of infected milk cells in HIV transmission during long-term breastfeeding. The HIV innoculum in milk needs to be precisely defined to better target the HIV forms that have to be eliminated by drugs or vaccination in order to lower the transmission rate. We suggest to identify productively infected milk-borne cells capable of transferring infectious HIV to the child's oral and GI tissues, and correlate their frequency and cell type with maternal transmission rate. Elimination of those forms could be then targeted in drug-efficacy and vaccination studies.

7. Milk-Saliva Interactions and Their Possible Role in HIV Transmission

During breastfeeding, milk is mixed with saliva and therefore oral HIV transmission may be actually mediated by the resulting milk-saliva mixture. To model changes in cell-associated infectivity which could be triggered during ingestion of infected milk, we investigated the interaction of normal or HIV/HTLV infected milk cells with normal saliva.[3] When milk was mixed with saliva, milk cells became rapidly disrupted, or bound and endocytized by salivary epithelial cells. The salivary epithelial cells remained intact. While high levels of viral antigens were detected in many salivary epithelial cells, these cells did not appear to support productive infection. On the other hand, HIV-1 RNA was detected within the T cells that were enfolded inside salivary epithelial cells suggesting that the enfolded infected cells could have remained infectious.

It is likely that the cellular events visualized upon mixing milk and saliva reflect a complex homeostatic process occurring continuously during ingestion, possibly playing a role in immune surveillance of the oral environment, and inadvertently mediating transmission of HIV or other oral pathogens.

8. Summary of Results and Hypotheses

During long-term breastfeeding milk continuously contains high concentrations of viable maternal cells including memory T cells, macrophages and epithelial cells. These cells have the potential to transfer beneficial properties, such as immunological help, as well as injurious properties, such as infectious retroviruses, to the breastfed child. Several constituent cell types in milk can be productively infected with HIV and HTLV, suggesting that milk-borne cells can transfer the viruses in all stages of lactation. Mixing milk and saliva produces fundamental changes in the cellular composition of milk, suggesting that ingestion may significantly impact cell-mediated transmission of HIV and HTLV. In HIV and HTLV infected people, the mammary epithelium can contain productively infected epithelial cells and leukocytes, as well as a much larger number of cells displaying high levels of stress activated proteins. This suggests that the breast can be a retroviral reservoir contributing to the infectious innoculum, and to the anti-retroviral constituents in milk. Our observations offer a new insight into the cellular and molecular sequence of events that occur during milk-borne transmission of HIV and HTLV. Considering the similarity between the maternal and sexual routes of transmission, and the relatedness between HTLV and HIV, our findings may have general applications for the study of human retroviral infections.

9. Future Research Goals

In the recent years, a large international effort generated many fundamental insights into the molecular biology and epidemiology of the HIV and HTLV infections. However, more information is urgently needed for designing and applying efficient anti-HIV vaccines. We propose to further investigate the inter-linked mechanisms of HIV transmission and anti-HIV protection during long-term breastfeeding, in the directions suggested by our previous results.

First, the HIV innoculum in milk needs to be better defined to provide the basis for specific targeting of drug therapy and vaccination. It is necessary to isolate and characterize milk-borne forms of HIV to evaluate the relationship between the population of viral genomes found in milk and those found in blood and genital secretions, since it is possible that HIV compartmentalization occurs in breast.[3] It needs to be determined which milk-borne cells have the capacity to transfer HIV infectivity to the child's oral and intestinal tissues. This issue needs to be investigated separately in the colostrum/early milk and in the mature milk, since the cellular composition of milk is different in those stages of lactation. Mastitis should be studied separately since inflammation in breast is likely to produce changes in the cellular content of milk and the composition of the milk-borne HIV innoculum. What changes to the HIV innoculum happen as the result of milk-saliva interactions, and other events during ingestion?

Second, it is urgently needed to evaluate whether milk-borne cells contribute to the immunological defense of the breastfed children against HIV. Specifically, it needs to be determined whether HIV-specific memory T cells are present in milk, and whether they appear to play protective role in breastfed children. It is also important to investigate whether milk from HIV seropositive mothers has vaccine-like properties for newborns. The natural immunological protection against HIV should be determined during long-term exclusive breastfeeding, when the immunological relationship between the mother and her child can become fully developed. If beneficial, the presence of HIV-specific memory T cells in milk could possibly be augmented by targeted maternal vaccination. If milk from HIV infected mothers had vaccine-like properties for the newborns, then milk could provide clues for vaccine design. Looking beyond childhood, it needs to be determined whether breastfeeding can protect against sexually transmitted HIV. Specifically, to determine the rate of the sexual transmission of HIV, and clinical AIDS progression in exclusively breastfed, AIDS-free children of HIV positive mothers. This research will be possible when cohorts of such children reach the appropriate age. Exclusively breastfed newborns receive about 10^7 - 10^9 HIV infected cells during the first 3 months, and have a very low risk of milk-borne HIV transmission.[16] In contrast, about 10^2 - 10^4 HIV infected cells are transferred in one ejaculate,[17] and the risk of HIV transmission is very high. Therefore it seems that breastfeeding can teach us best about the natural protection against HIV. All the natural protective mechanisms at work during breastfeeding should be discovered and adapted for general use in the fight against AIDS and other retroviral diseases.

Breastfeeding is the defining behavior of mammals. It has a fundamental beneficial role for children. In some situations, when the mother is ill, the beneficial outcome of breastfeeding can be diminished and even overturned by potentially injurious cells and molecules that become a part of milk, as in HIV and HTLV infections and perhaps some autoimmune diseases. The balance between benefit and injury can become overturned. When we better understand the nature of the different properties of milk, we can try to keep the balance tipped towards benefiting the breastfed children. As we work on further illuminating the beneficial role of long-term exclusive breastfeeding, we should also work on providing to every mother the possibility to breastfeed her child, as a human rights issue, and in the interest of improving the public health.

Acknowledgements

This work has been inspired by love and care for our children, Derek and Carolyn, and all the children in the world. The research on HTLV and HIV described here was funded in part by the Minnesota Medical Foundation.

References

1. S.O. Southern and P.J. Southern, Persistent HTLV-I infection of breast luminal epithelial cells: a role in HTLV-I transmission?, Virology 241(2):200-215 (1998).
2. R.J. LeVasseur, S.O. Southern, and P.J., Southern, Mammary epithelial cells support and transfer productive HTLV infection, J. Human Virol. 1(3):214-223 (1998).

3. S.O. Southern, Milk-borne transmission of HIV. Characterization of productively infected cells in breast milk and interactions between milk and saliva, J. Human Virol. 1(5):328-337 (1998).
4. P. Loureiro, S.O. Southern, and M.S.P., Oliveira, Adult T cell leukemia associated with breast disease in HTLV-I patient: clinical, immunopathological and virological studies, Am. J. Hematol. 2000 (in press).
5. F.L. Black, R.J. Biggar, J.V. Neel, E.M. Maloney, and D.J., Waters, Endemic transmission of HTLV type II among Kayapo Indians of Brazil, AIDS Res. Hum. Retroviruses 10(9):1165-1171 (1994).
6. J. Goodall, *The chimpanzees of Gombe: patterns of behavior* (Harvard University Press, Cambridge, MA), (1986).
7. R.S. Wells et al., The social structure of free-ranging bottlenose dolphins, in: *Current Mammalogy,* edited by H. Genoways (Plenum Press, New York), pp. 247-305 (1987).
8. J.A. Mohr, The possible induction and/or acquisition of cellular hypersensitivity associated with ingestion of colostrum, J. Pediatr. 82(6):1062-1064 (1973).
9. S.S. Ogra, D. Weintraub, and P.L. Ogra, Immunologic aspects of human colostrum and milk, J. Immunol. 119(1):245-248 (1977).
10. L.L. Seelig and R.E. Bilingham, Concerning the natural transplantation of maternal lymphocytes via milk, Transplant. Proc. 13:1245-1249 (1981).
11. K.L. Schnorr and L.D. Pearson, Intestinal absorption of maternal leukocytes by newborn lambs, J. Reprod. Immunol. 6(5):329-337 (1984).
12. S. Nakano, Y. Ando, K. Saito, I. Moriyama, M. Ichijo, T. Toyama, K. Sugamura, J. Imai, and Y. Hinuma, Primary infection of Japanese infants with adult T-cell leukaemia-associated retrovirus (ATLV): evidence for viral transmission from mothers to children, J. Infect. 12(3):205-212 (1986).
13. M. Sarzotti, D.S. Robbins, and P.M. Hoffman, Induction of protective CTL responses in newborn mice by murine retrovirus, Science 271(5256):1726-1728 (1996).
14. U. Zügel and S.H. Kaufmann, Role of heat shock proteins in protection from and pathogenesis of infectious diseases, Clin. Microbiol. Rev. 12(1):19-39 (1999).
15. A. Toniolo, C. Serra, P.G. Conaldi, F. Basolo, V. Falcone, and A. Dolei, Productive HIV-1 infection of normal human mammary epithelial cells, AIDS 9(8):859-866 (1995).
16. J. Quayle, C. Xu, K.H. Mayer, and D.J. Anderson, T lymphocytes and macrophages, but not motile spermatozoa, are a significant source of human immunodeficiency virus in semen, J. Infect. Dis. 176(4):960-968 (1997).
17. A. Coutsoudis, K. Pillay, E. Spooner, L. Kuhn, and H.M. Coovadia, Influence of infant-feeding patterns on early mother-to-child transmission of HIV-1 in Durban, South Africa: a prospective cohort study, South African Vitamin A Study Group, Lancet 354(9177):471-476 (1999).

PREVENTION OF VERTICAL TRANSMISSION OF HIV: STRATEGIES, SUCCESSES, AND FAILURES

Philippe Van de Perre*

1. INTRODUCTION

In this presentation, I would like to review what is working and **what** is apparently not working as tools to combat Mother-to-Child Transmission (MTCT) of HIV. With only 10% of the world population, Africa **has** more than 60% of the world's people living with HIV and more than 90% of children with the virus are living in Sub Saharan Africa. It **would** take 8 years in the **United States** to reach the number of infants and children infected with HIV each day in developing countries.

2. MOTHER-TO-CHILD TRANSMISSION OF HIV

In the absence of treatment, 12 to 40% of HIV-1-infected pregnant women transmit the virus to their offspring. This rate of transmission is much lower for HIV-2. In order to design appropriate preventive interventions, description of the timing of transmission is critical. While MTCT of HIV-1 in artificially fed infants occurs during the last trimester of pregnancy and the other two thirds during labor and delivery, in breastfed infants, breastfeeding is an important contributor, representing one to two thirds of overall transmissions.[1] Recent studies suggest that the contribution of early postnatal transmission (before six months of age) may be more important than previously thought.[2] In addition, an international pooled analysis **has** recently estimated the incidence of late postnatal transmission at 3.2 per 100 child-year of breastfeeding.[3]

*Philippe Van de Perre, Centre Muraz/OCCGE, 01 BP 153 Bobo-Dioulasso, Burkina Faso

3. POTENTIAL INTERVENTIONS VITAMIN A SUPPLEMENTATION

Let's now have a look **at** potential interventions aimed at reducing MTCT of HIV. It should be kept in mind that indirect interventions are presently most relevant to the developing world. First, is the prevention of vulnerable young women from becoming infected by HIV. Second, is to provide support and health services to already infected women who choose not to become pregnant. Potential direct interventions to reduce MTCT of HIV-1 are numerous and I will further review the use vitamin A supplementation, obstetrical interventions, breast-feeding alternatives and antiretrovirals. An observational study conducted in Malawi, suggested that maternal vitamin A deficiency is associated with an increased risk of MTCT of HIV-1. A clinical **trial** conducted by Richard Semba and co-workers in Malawi **has** evaluated the efficacy of a daily maternal supplementation with 10.000 IU of retinol palmitate during the last trimester of pregnancy, versus placebo. No significant effect of Vitamin A supplementation on HIV transmission was observed (R. Semba, presentation at the 3rd International Symposium "Global Strategies to Prevent Perinatal HIV Transmission"; Valencia, Spain, 9-11 November 1998). Other trials have confirmed these results in other African settings .[4] However, the effects of vitamin A supplementation on maternal and child morbidity and mortality has still to be analysed in this study **and** that may justify this simple and inexpensive intervention.

4. OBSTETRIC INTERVENTIONS

Recently, a multicentric European study **has randomised** HIV-infected pregnant women according to the mode of delivery. A dramatic reduction of 80% of the risk of vertical transmission was attributed to elective caesarean section.[5] In addition, Laurent Mandelbrot and co-workers have recently demonstrated a synergistic effect of elective caesarean section with the use of antiretrovirals, showing transmission rates as low as less than 1%.[6] Although impressive, these results have only limited relevance to developing countries where severe maternal complications are often observed as a consequence of caesarean section, especially in HIV-infected women. An African study **has** shown maternal mortality in HIV-infected women to be 5 times higher after caesarean section than after vaginal delivery.

5. INFANT FEEDING PRACTICES

Infant feeding practices, **alone or as part of a total package,** are also the focus of potential interventions to reduce MTCT of HIV. A few months ago, the results of the first randomised controlled trial of formula versus breastfeeding was reported from Nairobi, Kenya.[7] This trial screened over 16,000 mothers among **whom** 14% were HIV-1-infected. Relatively few women returned for the HIV test result and only 425 met the study eligibility criteria. **At** 32 weeks of gestation **this latter group was divided between** the breastfeeding arm (n=212) **and** the formula feeding arm (n=213) **of the**

study. In the breastfed group there was a 96% compliance with "any breastfeeding" but only 56% exclusively breastfed for at least 3 months. In the formula-fed group, only 70% of women completely avoided breastfeeding. At 24 months of age, there was a significant 44% reduction of the risk of MTCT in the formula-fed arm as compared with the breastfed arm (from 36.7 to 20.5%). Although overall mortality was not significantly different among the two arms (around 20%), HIV-free survival at 2 years was significantly higher in the formula-fed group (70% versus 58%). This suggests that at least in the environment and the context of the study, formula feeding may be a safe alternative to mixed feeding.[7]

Different breastfeeding practices are likely to be associated with different risks of transmission. In an observational study, Anna Coutsoudis and coworkers have recently measured a lower transmission rate in exclusively breastfed infants **during the first three months of life** (14.6%) as compared with mixed-fed infants (24.1%).[8] This risk reduction of 48% is plausible because of the protective immunologic painting of **the infant's** mucosal surfaces conferred by exclusive breastfeeding. Although mothers choosing breastfeeding are already advised to exclusively breastfeed, confirmation of these results would have important implications for safer infant feeding strategies.

Richard Semba and colleagues recently reported on the impact of mastitis on HIV viral load and transmission.[9] HIV viral load was detectable in breast milk in 75% of women with sub-clinical mastitis versus 33% in women without mastitis and was associated with MTCT. Mastitis may therefore represent a major risk factor for postnatal transmission of HIV-1 by breastfeeding that could be amenable to innovative interventions such as chemoprophylaxis.

6. ANTIRETROVIRALS

Today, the direct intervention that offers **the** most **promise** is antiretroviral prophylaxis. Six years ago the results of the AIDS Clinical Trial Group 076 were published that revolutionised antenatal care of HIV infected women.[10] In this randomised controlled trial conducted in the US and France, a complex regimen which included oral administration of Zidovudine (ZDV) to women beginning 14 to 34 weeks **after** amenorrhea, intravenous administration during labour and delivery **and** oral administration to the neonate until six weeks of life, was shown to reduced MTCT by nearly two thirds. However, this regimen **cannot be implemented** in many developing countries for reasons of social and cultural acceptability, cost, logistics and structure of health systems.

Within the past few months, a first analysis of clinical trial **data** comparing a short ZDV treatment to placebo in non breast-feeding mothers **has** been performed in Bangkok, Thailand.[11] The treatment regimen consisted of maternal administration of 300 mg of **ZDV** two times daily from the 36th week of amenorrhea plus 300mg every three hours during labour and delivery. After a median **treatment** duration of 24 days, ZDV-treated mothers had a 51% reduced risk of transmission of HIV to their babies than placebo-treated mothers.

Recently, in several clinical trials performed in African **countries**, a short antiretroviral treatment has been shown to significantly reduce the risk of **HIV** transmission, **during the first three to six months of** breastfeeding. The ANRS 049 trial conducted in Bobo-Dioulasso, Burkina Faso and Abidjan, Côte d'Ivoire used a ZDV regimen that was very similar to the Thai study.[12]

At six month follow up, while more than 80% of children were still breastfed, transmission of HIV-1 was reduced by 38% in the treated group compared with placebo. Another zidovudine trial performed by the RETROCI project in Abidjan and using **a** very similar treatment regimen, **also showed a** 38% reduction in **HIV** transmission attributed to the treatment.[13]

Laura Guay and co-workers reported on the results of a randomised controlled trial in Uganda, evaluating the efficacy of nevirapine, a simple and inexpensive **nonnucleosidic** reverse transcriptase inhibitor.[14] It was given as a single 200 mg oral dose to the mother at the beginning of labor and to the infant in the first 72 hours after birth. The comparison regimen was an equivalent **intrapartum** and post partum regimen of ZDV. **The nevirapine** regimen reduced MTCT by 48% at 14 to 16 weeks of life in comparison to extra-short ZDV treatment (which is itself of unknown efficacy). At that time nearly all babies were still breastfed. The cost of this treatment is quite low, **$4** US per treatment, in comparison with **$50** US for short ZDV treatment. A cost-effectiveness analysis of this treatment has been performed by Elliot Marseille.[15] At 30% HIV prevalence, a targeted intervention using nevirapine would cost **$298** US per case of pediatric HIV infection averted **compared to $3000-6000** US for short ZDV treatments. Because of the low cost of treatment relative to **HIV** testing, universal treatment without testing has been suggested because of an even better cost-efficacy **of $138** US per case averted. However, there are **many other reasons** why voluntary counselling and testing might be favored despite the cost-effectiveness arguments. **WHAT ARE THEY?**

It should be noted, however, that although these short antiretroviral regimens seem effective in trials conducted in developing countries, transmission rates calculated in the treated arms are consistently higher than those recorded in clinical trial and in clinical practice in North America and Europe. This suggests that **the** performance of antiretrovirals could be further increased by improving the management of pregnancy and by avoiding **exposing infants** to HIV-1 by breastfeeding.

Follow up of therapeutic cohorts **has** shown that in **nonbreastfed** infants, abbreviated antiretroviral regimens including a simple oral administration of antiretrovirals to the neonate within 48 hours of life may still be effective in reducing vertical transmission.[16] This observation suggested that simple and short antiretroviral **regimens** could be translated into public health interventions even if the first contact of mothers with health care services occurs after delivery.

7. POTENTIAL OBSTACLES

All these encouraging results should be rapidly translated into pilot programmes in order to **expand** these procedures on a large scale and **institute** innovative health policies. Several African countries have already committed **to** this **process,** such as Côte d'Ivoire, Zimbabwe and Rwanda. However, **a number of** obstacles **remain**. There are at least two major concerns regarding the use of antiretrovirals **for** prophylaxis of MTCT of HIV in developing countries: a catchment effect due to continuous exposure of infants to HIV through breastfeeding and emergence of antiretroviral resistance.

A few weeks ago, during the Durban Conference, the results of a pooled analysis of the long term efficacy of a short ZDV regimen was reported on 641 mother-child pairs (RETROCI, Abidjan and ANRS 049, Bobo-Dioulasso and Abidjan).[17] Probability of postnatal transmission in infants was rigorously comparable from ZDV- versus placebo-treated mothers. In these trials, although the efficacy of treatment was reduced over time, due to postnatal transmission through breastfeeding, at 24 months of age, the reduction of MTCT conferred by maternal ZDV treatment during gestation and delivery remained significant (27.7%). This suggests that, at least when using ZDV monotherapy, breastfeeding does not erase completely the benefit of maternal treatment.

Long term analysis of another trial **however,** showed completely opposite results. The PETRA study which was conducted in Uganda, Tanzania and South Africa with support of UNAIDS has evaluated the efficacy of 3 different regimens of a combination of zidovudine and 3TC.18 A 50% reduction of transmission was measured for this combination treatment applied pre-, intra- and post partum, 37% for the same combination applied during labour and delivery and post partum and no effect for the intra partum component only. When combining transmission rates and mortality at 24 months, however, there was no significant efficacy in any of the treatment arms, suggesting that breastfeeding exposure had definitely reversed the benefit of earlier maternal treatment.[18]

The other concern is related to the emergence of resistance. A study conducted in the US, the Women and Infant Transmission Study, showed that in a group of 142 women treated with ZDV, 21% acquired a ZDV-resistant HIV genotype.[19] Intriguingly, this resistant genotype was independently associated with vertical transmission of HIV. In addition, in the above-mentioned HIVNET study, 3 out of 14 women had a nevirapine resistant HIV genotype after a single dose of 200 mg of nevirapine.[20] This observation was confirmed and extended in both mothers and infants in further analyses of the same study.[21] The implications of such resistance - for example whether these are reversible or not - are presently unpredictable. As nevirapine was given **to** mother and neonate at delivery/birth, nevirapine-resistant HIV strains may well be transmitted by breastfeeding to infants or introduced into the community by **the** heterosexual route.

However, the **major** obstacles to large scale use of antiretroviral prophylaxis **to prevent** MTCT of HIV **are** merely operational. The **primary** obstacle is certainly not **accessibility,** acceptability of voluntary counselling **or** testing which is the entry point into interventions. In an open cohort of HIV-infected women offered ZDV as prophylaxis in Bobo-Dioulasso and Abidjan, 80% accepted the principle of HIV voluntary counselling and testing, but only 50% effectively return for the test result. **This** clearly suggests that fear of discrimination and of adverse social and familial consequences of test result disclosure is still present in many women in developing countries.

CONCLUSIONS

After more than a decade of short-lived hopes, today the objective of reducing vertical **HIV** transmission in developing countries seems **attainable** providing a stronger

scientific, societal, creative and political commitment **is provided**. Indeed, with the extremely encouraging results obtained **in** clinical trials **using** antiretroviral prophylaxis, the battle against MTCT of HIV should enter a new era by rendering voluntary **counselling, testing,** antiretroviral prophylaxis and alternatives to breast-feeding, accessible to an increasing proportion of women in developing countries.

We know what works although we do not know clearly how to implement it optimally in the real **world**. The developing world is not homogenous. It is quite clear that in some countries implementing pilot programmes using short antiretroviral treatment is an achievable goal. However, in many other settings basic prerequisites of health services are not **available**. In these settings, **alternative approaches must be used.**

First, is to increase **the** accessibility and quality of health services in antenatal care. It is unacceptable that in many African countries today a proportion as high as 50 to 70% of pregnant women deliver without having had a single contact with a health professional, and that for many of the remaining others who had this chance, resources are so scarce that the minimum of care is not provided. Second, is to offer access to emergency obstetrical services of acceptable standard. This simple service is **nonexistent** in many settings **leading to** maternal mortality **rate** reaching one maternal death per hundred life births. Third, is to prevent sexual transmission of HIV in young women. Recent data from seroprevalence studies in Uganda suggest that this may not be a **chimera anymore**. **WHAT DOES THIS MEAN?** Fourth, is to increase **the** accessibility and quality of family planning and, more generally of reproductive health services for HIV-infected women. This seems extremely relevant as diagnosis of HIV infection, both in women from industrialized and developing countries, has a substantial impact on their willingness to reduce conception that should be satisfied by services of **an** acceptable standard.

The challenges are ahead, those which will really make the difference. Having the tools in **hand, we will not be forgiven if we do not use them.**

REFERENCES

1. Simonon A, Lepage P, Karita E, et al. An assessment of the timing of mother-to-child tranmission of human immunodeficiency virus type 1 by means of polymerase chain reaction. J Acquir Immune Def Syndr 1994; 7: 952-7.
2. Miotti PG, Taha TE, Kumwenda NI, Broadhead R, Mtimavalye LA, Van der Hoeven L, Chiphangwi JD, Liomba G, Biggar RJ. HIV transmission through breastfeeding: a study in Malawi. JAMA 1999;282:744-9.
3. Leroy V, Newell ML, Dabis F, Peckham C, Van de Perre P, Bulterys M, Kind C, Simonds RJ, Wiktor S, Msellati P, for the Ghent International Working Group on Mother-to-Child Transmission of HIV. International multicentre pooled analysis of late postnatal mother-to-child transmission of HIV-1 infection. Lancet 1998; 352 : 597-600.
4. Coutsoudis A., Pillay K., Spooner E., Kuhn L., Coovadia H.M. Randomized trial testing the effect of vitamin A supplementation on pregnancy outcome and early mother-to-child transmission in Durban, South Africa. South African Vitamin A Study Group. AIDS 1999; 13: 1517-24.
5 European Mode of Delivery Collaborative Study. Elective caesarean –section versus vaginal delivery in prevention of vertical HIV-1 transmission: a randomised clinical trial. Lancet 1999; 353: 1035-9.
6. Mandelbrot L., Le Chenadec J., Berrebi A., et al. Perinatal HIV-1 transmission: interaction between zidovudine prophylaxis and mode of delivery in the French Perinatal Cohort. JAMA 1998; 280: 55-60.

7. Nduati R. Clinical studies of breast versus formula feeding. Second Conference on Global Strategies for the Prevention of HIV Transmission from Mother to Child, Montreal, Canada, septembre 1999. Presentation # 047.
8. Coutsoudis A, Pillay K, Spooner E, Kuhn L, Coovadia HM for the South African Vitamin A Study Group. Influence of infant feeding patterns on early mother-to-child transmission of HIV-1 in Durban, South Africa. Lancet 1999; 353: 354:471-76.
9. Semba RD, Kumwenda N, Hoover DR, Taha TE, Quinn TC, Mtimavalye L, Biggar RJ, Broadhead R, Miotti PG, Sokoll LJ, van der Hoeven L, Chiphangwi JD. Human immunodeficiency virus load in breast milk, mastitis, and mother-to-child transmission of human immunodeficiency virus type 1. J Infect Dis 1999; 180: 93-8.
10. Connor ED, Sperling RS, Gelber R, et al. Reduction of maternal-infant transmission of human immunodeficiency virus type 1 with ZDV treatment. N Engl J Med 1994 ; 331: 1173-80.
11. Shaffer N., Chuachoowong R., Mock P.A., et al. Short-course zidovudine for perinatal HIV-1 transmission in Bangkok, Thailand: a randomised controlled trial. Lancet 1999; 353: 773-80.
12. Dabis F, Msellati P, Meda N, Welffens-Ekra C, You B, Manigart O, et al. Six months efficacy, tolerance and acceptability of a short regimen of oral zidovudine in reducing vertical transmission of HIV in breast-fed children. A double blind placebo controlled multicentre trial, ANRS049a, Côte d'Ivoire and Burkina Faso. Lancet 1999;353:786-92.
13. Wiktor S, Ekpini E, Karon J, Nkengasong J, Maurice C, Severin S, et al. Short-course oral zidovudine for prevention of mother-to-child transmission of HIV-1 in Abidjan, Côte d'Ivoire: a randomized trial. Lancet 1999;353:781-5.
14. Guay LA, Musoke P, Fleming T, Bagenda D, Allen M, Nakabiito C, Sherman J, Bakaki P, Ducar C, Deseyve M, Emel L, Mirochnick M, Fowler MG, Mofenson L, Miotti P, Dransfield K, Bray D, Mmiro F, Jackson JB. Intrapartum and neonatal single-dose nevirapine compared with zidovudine for prevention of mother-to-child transmission of HIV-1 in Kampala, Uganda: HIVNET 012 randomised trial. Lancet 1999;354:795-802.
15. Marseille E., Kahn J., Mmiro F., et al. Cost effectiveness of single-dose nevirapine regimen for mothers and babies to decrease vertical HIV-1 transmission in sub-Saharan Africa. Lancet 1999;354:803-9.
16. Wade NA, Birkhead GS, Warren BL, et al.Abbreviated regimens of zidovudine prophylaxis and perinatal transmission of the human immunodeficiency virus. N Engl J Med 1998; 339: 1409-14.
17. Wiktor S.Z., Leroy V., Ekpini E.R., et al. 24-month efficacy of short-course maternal zidovudine for the prevention of mother-to-child HIV-1 transmission in a breast feeding population. A pooled analysis of two randomized clinical trials in West Africa. 13th International AIDS Conference, Durban, South Africa, 9-14 July 2000. Abstract TuOrB354.
18. Gray G. for the PETRA Trial Management Committee. The Petra study: early and late efficacy of three short ZDV/3TC combination regimens to prevent mother-to-child transmission of HIV-1. 13th International AIDS Conference, Durban, South Africa, 9-14 July 2000. Abstract LbOr5.
19. Welles S.L., Pitt J., Colgrove R. et al. HIV-1 genotypic zidovudine drug resistance and the risk of maternal-infant transmission in the women and infants transmission study. The Women and Infants Transmission Study Group. AIDS 2000; 14: 263-71.
20. Beker-Pergola G., Guay L., Mmiro F., et al. Selection of the K103N Nevirapine resistance mutation in Ugandan women receiving NVP prophylaxis to prevent HIV-1 vertical transmission (HIVNET-006). 7th Conference on Retroviruses and Opportunistic Infections, San Francisco, Jan 30-Feb 2 2000.
21. Jackson J.B., Mracna M., Guay L., et al. Selection of Nevirapine (NVP) resistance mutations in Ugandan women and infants receiving NVP prophylaxis to prevent HIV-1 vertical transmission (HIVNET-012). 13th International AIDS Conference, Durban, South Africa, 9-14 July 2000.

LACTATIONAL ENDOCRINOLOGY: THE BIOLOGY OF LAM

Alan S. McNeilly*

1. INTRODUCTION

The scepticism shown by many that breastfeeding suppresses fertility is not surprising when one considers that the major stimulus that governs the success of inhibiting reproduction is the suckling stimulus from the baby. This is a fickle controller, and it is not always, if ever, possible to control how much of a stimulus will be supplied. The rates at which babies suckle to achieve a satisfying intake of milk depends not only on the strength of the suckling stimulus, but also on the efficiency of milk let-down by the mother — the milk-ejection reflex related to the release of oxytocin (McNeilly et al., 1983). A further complication is a consistent definition of breastfeeding. We have shown over a number of studies that both frequency and duration of suckling play a major part in maintaining infertility. However, the parameters that we have observed, while adequate for many in the Western world, do not provide useful guidelines for others where the pattern of breastfeeding, and introduction of supplements varies from those commonly found in our study groups.

In spite of these confounding issues related to the strength of the suckling input, we now have a reasonably clear picture of the essential components of breastfeeding that result in infertility (Kennedy et al., 1989; WHO 1998a,b). From this basis we can begin to examine ways of enhancing the effectiveness of breastfeeding as a fertility regulator. However, we are now faced with an ethical brick wall beyond which we may not be able to proceed in our quest to understand the precise endocrine mechanisms through which the suckling stimulus in women suppresses fertility. While appropriate drugs may be available to dissect the pathways involved, it would not be ethical to use many of these in studies of normal mothers and babies, since many of the drugs will pass through into the breast milk and be consumed by the baby.

*Alan S. McNeilly, MRC Human Reproductive Sciences Unit, University of Edinburgh Centre for Reproductive Biology, 37 Chalmers Street, Edinburgh EH3 9ET, Scotland, UK

This brief paper will give an overview of the present state of knowledge of the changes in endocrine function that lead to and maintain infertility during breastfeeding. Several more comprehensive reviews are available (McNeilly, 1994;1997).

2. Ovarian Function during Lactation

Assessment of ovarian function by routine measurement of the secretion patterns of ovarian steroids, estradiol from the follicle and progesterone from the corpus luteum, during lactation in women, indicate that the ovaries are almost non-functional for prolonged periods from birth. If suckling declines slowly over time then there is a resumption of estradiol production indicating a resumption of the growth of follicles, but there is a variable time during which follicle growth is not accompanied by any, or only limited secretion of progesterone. This indicates that for a variable time ovulation either does not take place, or that the function of the corpus luteum that forms from the follicle after ovulation is compromised, and would almost certainly not support a pregnancy (McNeilly et al., 1983). Thereafter, when the suckling stimulus declines further, normal menstrual cycles can resume with normal follicle growth and corpus luteum function and potential to carry a pregnancy. Difficulties arise in defining a suckling parameter(s) which will maintain each of these phases, but full breastfeeding in the absence of high calorie supplements given to the baby is usually associated with complete suppression of ovarian activity, and amenorrheoa (McNeilly 1997).

Monitoring ovarian follicle growth using ultrasound has confirmed that during a large part of lactational amenorrhoea, follicle growth is limited (Flynn et al., 1991; Perheentupa et al., 2000). In some instances, particularly during the later stages of amenorrhoea, medium to large, normal follicles are observed, which may or may not produce estradiol (Flynn et al., 1991). However, most studies would indicate that during amenorrhoea functional ovarian activity is suppressed indicating a state of reduced or absent fertility (Kennedy et al., 1989). Even when menses resumes, this is often preceded by inadequate luteal function and infertility (McNeilly 1997).

Recent monitoring of ovarian follicle function by the more sensitive measurement of inhibin B produced by small growing follicles confirms the lack of ovarian activity during the prolonged periods of lactational amenorrhoea (Perheentupa et al., 2000).

Since ovarian follicle growth can easily be initiated by treatment with GnRH-induced gonadotropins (Glasier et al., 1986) it is clear that the suppression of ovarian activity is directly related to an alteration in the pattern of gonadotropin secretion induced by the suckling stimulus, and not by any intrinsic insensitivity of the ovaries to gonadotropin stimulation.

3. The Pattern of Gonadotropin Secretion

During the normal menstrual cycle the final stages of ovarian follicle growth is stimulated by follicle stimulating hormone (FSH) while the secretion of steroids is stimulated by luteinising hormone (LH). In turn the release of FSH and LH are regulated by gonadotropin-releasing hormone (GnRH) released in pulses every hour in the follicular phase of the cycle, and at a slower frequency during the luteal phase of the cycle. Each pulse of GnRH releases a pulse of LH but the release of FSH, while dependent on GnRH, is not pulsatile, but is regulated by feedback of both estradiol and inhibin released from the developing preovulatory follicle. The co-ordinated release of FSH which increases at menses and increase frequency of GnRH/LH pulses is required for normal follicle growth and formation of a normal corpus luteum after ovulation.

During lactation FSH levels increase to near normal levels within 4 weeks of delivery, but pulsatile release of LH is almost entirely absent in the first 4-6 weeks postpartum (Glasier et al., 1984; Nunley et al., 1991; Tay et al., 1992). Normal pulses of LH reappear around 6-8 weeks postpartum, but the frequency is low, and the pattern of pulsatile secretion is extremely erratic, with pulses occurring randomly during 24h (Tay et al., 1992). This pattern is unlike that during puberty when there is a co-ordinated pulsatile release of LH during the period of sleep, associated with a rise in prolactin (Wu et al., 1991). A more normal 24h pattern of pulsatile LH secretion is observed around the time of consistent increases in ovarian estradiol secretion strongly suggesting that it is the abnormal pattern of GnRH/LH release that is responsible for the absence of normal follicle development (McNeilly 1994).

Indeed when a normal pattern of GnRH is given by pulsatile pump in breastfeeding women, there is an immediate stimulation of follicle growth and ovulation (Glasier et al., 1986; Zinaman et al., 1995). Thus it is clear that the disrupted pattern of LH and FSH release from the pituitary results in the failure of normal ovarian follicle growth, and the infertility related to breastfeeding. It is also clear that this aberrant pattern of LH and FSH release is related to a reduced frequency of GnRH pulsatile secretion from the hypothalamus caused by the suckling stimulus. This pattern of changes in gonadotropin secretion is similar situation in all species in which there is adequate information on the patterns of gonadotropin secretion during lactation (McNeilly 1994).

4. Factors Regulating GnRH Release

It is clear that the key effects of suckling in regulating fertility during lactation are mediated through an alteration in the frequency of pulsatile secretion of GnRH from the hypothalamus. The secretion of FSH is dependent on GnRH, but the major regulation of FSH secretion is a direct effect of estradiol and inhibin at the pituitary. A number of factors could be involved in the suckling-related suppression of GnRH, and a number of studies have addressed the involvement of opioids, dopamine, leptin, prolactin and steroids in women.

The generation of the preovulatory LH surge is dependent on the increase in plasma levels of estradiol released from the preovulatory follicle. In breastfeeding women this so-called positive feedback effect of estradiol was absent, and there was an increase in the sensitivity of the GnRH system to the negative effects of estradiol which resulted in an inhibition of GnRH/LH release (Baird et al., 1979). We have recently shown that this enhanced negative effect of estradiol suppresses the pulsatile secretion of GnRH assessed by measurement of the pulsatile release of LH (Illingworth et al., 1995). There was no effect on the sensitivity of the pituitary to GnRH indicating an effect only on GnRH release from the hypothalamus. There did not appear to be any involvement of the opioid system within the hypothalamus since treatment with opioid blockers did not affect the effect of estradiol (Illingworth et al., 1995).

Subsequently we have shown that this enhanced sensitivity to low doses of estradiol can persist for a prolonged period during lactation, resulting in a suppression of the basal levels of both FSH and LH and a complete inhibition of functional ovarian activity, as monitored by ultrasound, estradiol, and inhibin B measurements (Perheentupa et al., 2000). However, it is still not clear how these effects of estradiol are mediated since GnRH neurones in the hypothalamus do not express estrogen receptors (Scott et al., 2000). The effect is related solely to estradiol since the progestagen mini-pill does not affect gonadotropin secretion, and must act as a contraceptive entirely by non-gonadotropin effects on the reproductive tract during lactation.

As mentioned above, the opioids do not appear to play any major role in regulating GnRH release during lactation. Treatment with the opioid blocker naloxone did not influence LH or FSH secretion (Illingworth et al., 1995; Tay et al., 1993). Furthermore, dopamine also does not appear to be involved since treatment with the dopaminergic blocker metoclopramide did not affect LH or FSH secretion, but did result in a massive release of prolactin (Tay et al., 1993).

Leptin is secreted by adipose tissue and may affect fertility (Clarke and Henry 1999). Furthermore, leptin and leptin receptors have been identified on gonadotropes in the pituitary (Iqbal et al 2000; Vidal et al., 2000). However, plasma levels of leptin are within normal ranges in women during lactation (Lage et al., 1999). This area of the influence of nutrition and obesity on lactation will be discussed elsewhere in this issue.

Finally the textbook parameter for suppressing gonadotropin secretion in women is prolactin, since pituitary tumor-related hyperprolactinaemia in women is associated with infertility, and suppression of prolactin leads to a resumption of menstrual cyclicity and fertility. Lactational hyperprolactinaemia is related to the duration of ammenorhoea (Duchen and McNeilly 1980).

The production of milk in women is dependent on prolactin, and each suckling episode causes the release of substantial amounts of prolactin (Tay et al., 1992, 1996). In our own studies relatively slow frequencies of suckling plasma concentrations of prolactin increase dramatically, whereas high frequency suckling is associated with maintained high plasma concentrations of prolactin with much smaller increases in prolactin during each suckling episode (Tay et al., 1996). In these studies we found no relationship between the duration of infertility, or the pulsatile pattern of secretion of LH and the overall levels of prolactin over 24h periods at different stages of lactation. However, in other studies longer durations of amenorrhoea were associated with a greater release of prolactin at each suckling episode (Campino et al., 1999; Stallings et al., 1998) and the biological activity of prolactin is enhanced in women with prolonged infertility during lactation (Campino et al., 1999). However, it is not clear where this prolactin would act within the hypothalamus to affect GnRH secretion, although prolactin receptors have been located within the brain.

It is difficult to present a clear picture regarding the role of prolactin in the suppression of GnRH release during lactation in women and there is conflicting information on the effects of prolactin on reproductive function in various species (McNeilly 1994). Indeed, even in primate species there is some confusion since there is little or no delay in the resumption of fertile cycles postpartum in lactating marmosets even though lactation is associated with hyperprolactineamia, (McNeilly et al., 1981). Nevertheless, the close association between high plasma levels of prolactin and the duration of infertility requires that we continue investigations into the potential role of prolactin in suppressing GnRH release. Although found in the sheep (Tortonese et al., 1999) we have recently failed to show the presence of prolactin receptors in the human pituitary gonadotropes suggesting that effects of prolactin, if any, are mediated through the hypothalamus.

Conclusions

There is no doubt that suckling suppresses ovarian activity postpartum in women. During the period of amenorrhoea there is limited ovarian follicle growth associated with low estradiol and inhibin B production, and erratic pulsatile GnRH/LH secretion. Treatment with regular pulses of GnRH induces a normal pattern of LH release with subsequent follicle growth and ovulation. Since there is no apparent effect on the sensitivity of the ovaries to gonadotropins, or the pituitary to GnRH, the effect of suckling must be mediated through an effect on the pulsatile secretion of GnRH release from the hypothalamus.

The factors associated with suckling in women that affect GnRH release are unclear. There is little evidence to support a role for opioids, dopamine or leptin. However, there is some evidence to link the high plasma concentrations of prolactin although the mechanisms are still not clear. Certainly suckling does increase the sensitivity of the hypothalamus to the negative feedback effects of estradiol resulting in a suppression of pulsatile GnRH/LH release and loss of the positive feedback effects of estradiol. There is no doubt that lactation and the suckling stimulus suppresses ovarian activity and fertility in breastfeeding women, and the endocrine changes observed during this provides a sound basis for LAM.

References

Baird, D.T., McNeilly, A.S., Sawers, R.S., and Sharpe, R. M., Failure of estrogen-induced discharge of luteinizing hormone in lactating women, J. Clin. Endocrinol. Metab. 49:500-506 (1979).

Campino, C., Torres, C., Ampuero, S., Diaz, S., Gonzalez, G.B., and Seron-Ferre, M., Bioactivity of prolactin isoforms: lactation and recovery of menses in nursing women, Human Reprod. 14:898-905 (1999).

Clarke, I.J. and Henry, B.A., Leptin and Reproduction, Rev. Reprod. 4:48-55 (1999).

Duchen, M.R., and McNeilly, A.S., Hyperprolactinaemia and long-term lactational amenorrhoea, Clin. Endocrinol. 12:621-627 (1980).
Flynn, A., Docker, M., Brown, J.B., and Kennedy, K.I., Ultrasonographic patterns of ovarian activity during breastfeeding, Amer. J. Obstet. Gynecol. 165:2027-2031 (1991).
Glasier, A., McNeilly, A.S., and Baird, D.T. Induction of ovarian activity by pulsatile infusion of LHRH in women with lactational amenorrhoea, Clin. Endocrinol. 24:243-252 (1986).
Glasier, A.F., McNeilly, A.S., and Howie, P.W., Pulsatile secretion of LH in relation to the resumption of ovarian activity post partum, Clin. Endocrinol._20:415-426 (1984).
Illingworth, P.J., Seaton, J.E.V., McKinlay, C., Reid-Thomas, V., and McNeilly, A.S., Low dose transdermal oestradiol supresses gonadotrophin secretion in breastfeeding women, Human Reprod. 10:1671-1677 (1995).
Iqbal J., Pompolo S., Considine R.V., and Clarke, I.J., Localization of leptin receptor-like immunoreactivity in the corticotropes, somataotropes, and gonadotropes in the ovine anterior pituitary, Endocrinology. 141:1515-1520 (2000).
Kennedy, K.I., Rivero, R., and McNeilly, A.S., Consensus statement on the use of breastfeeding as a family planning method, Contraception. 39:477-496 (1989).
Lage, M., Garcia-Mayor, R.V., Tome, M.A., Cordido, F., Valle-Inclan, F., Considine, R.V., Caro, J.F., Dieguez, C., and Casanueva, F.F., Serum leptin leves in women throughout pregnancy and the postpartum period and in women suffering spontaneous abortion, Clin. Endocrinol. 50:211-216 (1999).
McNeilly, A.S., Suckling and the control of gonadotropin secretion, in: *The Physiology of Reproduction*, E. Knobil and J.D. Neill, eds., Raven Press, New York, pp 1179-1212 (1994).
McNeilly, A.S., Lactation and fertility, J. Mamm. Gland. Biol. Neoplasia 2:291-298 (1997).
McNeilly, A.S., Abbott, D.H., Lunn, S.F., Chambers, P.C., and Hearn, J.P., Plasma prolactin concentrations during the ovarian cycle and lactation and their relationship to return of fertility post partum in the common marmoset (*Callithrix jacchus*), J. Reprod. Fertil. 62:353-360 (1981).
McNeilly, A.S., Robinson, I.C.A.F., Houston, M.J., and Howie, P.W., Release of oxytocin and prolactin in response to suckling, Brit. Med. J. 286:257-259 (1983a).
McNeilly, A.S., Glasier, A.F., Howie, P.W., Houston, M.J., Cook, A., and Boyle, H., Fertility after childbirth: pregnancy associated with breast feeding, Clin. Endocrinol. 18:167-173 (1983b).
Nunley, W.C., Urban, R.J., and Evans, W.S., Preservation of pulsatile luteinizing hormone release during postpartum lactational amennorhoea, J. Clin. Endocrinol. Metab. 73:629-636 (1991).
Peerhentupa, A., Chritchley, H.O.D., Illingworth, P.J., and McNeilly, A.S., Enhanced sensitivity to steroid negative feedback during breast-feeding: Low dose estradiol (Transdermal Estradiol Supplementation) suppresses gonadotropins and ovarian activity assessed by inhibin B, J. Clin. Endocrinol. Metab., in press, (2000).
Scott, C.J., Tilbrook, A.J., Simmons, D.M., Rawson, J.A., Chu, S., Fuller, P.J., Ing, N.H., and Clarke, I.J., The distribution of cells containing estrogen receptor-α (ERα) and ERß messenger ribonucleic acid in the preoptic area and hypothalamus of the sheep: comparison of males and females, Endocrinology 141:2951-2962 (2000).
Stallings, J.F., Worthman, C.M., and Panter-Brick, C., Biological and behavioural factors influence group differences in prolactin levels among breasfeeding Nepali women, Amer. J. Human Biol. 10:191-210 (1998).
Tay, C.C.K., Glasier, A.F., and McNeilly, A.S., The twenty-four hour pattern of pulsatile luteinizing hormone, follicle stimulating hormone and prolactin release during the first eight weeks of lactational amennorhoea in breastfeeding women, Human Reprod. 7:951-958 (1992).
Tay, C.C.K., Glasier, A.F., and McNeilly, A.S., Effect of antagonists of dopamine and opiates on the basal and GnRH-induced secretion of luteinizing hormone, follicle stimulating hormone and prolactin during lactational amenorrhoea in breastfeeding women, Human Reprod. 8:532-539 (1993).
Tay, C.C.K., Glasier, A.F., and McNeilly, A.S., Twenty-four hour patterns of prolactin secretion during lactation and the relation to suckling and the resumption of fertility in breastfeeding women, Human Reprod. 11:950-955 (1996).
Tortonese, D., Brooks, J., Ingleton, P.M., and McNeilly, A.S., Detection of prolactin receptor gene expression in the sheep pituitary gland and visualisation of the specific translation of the signal in gonadotrophs, Endocrinology 139:5212-5223 (1998).
Vidal, S., Cohen, S.M., Horvath, E., Kovacs, K., Scheithauer, B.W., Burguera, B.G., and Lloyd, R.V., Subcellular localization of leptin in non-tumorous and adenomatous human pituitaries. An immuno-ultrastructural study, J. Histochem. Cytochem. 48:1147-1152 (2000).
WHO: The World Health Organization multinational study of breastfeeding and lactational amenorrhoea. I. Description of infant feeding patterns and of the return of menses, Fertil. Steril. 70:448-460 (1998a).

WHO: The World Health Organization multinational study of breastfeeding and lactational amenorrhoea. II. Factors associated with the length of amenorrhoea, Fertil. Steril. 70:461-471 (1998b).
Wu, F.C.W., Butler, G.E., Kelnar, C.H.J., Stirling, H.F., and Huhtaniemi, I., Patterns of pulsatile luteinizing and follicle stimulating hormone secretion in prepubertal (midchildhood) boys and girls and patients with idiopathic hypogonadotrophic hypogonadism (Kallman's syndrome): a study using an ultrasensitive time-resolved immunofluorometric assay, J. Clin. Endocrinol. Metab. 72:1229-1237 (1991).
Zinaman, M..J., Cartledge, T., Tomai, T., Tippett, P., and Merriam, G.R., Pulsatile GnRH stimulates normal cycle ovarian function in amenorrhoeic lactating postpartum women, J Clin Endocrinol Metab 80:2088-2093 (1995).

EFFICACY AND EFFECTIVENESS OF LAM

Kathy I. Kennedy*

1. ABSTRACT

Two major protocols of non-randomized clinical trials of the efficacy of the Lactational Amenorrhea Method of contraception (LAM) were performed in the 1990s that suggested LAM to be a highly effective albeit temporary method of contraception. Data from a longitudinal study of over 4000 breastfeeding women performed by the World Health Organization provide supporting information as to the efficacy of LAM. Effectiveness data are scarce, as is information on the use of LAM in industrialized countries. Issues surrounding LAM efficacy and effectiveness are reviewed, and the existing information from industrialized countries is highlighted.

2. BACKGROUND

Population scientists have long observed that breastfeeding is associated with infecundity. In the 1970s and 1980s reproduction scientists became interested in understanding the mechanisms of action responsible for lactational infertility, as eloquently described in the previous chapter. These efforts were directed toward learning how to harness the natural contraceptive effect of lactation so that it could be used pro-actively for family planning. By the end of the 1980s, a body of basic clinical research had been amassed on the behavior of the hypothalamic-pituitary-ovarian axis during lactation. Researchers around the globe who had conducted these studies came together to see whether they could create a consensus on the conditions under which an individual woman would be protected from pregnancy due to breastfeeding. Their conclusion is published as the "Bellagio Consensus" which states that:

*Kathy I. Kennedy, 2201 South Fillmore Street, Denver, CO 80210, US kkennedy@du.edu, Associate Clinical Professor of Preventive Medicine, University of Colorado Health Sciences Center, Director, Regional Institute for Health and Environmental Leadership, University of Denver

Integrating Population Outcomes, Biological Mechanisms and Research Methods in the Study of Human Milk and Lactation
Edited by Davis *et al.*, Kluwer Academic/Plenum Publishers, 2002

> "During amenorrhea and full (or nearly full) breastfeeding,
> the chance of pregnancy is less than 2% in the first six months postpartum.".[1]

The consensus group also recognized the need to test this statement with clinical trials of breastfeeding women who were actually using this algorithm to delay or avoid a subsequent pregnancy. Thus began the research of the early 1990s to determine the efficacy and the effectiveness of the Bellagio Consensus, which, when applied, became known as the Lactational Amenorrhea Method (LAM) of family planning. The consensus group also agreed to reconvene after the next generation of studies was completed.

The Bellagio Consensus scientists made their prediction on the basis of small biomedical studies that observed the physiological correlates of the return of fertility in lactating women around the world. Yet their prediction should have come as no surprise to demographers and family planning specialists since it is consistent with a large number of earlier surveys. Reviews of the literature in the 1970s and 1980s concluded that about 5% (3%-10%) of breastfeeding women conceive *during amenorrhea* in the *first year* postpartum.[2-4] The Bellagio Consensus scientists recognized from their basic studies that the rate of return of fertility increased with time, and that the return of fully fertile cycles during amenorrhea was not the norm, although it was more likely to occur with each passing month. They also realized that it is hard to quantify breastfeeding, but that whatever amount of breastfeeding results in amenorrhea is largely (98% of the time) enough to prevent pregnancy in the first six months of infancy.

When studying the ability of a device, drug or method to prevent pregnancy, two kinds of success are studied. Both efficacy and effectiveness are valuable and important to measure, but subtly different. **Efficacy** is the extent to which a specific regimen produces a beneficial result under *ideal conditions*. Ideally, the determination of efficacy is based on the results of randomized controlled trials. **Effectiveness** is the extent to which a specific regimen, when deployed in the field in *routine circumstances*, does what it is intended to do.[5] Both the efficacy and the effectiveness of LAM have been studied, and that research is reviewed here.

3. Studies of LAM Efficacy

A set of non-comparative contraceptive efficacy studies of LAM was conducted using a common protocol, but with large enough numbers for each study to be evaluated separately. These studies were conducted in Pakistan[6] and the Philippines.[7] A health educator who had previously worked in both countries, and who knew both Urdu and Tagalog, created the LAM education regimen and trained the staff at the family planning clinics involved about LAM and the research protocol. The details of the study designs can be read elsewhere, but some aspects of the protocol are important enough to mention here.

The volunteers in the studies chose LAM as their postpartum family planning method from among the array of methods offered by the maternity service or family planning clinic involved. Learning the method involved learning a jingle that described the LAM algorithm and also learning about the breastfeeding practices that simultaneously maximize milk production and lactational infertility. The mothers (and infants) were normal and healthy, and expected to be sexually active in the ensuing months. All women had previously breastfed a child for at least one year, planned to fully breastfeed the current infant for 4-6 months and generally expected not to be separated from child. The characteristics of the mothers are seen in Table 1. They were largely parous, urban women with some but not much family planning experience. The Philippine women were well educated, and resumed sexual relations somewhat later than the Pakistani volunteers.

Table 1. Characteristics of the Volunteers in the Efficacy Trials

Pakistan (Kazi et al, 1995)

N = 391

- mean age 27
- mean number of living children 4.4
- 75% urban
- 38% literate
- 6% ever used family planning
- 80% sexually active by second month
- 59% fully breastfeeding through 6th month
- median duration amenorrhea 10th month

Philippines (Ramos et al, 1996)

N = 485

- mean age 27
- mean number of living children 2.9
- 100% urban, low income
- mean education 8.8 years
- 21% ever used modern family planning
- 75% sexually active by third month
- 62% fully breastfeeding through 6th month
- median duration amenorrhea 10th month

The Pakistani data are reprinted by permission from the American Society for Reproductive Medicine (***Fertility and Sterility*** 1995, 64(4):717-723). The Philippine data are reprinted by permission of the BMJ Publishing Group (***British Medical Journal*** 1996, 313:909-912.)

The six-month lifetable rates of pregnancy during correct use of the method (i.e. during lactational amenorrhea and full/nearly full breastfeeding) are seen in Table 2. During correct use, there was one pregnancy in Pakistan for a rate of 0.58 under the Bellagio Consensus criteria and in the Philippines there was a pregnancy in the fourth month and one in the fifth month, for a rate of 0.97.

Table 2. Pregnancy During Correct Use of LAM – Efficacy Trials

Month Postpartum	woman-mo exposure	number of pregnancies	cum failure rate (95% ci)
A. Pakistan (N = 391)			
1	61	0	0.00
2	276	0	0.00
3	291	0	0.00
4	278	0	0.00
5	234	0	0.00
6	173	1	**0.58 (0.00 - 1.58)**
B. Philippines (N = 485)			
1	60	0	0.00
2	189	0	0.00
3	218	0	0.00
4	217	**1**	**0.46 (0.00-1.47)**
5	197	**1**	**0.97 (0.00-2.41)**
6	150	**0**	**0.97 (0.00-2.43)**

The Pakistani data are reprinted by permission from the American Society for Reproductive Medicine (***Fertility and Sterility*** 1995, 64(4):717-723). The Philippine data are reprinted by permission of the BMJ Publishing Group (***British Medical Journal*** 1996, 313:909-912.)

Table 3 shows the six-month lifetable pregnancy rates during *any* use of the method, correct or incorrect, i.e. regardless of whether the women were amenorrheic or fully/nearly fully breastfeeding. Under these conditions, four more pregnancies occurred in Pakistan and two more in the Philippines, for six-month pregnancy rates of 1.65 and 1.53, respectively. Thus even when the Bellagio Consensus criteria did not apply, these women still had pregnancy rates of less than 2% in the first six months postpartum.

Table 3. Pregnancy During Correct and Incorrect Use of LAM – Efficacy Trials

month postpartum	woman-mo exposure	number of pregnancies	cum failure rate (95% ci)
A. Pakistan (N = 391)			
1	61	0	0.00
2	283	0	0.00
3	317	0	0.00
4	329	0	0.00
5	312	**3**	**0.96 (0.00-2.15)**
6	287	**2**	**1.65 (0.12-3.18)**
B. Philippines (N = 485)			
1	61	0	0.00
2	204	0	0.00
3	251	0	0.00
4	273	**1**	**0.37 (0.00-1.21)**
5	266	**2**	**1.11 (0.00-2.52)**
6	243	**1**	**1.52 (0.12-3.14)**

The Pakistani data are reprinted by permission from the American Society for Reproductive Medicine (***Fertility and Sterility*** 1995, 64(4):717-723). The Philippine data are reprinted by permission of the BMJ Publishing Group (***British Medical Journal*** 1996, 313:909-912.)

Table 4 presents the twelve-month pregnancy rates *during amenorrhea*, i.e. regardless of the amount of breastfeeding going on. This is of interest in determining whether the LAM criteria can be simplified to only the amenorrhea criteria, since breastfeeding is difficult to quantify, and since infants should not remain fully breastfed after about the sixth month anyway. Twelve-month pregnancy rates of 1.12 and 2.56 were found in Pakistan and the Philippines, respectively. These rates of pregnancy during amenorrhea in the first year postpartum are better (lower) than in the reviews of the literature mentioned above, but the women in the efficacy trials were informed as to the best breastfeeding practices to simultaneously maximize milk production and infertility.

The breastfeeding requirements of the protocol were requirements for the research and not requirements for using LAM, yet the results may have been affected by these selection criteria. Experienced breastfeeders were sought because 12 month pregnancy rates during breastfeeding were sought, and if the studies had included large numbers of women who ceased breastfeeding earlier in the first year, there might not have been enough power to draw conclusions with confidence. This demonstrates the need for effectiveness trials to see how LAM performs among all kinds of breastfeeding women. The results from the efficacy trials may indeed have elicited data as to the most protection that a group of women could expect.

An important strength of the efficacy studies is that they collected and analyzed contraceptive clinical trial data in the best possible way: The lifetable rates were left and right censored for the use of other contraceptive methods, and for sexual activity. (Women were only included in the lifetable analyses during months in which they were sexually active and not using another method.) Data on contraceptive use and sexual activity were collected each month, and the women's status in the analysis adjusted accordingly. These procedures for collecting and analyzing contraceptive efficacy data are now used in the general practice of contraceptive clinical trials.

Table 4. Pregnancies During Amenorrhea – Efficacy Trials

month postpartum	woman-mo exposure	number of pregnancies	cum failure rate (95% ci)
A. Pakistan (N = 391)			
6	244	1	0.41 (0.00-1.21)
7	213	0	0.41 (0.00-1.22)
8	182	0	0.41 (0.00-1.23)
9	164	0	0.41 (0.00-1.25)
10	139	1	1.12 (0.00-2.53)
11	120	0	1.12 (0.00-2.56)
12	103	0	1.12 (0.00-2.59)
B. Philippines (N = 485)			
4	233	1	0.43 (0.00-1.38)
5	221	1	0.88 (0.00-2.20)
6	194	0	0.88 (0.00-2.20)
7	148	0	0.88 (0.00-2.22)
8	116	1	1.72 (0.00-3.65)
9	117	1	2.56 (0.18-4.94)
10	97	0	2.56 (0.13-4.99)
11	84	0	2.56 (0.07-5.05)
12	73	0	2.56 (0.02-5.11)

The Pakistani data are reprinted by permission from the American Society for Reproductive Medicine (***Fertility and Sterility*** 1995, 64(4):717-723). The Philippine data are reprinted by permission of the BMJ Publishing Group (***British Medical Journal*** 1996, 313:909-912.)

The efficacy trials in Pakistan and the Philippines conclude that LAM is greater than 98% effective when used correctly, that the protection afforded by LAM is not due to postpartum abstinence, and that the method is forgiving of incorrect use. The trials also suggest that amenorrhea alone is associated with a large amount of protection through the first 12 months of breastfeeding.

4. Studies of LAM Effectiveness

A non-comparative multicenter study of the effectiveness of LAM was conducted in ten centers of nine countries. Table 5 shows the number of volunteers from each center and some important characteristics of the total group. A detailed description of the study design, protocol, and results is available.[8] Pregnancy rates during LAM use were determined in women whose breastfeeding practices were more representative of normal conditions. Both primiparas and multiparas were included in the study, the women needed to express no particular breastfeeding intentions, and no restrictions were placed on separations from the infant. Unfortunately, there was no censoring for coitus or family planning use in the published analysis, but this was not a conventional feature of lifetable data analysis at the time.

The six-month lifetable pregnancy rate during correct LAM use was 1.5 in the

Table 5. Characteristics of the Volunteers in the Effectiveness Study

Number of Volunteers by Center

Egypt	59	• ever used family planning	16% to 100%
Indonesia	61	• mean age	23 to 32 yrs
Mexico	50	• mean parity	1.5 to 3..2
Nigeria, Jos	60	• primiparas	2% to 64%
Nigeria, Sagamu	47	• mean education	4 to 17 years
Philippines	47	• Catholic	0% to 94%
Germany/Italy	47	• work at 6 months	0% to 51%
Sweden	51		
United Kingdom	49		
United States	48		
TOTAL	**519**		

Reprinted by Permission. ***Contraception*** 1996, 55, 327-336.

pooled dataset. There were no pregnancies in most centers, high pregnancy rates in some centers, and no single center has enough cases for analysis by itself. The pregnancy rate during amenorrhea (i.e. regardless of breastfeeding behavior) was 2.0% (0.7% SE) at six months and 8.8% (2.0% SE) at twelve months. Thus, as might be expected, these effectiveness results are not quite as strong as the efficacy results, but clearly uphold the Bellagio Consensus and are consistent with the early surveys on pregnancy during amenorrhea in the first year. The particular value of these results is that they were obtained under more field-like conditions than the efficacy study results.

Table 6. Pregnancy During Correct Use of LAM – Effectiveness Study (N=519,10 centers)

Study center	cum woman-mo exposure	number of pregnancies	cum 6 mo failure rate (SE)
Egypt	330	1	2.0 (2.0%)
Indonesia	318	1	2.6 (1.6%)
Mexico	249	2	7.5 (5.1%)
Nigeria, Jos	352	0	0.0
Nigeria, Sagamu	245	1	4.2 (4.1%)
Philippines	236	0	0.0
Germany/Italy	237	0	0.0
Sweden	261	0	0.0
United Kingdom	250	0	0.0
United States	240	0	0.0
TOTAL	2718	5	**1.5 (0.7%)**

Reprinted by Permission. ***Contraception*** 1996, 55, 327-336.

5. Other Studies Related to LAM Efficacy/Effectiveness

There are several additional studies related to LAM efficacy/effectiveness that are not controlled trials. For example, a study was conducted in Ecuador during the integration of LAM into an existing family planning service.[9] LAM effectiveness was found to be consistent with the Bellagio Consensus, but the numbers involved were small. A study of LAM extended for nine months was begun in Rwanda[10], but the civil war resulted in most cases being lost to follow-up. Among the small percent of cases that could be found, the pregnancy rate was in the expected range. More recently, a study of 40 women in Italy observed no pregnancies during amenorrhea.[11] Finally, secondary analyses of the occurrence of pregnancy in breastfeeding women (i.e., the women were not using LAM), were conducted by numerous authors.[12-18] In all cases, the observed rate of pregnancy was consistent with the Bellagio Consensus.

All of these latter secondary analyses of existing data sets are similar to the World Health Organization (WHO) Infant Feeding Study.[19] The WHO studied more than 4,000 women in seven countries to determine the relationship between infant feeding behaviors and the return of menses in breastfeeding women, with particular emphasis on whether women from diverse gene pools and cultures who ostensibly breastfeed the same amount obtain the same amount of protection from pregnancy. To facilitate the research protocol, the mothers in the study needed to be aged 20-37, normal and healthy, to have breastfed at least 1 infant for at least 4 months and intend to breastfeed the current infant for at least 6 months, be literate, not planning to use a hormonal method of contraception but expecting to be sexually active. Table 7 shows the number of pregnancies observed over the entire duration of the study as long as the mothers were still breastfeeding and not using another contraceptive method (i.e. they may or may not have been amenorrheic).

Table 7. Number of Pregnancies in the WHO Infant Feeding Study in Non-contracepting Breastfeeding Women

Center	Number of Pregnancies
Chengdu, China	20
Guatemala City	9
Melbourne/Sydney	5
New Delhi, India	5
Sagamu, Nigeria	0
Santiago, Chile	2
Uppsala, Sweden	5

Reprinted by permission of the American Society for Reproductive Medicine (***Fertility and Sterility*** 1999, 72(3):431-440.)

The WHO study collected so much information on postpartum menses, that four different definitions of the return of menses could be created. The first menses that was known to be part of a cycle (since it was followed by a second menses or pregnancy) was called the "confirmed first menses"; the first menses whether or not there was evidence of cyclicity was called "any first menses", the first bleeding episode that the woman believed was the return of menses was called "woman's perception"; and "all first bleeding" referred to bleeding whether or not it was long enough in amount and duration to qualify as a bleeding episode. Table 8 shows that regardless of how the return of menses was measured, the six-month pregnancy rate during full/nearly full breastfeeding and amenorrhea was 0.9 to 1.2%, consistent with the Bellagio Consensus. In the lifetable pregnancy rates, the WHO was able to censor for sexual activity and the use of other contraceptives.

Table 8. Six-month Pregnancy Rate During "Correct Use of LAM " by Definition of the First Menses (i.e. of the End of Amenorrhea)

Menses Definition	cum 6 mo woman-mo exposure	cum 6 mo failure rate
confirmed first	2969	**1.2** (0.0-2.4)
any first	2963	**0.9** (0.0-2.0)
woman's perception	2939	**0.9** (0.0-2.0)
first bleeding	2831	**1.0** (0.0-2.1)

n=4118 in 7 countries

Reprinted by permission of the American Society for Reproductive Medicine (***Fertility and Sterility*** 1999, 72(3):431-440.)

Table 9 shows that as long as the women were breastfeeding, in the first year postpartum, the pregnancy rate during amenorrhea was 3.7 to 5.2%, consistent with the early reviews of the literature on pregnancy during amenorrhea.

Table 9. Pregnancy During Amenorrhea and Breastfeeding

Menses Definition	cum 12 mo woman-mo exposure	cum 12 mo failure rate
confirmed first	8033	**4.4** (2.5-6.3)
any first	7979	**3.7** (1.9-5.5)
woman's perception	7869	**5.2** (3.1-7.4)
first bleeding	7407	**4.1** (2.1-6.1)

n=4118 in 7 countries

Reprinted by permission of the American Society for Reproductive Medicine (***Fertility and Sterility*** 1999, 72(3):431-440.)

Thus although the WHO study was not designed to test the Bellagio Consensus, it is a very large dataset onto which the LAM criteria could be applied. Like the other studies mentioned in this section that were not prospective LAM contraceptive efficacy trials, the results are consistent with the Bellagio Consensus. I have been unable to locate published studies that conflict with the Bellagio Consensus, but presumably some studies, especially studies of small numbers of women, or studies that include women who do not breastfeed well, could have poorer results than those mentioned here.

The breastfeeding practices that maximize milk production and infertility are likely to occur in developing countries, and access to, or use of, family planning is lower in developing countries. These generalities make it appear that LAM is better suited for non-industrialized populations. Yet it is of interest to know whether LAM works as well for "Western" women as for others. In the LAM Multi-Center effectiveness study there were no pregnancies in 988 woman-months of use by mothers in the U.S., the U.K.,

Germany/Italy or Sweden. In the WHO secondary analysis, there were no pregnancies in 1039 woman-months of use in Australia and Sweden.

Conclusion

The results of the efficacy and effectiveness studies reviewed here uphold the Bellagio Consensus, and concur with the early surveys which determined that about 5% (3%-10%) of breastfeeding women will conceive *during amenorrhea* in the *first year* postpartum.[2-4] Thus, when the second Bellagio Consensus Conference was held in 1995, after reviewing these and other data, the scientists concluded that (emphasis added):

> "The efficacy of LAM has now been well established..."
>
> and
>
> "(1) It is not possible to eliminate the amenorrhea criterion. ...
> (2) It *may* be possible to relax the requirement of full or nearly full breastfeeding. ...
> (3) It *may* be possible to extend LAM beyond 6 months postpartum. ...".[20]

Both programmatic and biomedical research needs were stated by the consensus crafters. Chief among the continuing questions about LAM is whether the achievable rates of contraceptive efficacy/effectiveness of lactational amenorrhea alone for up to a year postpartum are acceptably high. Even if they are, does LAM, even used for 12 months, provide enough protection from pregnancy to provide adequate child spacing for both maternal and child health? Assuming that it does not, can LAM be a conduit to the use of other family planning methods? Many of the women who chose to use LAM in the efficacy and effectiveness studies had little experience with modern contraceptive methods. Can the successful use of LAM be used to engender confidence in family planning, the use of another modern method after LAM protection expires, and result in adequate child spacing?

References

1. K.I. Kennedy, R. Rivera, and A. McNeilly, Consensus statement on the use of breastfeeding as a family planning method, Contracept. 39(5):477-496 (1989).
2. J.K. Van Ginnekin, Prolonged breastfeeding as a birth spacing method, Stud. Fam. Plann. 5:201-206 (1974).
3. R. Rolland, Bibliography (with review) on contraceptive effects of breastfeeding, Biblio. Reprod. 28(93):1-4 (1976).
4 M.Simpson-Hebert, and S.L. Huffman, The contraceptive effect of breastfeeding, Stud. Fam. Plann. 12:125-133 (1981).
5 J.M. Last, *A Dictionary of Epidemiology*, fourth edition, Oxford University Press, Oxford (2001).
6. A. Kazi, K.I. Kennedy, C.M. Visness, and T. Khan, Effectiveness of the lactational amenorrhea method in Pakistan, Fertil. Steril. 64(4):717-723 (1995).
7. R. Ramos, K.I. Kennedy, and C.M. Visness, Effectiveness of lactational amenorrhoea in prevention of pregnancy in Manila, the Philippines: non-comparative prospective trial, Brit. Med. J. 313:909-912 (1996).
8. M.H. Labbok, V. Hight-Laukaran, A.E. Peterson, V. Fletcher, H. von Hertzen, and P.F.A. Van Look, Multicenter study of the lactational amenorrhea method (lam): I. efficacy, duration, and implications for clinical application, Contracept. 55:327-336 (1996).
9. K.B. Wade, F. Sevilla, and M.H. Labbok, Integrating the lactational amenorrhea method into a family planning program in Ecuador, Stud. Fam. Plann. 25:162-175 (1994).

10. K. Cooney, Assessment of the nine month lactational amenorrhea method in Rwanda (MAMA-9), Stud. Fam. Plan. 27:162-171 (1996).
11. G.A. Thommaselli, M. Guida, S. Palomba, and M. Barbata, Using complete breastfeeding and lactational amenorrhea as birth spacing methods, Contracept., 61:253-257 (2000).
12. M. Bracher, Breastfeeding, lactational infecundity, contraception and the spacing of births: implications of the Bellagio Consensus Statement, Health Transit. Rev. 2(1):19-47 (1992).
13. K.I. Kennedy, S. Parenteau-Carreau, A. Flynn, et al., The Natural family planning -- lactational amenorrhea method interface. Observations from a prospective study of breastfeeding users of natural family planning, Am. J. Obstet. Gynecol. 165(6) part 2: 2020-2026 (1991).
14. K.I. Kennedy, B.A. Gross, S. Parenteau-Carreau, et al., Breastfeeding and the symptothermal method, Stud. Fam. Plann. 26(2):107-115 (1995).
15. K.I .Kennedy, and C.M. Visness, Contraceptive efficacy of lactational amenorrhea. *Lancet* 339:227-230.
16. B. Rojnik, K. Kosmelj, and L. Andolsek-Jeras, Initiation of contraception postpartum, Contracept. 51:75-81 (1995).
17. R.V. Short, P.R. Lewis, M.B. Renfree et al., Contraceptive effects of extended periods of lactational amenorrhea beyond the Bellagio consensus, Lancet 337:715-717(1991).
18. P. Weis, The contraceptive potential of breastfeeding in Bangladesh, Stud. Fam. Plann. 24(2):100-108 (1993).
19. World Health Organization Task Force on Methods for the Natural Regulation of Fertility, The WHO multinational study of breastfeeding and lactational amenorrhoea. iii. Pregnancy during breastfeeding, Fertil. Steril. 72:431-440 (1999).
20. K.I. Kennedy, M.H. Labbok, and P.F.A. Van Look, Consensus statement -- lactational amenorrhea method for family planning, Int. J. Gynecol. Obstet. 54(1):55-57 (1996).

OBESITY AS A RISK FACTOR FOR FAILURE TO INITIATE AND SUSTAIN LACTATION

Kathleen M. Rasmussen, Julie A. Hilson, and Chris L. Kjolhede*

INTRODUCTION

It is well documented that socioeconomic status, education, race/ethnicity and social support are determinants of which women attempt to breastfeed their newborn infants.[1] Although poor nutrition may compromise lactational performance,[2] evidence has only recently emerged that overnutrition may also compromise lactation. Investigators in Australia[3] studied women who had breastfed their infants for at least 2 wk and observed that those with a body mass index (BMI) value > 26 kg/m^2 at 1 mo postpartum had 1.5 times the risk of early cessation of breastfeeding (BF) compared to those whose BMI was below this value. These findings are of particular concern because a high proportion of women have BMI values this high so soon after delivery. In addition, the proportion of women with higher BMI values is likely to grow with the increasing rates of obesity that have been observed in American women.[4]

In our research with rats fed a high-fat diet before conception and throughout the reproductive period,[5] we found that initiation of lactation was so severely impaired that many pups died. In a subsequent experiment, we[6] found that the pups of the dams fed the high-fat diet that survived did not grow as well as pups reared by control dams. This appeared to be caused by reduced milk production by the dams.

Based on our findings in rats, we investigated the possibility that both initiation and duration of lactation would be negatively affected by maternal overnutrition. We conducted a review of medical records among white women who lived in a rural area of upstate New York.[7] In this population, at least 75% of women attempt to breastfeed. We showed that, among those who had attempted to breastfeed their newborns, the odds of failing to initiate BF successfully were significantly increased among women who were overweight (BMI 26.1-29.0 kg/m^2) or obese (BMI > 29 kg/m^2) before conception, 2.54- and 3.65- fold, respectively. In addition, both overweight and obesity before conception were significantly

*K.M. Rasmussen and J.A. Hilson, Division of Nutritional Sciences, Cornell University, Ithaca, NY 14853; C.L. Kjolhede, Bassett Research Institute, Cooperstown, NY 13326.

associated with shortened duration of exclusive BF as well as shortened duration of any BF. This investigation clearly established that both initiation and duration of BF could be negatively affected by maternal fatness. If this association were causal, we calculated that being overweight or obese accounted for 41% of the failure to initiate lactation in this population.

In this investigation we did not have adequate statistical power to investigate whether weight gain during pregnancy might modify the effect of preconceptual fatness on lactational performance. In addition, we were unable to explore whether this association resulted from psychosocial correlates of both obesity and BF behavior or whether it resulted from biological factors, as suggested by our prior work in rats.[5] To address the first issue, we conducted a much larger review of medical records and, to address the second, we conducted a prospective study. We report some of the results of both of these investigations here.

2. Does Gestational Weight Gain Modify the Association between Prepregnant bmi and Lactational Performance?

2.1 Subjects and Methods

We expanded our prior review of medical records[7] to encompass a 9-y period (January 1988 - December 1997). We identified 3,803 women 19-40 y old who delivered singleton infants at the Mary Imogene Bassett Hospital in Cooperstown, NY. We selected the 2,859 of these who had no contraindications for BF or gestational diabetes, who attempted to breastfeed at delivery and for whom complete data were available. For the analyses reported here, the 365 women who were underweight (BMI < 19.8 kg/m^2) at conception or lost weight during pregnancy were excluded.

We extracted data on feeding through 24 mo of age from the infant's record and data on maternal characteristics and events surrounding delivery from the mother's record. Maternal prepregnant BMI was calculated from reported prepregnant weight and measured height. Gestational weight gain was calculated as the difference between the mother's weight at delivery and her reported prepregnant weight. If weight at delivery was unavailable, we used the weight recorded at the mother's last clinic visit, which usually took place within a week before delivery.

In the analyses reported here, women were categorized as normal-weight, overweight or obese using the BMI cutoffs of the Institute of Medicine (IOM) described above.[8] Their weight gain during pregnancy was also categorized as below, within or above the cutoffs of the IOM for the woman's category of prepregnant BMI. Inasmuch as the IOM guidelines do not specify an upper limit of gain for obese women, we chose a cutoff of 9.1 kg--a value that has been used by others. This 2-way categorization produced 9 groups for the analysis. Normal-weight women who gained within the IOM guidelines were the reference group. Multiple logistic regression analysis was used to compare the risk of unsuccessful initiation of BF in the other 8 groups to this reference group. Proportional hazards regression analysis was used to compare the risk of discontinuation of BF over time in the other 8 groups to this reference group. Analyses for the duration of exclusive BF and any BF were conducted separately. All regression analyses were adjusted for known confounding factors.

2.2 Results

Among the 2,494 women in this analysis, those who gained more than the IOM recommendation had a higher prepregnant weight and BMI than women who gained within the recommendations. Those who gained more than the IOM guidelines were younger and less educated; a higher proportion participated in social service programs and were nulliparous; and exclusive BF and any BF lasted for less time than among those who gained within the IOM guidelines. A lower proportion of those who gained within and above the guidelines smoked than among those who gained less than the guidelines. As weight gain relative to the guidelines increased, so did birthweight and the proportion of macrosomic infants.

The odds of unsuccessful initiation of BF (defined as still BF at 4 d postpartum) were significantly increased compared to the reference group among normal-weight women who gained more than the guidelines (OR 1.74, 95% CI 1.05-2.91) and also among all of the groups of obese women (below the guidelines, OR 4.20, 95% CI 1.91-9.15; within, OR 3.33, 95% CI 1.37-6.84; and above, OR 2.40, 95% CI 1.38-4.13).

The relative risk of discontinuation of exclusive BF was significantly increased only among the 3 groups of obese women (below, RR 1.42, 95% CI 1.02-1.98; within, RR 1.69, 95% CI 1.21-2.35; and above, RR 1.58, 95% CI 1.30-1.93).

The relative risk of discontinuation of any BF also was significantly increased only among the 3 groups of obese women (below, RR 2.03, 95% CI 1.40-2.95; within, RR 1.60, 95% CI 1.12-2.28; and above, RR 1.92, 95% CI 1.54-2.38).

2.3 Discussion

Although our sample was certainly large, we nonetheless had less power than we had calculated was necessary among the overweight and obese subjects who gained below and within the IOM guidelines. This is because more than 70% of the overweight and obese women exceeded the IOM guidelines for weight gain during pregnancy. The lack of power was not a problem among the obese subjects, but it may explain the lack of observed effect among the overweight women.

It is noteworthy that this sample contained subjects with an extraordinarily wide range of values for BMI before conception (15-55 kg/m^2) as well as for weight change during pregnancy (from a loss of 12.7 kg to a gain of 43 kg). We did not evaluate postpartum weight retention, but the high proportion of women who gained in excess of the IOM guidelines is cause for concern.

These findings suggest that normal-weight women whose weight gain during pregnancy exceeds the IOM guidelines are at excess risk of lactation failure. Our observation that weight gain during pregnancy did not appreciably modify the excess risk of lactation failure already characteristic of obese women suggests that they may have reached a critical level of fatness for lactation failure before conception.

3. Do Psychosocial or Behavioral Factors Contribute to the Association between Prepregnant bmi and Lactational Performance?

3.1 Subjects and Methods

We conducted a prospective study among pregnant women 19-45 y old who were carrying a singleton fetus and expressed an intention to breastfeed. All received their prenatal care through Bassett Hospital in Cooperstown, NY. Among the 151 who gave their informed consent, 138 were actually eligible for the study and 117 provided complete data.

During pregnancy, women completed a questionnaire to document their demographic and psychosocial characteristics. We sought information on those psychosocial constructs related to success of BF (e.g. behavioral beliefs, knowledge, social learning and maternal confidence about BF and also social support). In the first 5 d after delivery, women were called daily or until their milk "came in" to assess the onset of copious milk secretion (lactogenesis II) using a series of questions developed by others (R. Perez-Escamilla and D. Chapman, pers. comm., 1999). Data on BF behaviors during the perinatal period were obtained from the Mother-Baby Assessment scores[9] in the hospital record that are assessed and recorded by trained nurses as a part of usual care at Bassett Hospital. The duration of BF was assessed by recall as part of a telephone interview conducted 12-18 mo postpartum.

Multiple logistic regression analysis was used to calculate the odds of lactogenesis II occurring earlier (< 72 h postpartum) or later (≥ 72 h postpartum). These analyses were done sequentially: first, we evaluated the association between obesity and the timing of lactogenesis II; next, we evaluated whether this association was mediated by the scores on the Mother-Baby Assessment; and, finally, we evaluated whether this association was modified by any of the psychosocial constructs. Cox proportional hazards regression analysis was used to determine if obesity was associated with the discontinuation of BF and whether this association was mediated by the timing of lactogenesis II.

3.2 Results

Compared to normal-weight and overweight women, obese women had a shorter planned duration of BF and were less satisfied with their appearance. Other than these differences, the 3 prepregnant BMI groups differed little in their psychosocial characteristics. Women among whom lactogenesis was delayed had a higher prepregnant BMI and a lower infant score on the Mother-Baby Assessment compared to women in whom lactogenesis II arrived earlier. There were no significant differences in psychosocial characteristics between women with earlier and later onset of lactogenesis II.

The risk of later onset of lactogenesis II was significantly associated with prepregnant BMI. This association was attenuated after adjustment for confounding factors, but it was not modified by BF behavior or psychosocial factors. When expressed as a categorical variable, prepregnant obesity interacted with knowledge of BF. Women with a BMI ≤ 26 kg/m^2 experienced a 25% reduction in the risk of later onset of lactogenesis II as knowledge increased; in contrast, women with a BMI > 26 kg/m^2 experienced a 90% increase in the odds of later onset as knowledge increased.

Once again, we observed that obesity was associated with early discontinuation of BF. Adjustment for psychosocial factors (planned duration of BF, plan to return to work or school, behavioral beliefs about BF, and satisfaction with appearance) and later onset of lactogenesis II substantially attenuated this association, but it remained statistically significant.

3.3 Discussion

In this investigation, we selected a group of women who were personally motivated to breastfeed their infants and who delivered in an environment that was supportive of this planned method of infant feeding. Nonetheless, their psychosocial characteristics and BF behaviors were not significant predictors of either lactogenesis II or the duration of BF. This suggests that biological factors, such as the woman's hormonal and metabolic milieu after delivery, may be important causes of this consistent relationship between maternal fatness and poor lactation performance. In particular, we provided evidence of an association between maternal prepregnant BMI and lactogenesis II, a biological event.

The negative interaction between maternal prepregnant obesity and knowledge of BF suggests that educational interventions to improve knowledge of BF among these women may not be effective. However, counseling women to expect a delay in the onset of copious milk secretion may be helpful. We calculate that the difference could be as much as 10 h between a woman with a BMI of 20 kg/m^2 and another with a BMI of 40 kg/m^2.

Conclusions

In this series of studies, all conducted within the same community but involving subjects who delivered over more than 10 y, we have consistently observed an association between high BMI values before conception and poor lactation performance. This association is magnified by high weight gain during pregnancy among normal-weight women. We did not have adequate statistical power to tell if this is also the case among overweight women. In contrast, obese women have an excess risk of poor lactation performance regardless of their weight gain during pregnancy. These findings will be important to consider when future recommendations are made for weight gain during pregnancy.

Although there were differences among the BMI groups in some psychosocial characteristics and adjusting for these factors attenuated the association between maternal obesity and lactational performance, the association remained statistically significant. These findings, when added to the data from rats, suggest that the association has a biological basis that should now be investigated.

Based on these findings, pregnant women could reasonably be advised to begin pregnancy with a healthy BMI value (i.e. < 26 kg/m^2), keep their weight gain during pregnancy within the IOM guidelines, and–if they are overweight or obese–to allow more time than normal-weight women for their milk to "come in".

References

1. Institute of Medicine (Subcommittee on Nutrition during Lactation, Committee on Nutritional Status During Pregnancy and Lactation, Food and Nutrition Board). Nutrition during Lactation (National Academy Press, Washington, DC, 1991).

2. M. Rasmussen. The influence of maternal nutrition on lactation. Annu. Rev. Nutr. 12:103-117 (1992).
3. H.E. Rutishauser and J.B. Carlin. Body mass index and duration of breast feeding: a survival analysis during the first six months of life. J. Epidemol. Comm. Health 46:559-565 (1992).
4. A. Galuska, M.K. Serdula, E. Pamuk,.P.S. Siegel, and T. Byers. Trends in overweight among US adults from 1987 to 1993: a multistate telephone survey. Am. J. Public Health 86:1729-1735 (1996).
5. A. Shaw, K.M. Rasmussen, and T.R. Myers. Consumption of a high-fat diet impairs reproductive performance in Sprague-Dawley rats. J. Nutr. 127:64-69 (1997).
6. M. Rasmussen, M.H. Wallace, and E. Gournis. A low-fat diet but not food restriction improves lactational performance in obese rats. in: Bioactive Substances in Human Milk, edited by D.S. Newberg (Plenum Press, New York, 2000), in press.
7. A. Hilson, K.M. Rasmussen, and C.L. Kjolhede. Maternal obesity and breastfeeding success in a rural population of Caucasian women. Am. J. Clin. Nutr. 66:1371-1378 (1997).
8. Institute of Medicine (Subcommittees on Nutritional Status and Weight Gain During Pregnancy and Dietary Intake and Nutrient Supplements During Pregnancy, Committee on Nutritional Status During Pregnancy and Lactation, Food and Nutrition Board). Nutrition During Pregnancy: Part I, Weight Gain; Part II, Nutrient Supplements (National Academy Press, Washington, DC, 1990).
9. C. Mulford. The mother-baby assessment (MBA): An "Apgar Score" for breastfeeding. J. Hum. Lact. 8:79-82 (1992).

SUB-OPTIMAL BREAST FEEDING PRACTICES:
Ethnographic approaches to building "Baby Friendly" communities

Daniel W. Sellen*

1. INTRODUCTION

Observations that exclusive breastfeeding in the early months of life, extended partial breastfeeding and transition to high quality non-breast milk foods have biological benefits are unsurprising from an evolutionary perspective. Nevertheless, in order to ascertain the optimal timing of components of the human weaning process, one must adopt a rigorous comparative biomedical approach. Most research on human milk and lactation therefore uses evidence-based methods and clinical outcomes to quantify the effects of different "doses" of breastfeeding ("none", "exclusive", "partial", "extended", etc.) and to identify the nutritional, immunological and endocrinological mechanisms mediating such effects. Comparisons among groups of children exposed to different patterns of breastfeeding and complementary feeding show that only some patterns maximize growth and development, energy and nutrient intakes in relation to estimated requirements, and survival for the normal, healthy child.

Health and nutrition policy makers use such data to define, recommend and promote a specific subset of the wide range of possible patterns of breastfeeding and complementary feeding practices. Randomized trials confirm that these patterns are indeed optimal for child outcomes across a range of settings, and evidence accumulates that the same patterns can also benefit the health and well being of mothers. Community-based studies have demonstrated the benefits of adoption of, or closer conformity with, the current, scientifically developed recommendations, and that these benefits are inversely related to the level of social and material deprivation mothers and children experience.

Yet despite enormous progress in our basic scientific understanding of lactation and its translation into useful policy recommendations, the extent to which modern populations deviate from clinically indicated (and presumably naturally selected) breast

*Departments of Anthropology and International Health, Emory University, Atlanta, GA 30022, USA

feeding patterns is surprising, frustrates attempts to improve maternal - child health and begs explanation. The focus of this chapter is on two crucial questions relevant to policy and the transfer on technical knowledge to the community: Why do caregivers often fail to adopt recommendations, and why are practices so often sub-optimal in the first place?

2. Socio-Cultural Influences on Young Child Care

2.1. Public Health Models

Efforts to investigate specific reasons caregivers fail to practice optimal feeding are few because strategies to identify the social determinants of health outcomes remain limited. One approach is to construct notional "key behaviors" based on biomedical models of the proximate determinants of health. The recently proposed UNICEF conceptual framework for understanding the causes of malnutrition, death and disability is an example (Figure 1). This model effectively captures the common etiology of these poor outcomes through proximate biological effects of inadequate dietary intake and prevalent disease. The socio-cultural and environmental roots of these insults are modeled as underlying causes that influence the supply of food, care and health to mothers and children. These in turn are assumed to be in some sense over-determined by so-called "basic causes", which include opportunities to gain knowledge and various aspects of infrastructure.

An important feature of the new UNICEF model is the central position of care. Care is conceptualized as based in the community, as encompassing both children and their mothers, and as having several components, of which one of the most crucial is breast feeding and complementary feeding. Nevertheless, the heuristic value of this model is limited because it only weakly incorporates cultural factors. A major flaw is that the framework relegates the influence of socio-cultural factors to a rather distal position in the flow of causality. Broad constructs such as "SES", "education", or "urban residence" are deployed as proxies for behaviors. The basic causes are vaguely assumed to be influenced by rather poorly specified constructs such as politics, ideology and economic relations. As a tool for analysis, it draws our attention away from the ways in which social interactions may directly constrain individuals.

Although criticized since the early 1980's,[1] this type of public health model is inevitable in the absence of a coherent theory of human social action. Indeed, many would argue that current public health approaches consistently fail to examine the social space between individuals and groups in research and policy-making. However, tensions between a health promotion focus on individuals as actors in a social vacuum and an epidemiological focus on population groups as homogenous units are forcing practitioners to confront this crisis of theory. There is renewed interest in the use of ethnographic approaches to fill this conceptual gap and measure the socio-cultural influences on health.

2.2. A challenge to Anthropologists

Understanding the cultural determinants of public health outcomes has proved a significant challenge for anthropologists interested in the interaction between social and

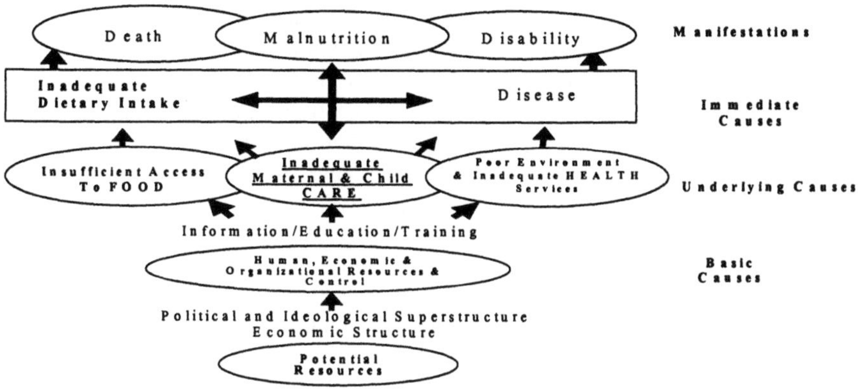

Figure 1. The recently proposed UNICEF conceptual framework for understanding the causes of malnutrition, death and disability (redrawn from.[2]

biological processes. Almost a half century ago, Benjamin Paul (then at the Harvard School of Public Health) reviewed the importance of including community perspectives in global health interventions and concluded that: "application of our available health knowledge is the weakest link in our chain of health protection".[3] His central claim was that for interventions to work well we needed to know more about local contexts. More recently, Robert Hahn observed that:

> "relevant anthropological knowledge and methods have grown substantially, but application remains a weak link".[4]

Although modern ethnographers have learned a considerable amount about local contexts, they have failed to share this knowledge effectively with health researchers and policy-makers. Thus, as we enter the new millennium, we continue to do a poor job of bringing together knowledge about the proximate determinants of human health and about how people in different cultural settings interact, cooperate and conflict with one another.

2.3. Components of Young Child Care

In particular, the precise role of care as an underlying cause of maternal and child health outcomes remains unclear. We do know that the quality of care is but one of many risk factors, and that care giving often trades off against other activities. We have recently learned that the non-material components of care are critical for child health because physical growth and psychological development interact.[5] However, we have yet to understand what the salient components really are.

Progress in understanding the most relevant *components* of care depends on development of methods to measure quality of care in appropriate currency. First, we need to quantify and compare the independent and combined impact of feeding, hygiene, and psychosocial stimulation in different settings. Second, we need to know the essential determinants of care quality. Only ethnographic techniques can generate this kind of information. Such approaches allow us to structure our investigations by simultaneously considering the provision of care at several levels.

The components of care that have received most research attention are of course breast feeding and complementary feeding. To date, most investigators have aimed to quantify the effects at different ages of exclusive, extended partial and no breast feeding in terms of functional development, nutrition status, health and survival of children and maternal reproductive health. Each category characterizes an "exposure" in the classic epidemiological sense, but it is important to remember that each can be enacted through a variety of maternal and/or care giver behaviors. Thus, just as infants in a community experience "partial breastfeeding" in different ways, many different sets of maternal behaviors may result in "exclusive breastfeeding", many enacted beyond the physical relationship that exists between the mother and child. We might expect that maternal choice, constraints and opportunities will operate as proximate determinants; information, material and social support as intermediate determinants; and attitudes and norms, and socio-economic context as underlying determinants.[6] The crucial point is that global policy recommendations shown biologically "optimal" from a child's perspective fail to offer solutions to the tradeoffs faced by caregivers.

2.4. Tradeoffs Faced by Care Givers

The gaps in information needed to understand the causes of sub-optimal breastfeeding practices for any particular community mainly concern the unexamined tradeoffs faced by caregivers trying to feed young children. Few data exist which allow us to examine conflicts between the needs of young child and those of a caregiver. For example, assumptions that the interests of the mother with usually coincide with those of the child remain untested. Similarly, in most contexts we lack information about conflicts of interest between individuals dependent on a single caregiver. Often, these individuals will compete directly or indirectly, actively or passively, for her care. Opportunity costs of time allocated to child care,[7] social costs of exerting control over feeding decisions,[8] and ideological concerns trigger by nutrition education and breast feeding promotion messages all generate conflicts for caregivers. The daily resolution of such conflicts must entail tradeoffs between currencies of caregiver utility, and knowledge transfer through breastfeeding promotion messages may not be sufficient to empower caregivers to privilege infant needs.

3. Current Policy and Practice

Policy statements made a decade ago evidence widespread recognition that the tradeoffs faced by mothers were contingent on factors often one-step removed from their control. Included in infant feeding recommendations were calls to remove "extrinsic barriers to optimal breast feeding" (Innocenti Declaration,[9] "inform all members of society of the benefits of breast feeding" (Convention on the Rights of the Child, 1991), to "empower all women to breast feed optimally" (World Summit for Children, 1991), and to "create environments supportive of breastfeeding" (UNICEFWHO Baby Friendly Hospital Initiative:[10] Thus, breast feeding was squarely framed as a social issue, a human rights issue, a health management issue and a feminist issue.

These policies have improved practices, as indicated by the marked increase in first and second-year breastfeeding continuation rates in developing countries[11] and limited evalution of promotion interventions in the West.[12] However, the current global situation suggests progress has been limited. Recent data suggest 85% of mothers do not conform to optimal recommendations.[13] Although 96% of children from developing countries who survive at least 3 days are breastfed , fewer than half of children under 4 mo are exclusively breast fed[11] and the median duration of any breast feeding is only 18 mo.[14]

The most likely causes of this mismatch between global policy and local practice are cultural factors that operate at different levels. First, the personal perceptions of policy makers, employers, and women may create strong and sometimes unconscious resistance to the optimal pattern. They may construct various kinds of cultural narratives about the high costs of achieving optimal practices and over-emphasize costs such as lost productivity, earnings, and health. For example, it would be useful to know whether the image of a "zero-sum game" in women's allocation to care among policy-makers parallels images of "milk insufficiency" among women. Second, shared local perceptions may reinforce the beliefs that exclusive, extended breast feeding is just not feasible. Local opinion-leaders often describe the sub-optimal status quo as "correct", "necessary" or "unavoidable". They may deploy various forms of folk reasoning to conclude that exclusive breastfeeding is not possible or not sensible. Conceptual dissonance between modern or scientific knowledge and traditional or local knowledge may mean that unmodified promotion messages lack credibility and foster feelings of inadequacy or hostility. Third, the classic public health focus on the maternal-child unit represents a particular cultural perspective, that of the community of people trained in the western biomedical tradition. This focus actively generates new barriers, and in fact undermines many of the policy initiatives. For example it usually leads to the exclusion of men in promotion and support group training and reinforces the erroneous view that protecting infant is solely the job of mother. It also diminishes the importance of a lack of maternal support systems at both household and community level.

We suggest that most so-called social, economic and political barriers in fact derive from these sorts of cultural factors. If superimposed on the UNICEF framework, such cultural factors would crosscut several levels of causality. For each community where infants are fed sub-optimally, support teams need to know who the caregivers are, what they do, and why. The following examples demonstrate how ethnographic approaches conducted at several levels can help us understand why feeding practices deviate from the ideal in a variety of contemporary cultural settings.

4. Examples of Ethnographic Approaches

4.1. Pastoralist Communities, Tanzania

The rural Datoga (Barabaig) follow a highly traditional lifestyle.[15, 16] Breast feeding is universally initiated and of long duration (median 22 months) and there is no evidence of recent declines.[17] Neverthelss, more than 50 % of infants are not offered the breast until more than 2 days after delivery, 45% are given prelacteals, and 80% do not receive colostrum. For many infants, non breast milk liquids are introduced too early and solid foods rather late. Only 25% are fully breast fed at 4 months and only 25% are regularly fed complementary solids between 6 and 9 months. These practices are of concern and probably related to the 20% infant mortality rate and the stunting of over one third of children under five.

Qualitative investigations reveal a complex set of underlying influences and suggest that the main barriers to optimal feeding are related to the organization of subsistence activities rather than maternal attitudes. Whereas focus groups of older women reported that weaning practices are related to seasonal availability of cow's milk, longitudinal observations showed that most mothers introduce non-breast milk foods and terminate breast feeding earlier when they had fewer helpers in the household. Although mothers often respond inappropriately to episodes of infant illness by withdrawing the breast, they express positive views about the value of breast feeding and mature breast milk. They use infant-centered cues to make decisions about the appropriate timing of milestones in the weaning process. Thus, household labor stress and a limited choice of foods appropriate to weaning placed major constraints on care and feeding practices.

4.2. Cane Cutting Communities, Brazil

Feeding practices in Sao Paulo have deteriorated precipitously in recent years, with the ever breast fed rate falling by over 50% between 1940-1975.[18] One community is characterized by almost universal use of prelacteal teas, bottle feeding, rejection of "dirty" colostrum, and perceived milk insufficiency after a few days post-partum.[19] Bottle-feeding is associated with a four-fold relative mortality risk and 30%of bottlefed infants are malnourished. Again, ethnographic observations shed light on how this situation came about.[20] These suggest that in the 1960's women used to have very positive views of breastfeeding and resisted formula. Present-day loss of faith in the value of breast feeding is linked to the severe impoverishment and disentitlement of the population and to women's extremely poor self efficacy and negative self-perception. Discourse analysis reveals the "unconscious sources" of women's rejection of breast feeding.[20, 21] Many women believe they are too "weak", "poor", or "wasted" to sustain the added burden of breast feeding and continually refer to their physical selves as "used up", "finished" and "no good". A strong belief that never breast fed, unbaptized babies go to heaven untainted exists as a coping response to extremely high mortality rates and miserable conditions. Ironically, the practice of formula feeding is reinforced by a new custom whereby new fathers provide large and expensive gifts of tins of formula to the mother to symbolize paternity and signal an intention to care for the infant.

4.3. Agrarian Communities in the Gambia

In a set of communities described by Samega-Janneh,[22] breast feeding is universal and of long duration, but again sub-optimal in its micro-structure. Nearly 40% of mothers do not initiate breast feeding in the first 24 hrs after delivery and only 1% exclusively breast feed at 4 mo. Ethnographic approaches were used in a community-based intervention to optimize practices. Village-based mother-to-mother support groups were elected in 9 locations, each with 5 women and 2 men. The training component emphasized linkages between breast feeding, maternal diet, weaning, environmental sanitation and personal hygiene. It was capped by an elaborate, locally organized graduation award ceremony attended by senior health administrators from the city government. Thus, the training mobilized local male support, elevated the status of participants, and signaled commitment from external power-holders.

An action component designed and initiated entirely by the teams working with their communities targeted all pregnant and lactating women and their spouses with messages about improved practices disseminated through songs, dances, gatherings, and home visits. Some communities invented a reciprocal social rule of 'community maternity leave', whereby new mothers rested while family and neighbors completed their work tasks. After examining the problem of infant separation while mothers worked, all communities built 'baby friendly rest houses' near the fields where infants were creched and breast fed at intervals during work breaks. Local linguistic shifts reinforced promotional messages, such as a newly coined term for colostrum meaning 'protective milk' and use a phrase meaning 'complete breastfeeding' for a password. Many mothers adopted a locally invented system for storage of expressed milk. Together, these interventions increased the 4 month exclusive breast feeding rate to 99.5%.

4.4. Refugee Communities in Britain

Young child diets are under threat in certain socially excluded group in industrial countries, such as recently arrived political asylum-seekers now living in London's east end.[23] Colleagues and I have worked among three refugee communities (Albanian, Somali, Colombian) to design a young child and family nutrition assessment tool to investigate the mechanisms by which cultural barriers may exacerbate the effects of poverty in these groups.[24] Family support for breast feeding and child care is challenged because many families are experiencing periods of homelessness, some family members have post-traumatic stress syndrome, and many families have been fragmented by deaths, disappearances and separation during migration. Most new arrivals lack competence in English, local knowledge, and transport. Those placed by government programs in temporary accommodation often have reduced access to cooking, cleaning and food storage facilities. Finally, a limited literature on communities moving to industrialized countries suggests that the erosion of traditional values and knowledge contributes to a decline in breast feeding.

Recently we assessed the salience of these suggested risk factors using preliminary data from a structured ethnographic survey among 30 families.[25] Findings are that refugee family diets conform to current health messages more closely than those of the local population on similar incomes do. Families rely on fresh foods, make little use of pre-packaged and cooked food outlets, and breastfeeding is common in all three communities. Income is the perceived limiting factor for the provision of a healthy diet, but access to a diverse range of food shops facilitates shopping and budgeting.

Thus, mothers surveyed maintain healthy dietary habits on low incomes by relying on sound cooking skills, tight budgeting, and help from friends and family (such as food lending and borrowing). Nevertheless, food insecurity is greatest at level of the mother, and inadequate kitchen facilities for cooking and storage, as well as lack of social contacts from whom to seek appropriate infant feeding advice undermine maternal confidence in adequate meal provision. Inferences are that policy should aim to prevent erosion of these coping mechanisms and that recent UK government proposals to introduce food vouchers in lieu of cash benefits and to disperse groups of refugees may compromise child diets.

5. Strategies to Build Baby Friendly Communities

Several general programmatic conclusions follow from this review of the cultural aspects of the child feeding process. First, ethnographic techniques reveal local opportunities to extend the concept care to the conditions of pregnant and lactating mothers. Widening the concept of care of the infant to include care of the mother, the nullipara, the girl-child, and the alternate care-giver should also help to increase the focus on women's health in settings where women's health problems are stigmatized. Second, ethnographic techniques can tell us what kinds of information should be included in health messages and how to target nutrition education and behavior change interventions beyond women and local health workers.

The concept of "baby friendly communities"[22] has potential to become a powerful catalyst for community change to make optimal child feeding practices a more feasible goal for more women and their families. Efforts to build such communities must include consideration of maternal diet and options for family planning, maternity protection for working mothers. They must also shift more responsibility to men, and emphasize the benefits of optimal breastfeeding to mothers and other family members. The local and household organization of labor can be modified to accommodate maternity leave, nursing breaks, and crèches even where employment is informal, such as agrarian communities. Existing reciprocal arrangements in neo-maternal support within and between houses can be reinforced and extended. The use of doula's should be encouraged at every opportunity. More effort could be made to educate about breast feeding benefits to mothers, which now include decreased postpartum bleeding, anemia, and mortality, and decreased premenopausal breast/ovarian cancer as well as increased child spacing, return to pre-pregnancy weight, bone re-mineralization, self confidence and satisfaction.

Community support can help build a stronger sense of self-efficacy among mothers. The ethnographic data suggest that non-biological breast milk insufficiency is a metaphor of maternal need. In many settings those who breast feed (in part because they cannot afford to do otherwise) often are plainly suffering and it should not be surprising that local people often attribute this to the act of breast feeding itself. For these reasons, promotion messages should address women as more than mothers, and explicitly link breastfeeding to long term goals such as education. Support to address social relations and self-perceptions may be as or more effective than technical training because mothers are, and need to feel, as important as their infants. Communities respond to the message that sophisticated but ordinary people breastfeed – not just especially poor or privileged women.

Finally, interventions should encourage stakeholding in the benefits of optimal breastfeeding at multiple levels. The often-remarkable involvement of men in the care of older children suggests they would become more involved in care of younger children if they could find ways to help. Ways may be found to involve male community and religious leaders as key participants in the breast feeding process. Where husbands are hostile or unsupportive (because they are insensitive to the benefits of breast feeding, they see conflicts with wage-earning, or they want wives and girl friends to return to work), messages about the economic value of breast feeding beyond the mother-infant dyad may be useful.

Conclusions

Studies focused on infant outcomes and the constraints on household food supply and hygiene usually fail to take into account the full suite of tradeoffs in time allocation, social agency and ideology that caretakers may face. Failure to set realistic behavioral goals for caretakers living under heavy constraints will limit the effectiveness of interventions and frustrate relations between policy makers and local people. Ethnographic information is crucial for the design and implementation of effective interventions to modify infant and young child feeding practices to improve maternal and child well being.

References

1. Mosley, W. and L. Chen. *An analytical framework for the study of child survival in developing countries*, in *Child survival: Strategies for research*, edited by W.H. Mosley and L.C. Chen, 1984) p. 25-45.
2. UNICEF. *The state of the world's children*, edited. (Oxford University Press, Oxford, 1997).
3. Paul, B. *Health, culture and community*. edited. (Russell Sage Foundation, New York, 1955).
4. Hahn, R. *Anthropology in Public Health: Bridging differences in culture and society*. edited. (Oxford University Press, New York, 1999).
5. Pelto, G., K. Dickin, and P. Engle. *A critical link: Interventions for physical growth and psychological development: A review*, edited. (Worl Health Organization, Geneva, 2000) 79.
6. ACC/SCN. *Breastfeeding and complementary feeding (Chapter 3)*, in *Fourth Report on the World Nutrition Situation*, edited. (United Nations Administrative Committee on Coordination Subcommittee on Nutrition in collaboration with International Food Policy Research Institute, Geneva, 2000) p. 33-41.
7. McGuire, J. and B. M. Popkin. *Beating the zero sum game: Women and nutrition in the third world*. United Nations: Administrative Committee on Coordination/Subcommitee on Nutrition: New York, 1990.
8. Maher, V. *Breast-Feeding in Cross-cultural Perspective: Paradoxes and Proposals*, in *The Anthropology of Breast-Feeding: Natural Law or Social Construct*, edited by V. Maher. (Berg Publishers Limited, Oxfrod University Press, Washington, D.C., 1992) p. 1-36.
9. Anon. Innocenti declaration on the protection, promotion and support of breast -feeding, Ecol. Food Nutr. 26:271-273 (1991).
10. WHO/UNICEF. *Protecting, promoting and supporting breast-feeding. The special role of maternity services. A joint WHO/UNICEF statement*, WHO/UNICEF: Geneva, (1989).
11. Haggerty, P. A. and S. O. Rutstein. *Breastfeeding and complementary infant feeding, and the postpartum effects of beastfeeding*. DHS Comparative Studies, edited by D.A.H. Surveys. Vol. 30. (Macro International, Inc, Calverton, MD, 1999) 282.
12. Tedstone, A., N. Dunce, M. Aviles, P. Shetty, and L. Daniels. *Effectiveness of interventions to promote healthy feeding in infants under one year of age: a review*, edited. (Health Education Authority, UK., London, 1998).
13. Obermeyer, C. M. and S. Castle. Back to nature? Historical and cross-cultural perspectives on barriers to optimal breastfeeding, Med. Anthropol. 17:39-63 (1997).
14. WHO, *Global data bank on breast-feeding*. World Health Organization Nutrition Unit (1996).
15. Sellen, D. W. Polygyny and child growth in a traditional pastoral society: the case of the Datoga of Tanzania, Human Nature: An Interdisciplinary Biosocial Journal. 10(4):329-371 (1999).
16. Sellen, D. W. Seasonal ecology and nutritional status of women and children in a Tanzanian pastoral community, Am. J. Hum. Biol. 12(6):758-781 (2000).
17. Sellen, D. W. Infant and young child feeding practices among African pastoralists: the Datoga of Tanzania, *J. Biosoc. Sci.* 30(4):481-499 (1998).
18. Victora, J. P. and E. All. Effects of bottle feeding on infant survival, Archives of Disease in Childhood 66:102-109 (1989).
19. Schepher-Hughes, N. Infant mortality and infant care: Cultural and economic constraints on nurturing in northest Brazil, Soc. Sci. Med. 19(5):535-546 (1984).
20. Schepher-Hughes, N. *Death without weeping: The violence of everyday life in Brazil*, edited. (University of California Press, Berkely, 1992).
21. Schepher-Hughes, N. Culture, scarcity and maternal thinking: Maternal detachment and infant survival in a Brazilian shantyton, Ethos. 13(4):291-317 (1985).
22. Samega-Janneh, I. J. *Breastfeeding: from biology to policy*. in *Challenges for the 21st century: A gender perspective on nutrition through the life cycle*. 1998. Oslo, Norway: United Nations.

23. Sellen, D. W. and A. Tedstone. Nutritional needs of refugee children in the UK, J. Roy. Soc. Med. 93:360-364 (2000).
24. Sellen, D. W. and A. Tedstone. *Assessing food security and nutritional well being of preschool refugee children in the United Kingdom*, in *Food in the migrant experience*, edited by A.J. Kershen and A. Penn. (Routledge, London, in press).
25. Sellen, D. W., A. Tedstone, and J. Frize. *Research Development: Young Refugee Children's Diets and Family Coping Strategies*. 2000, The Kings Fund: London. p. 15.

MATERNAL SMOKING HABITS AND FATTY ACID COMPOSITION OF HUMAN MILK

Carlo Agostoni, Francesca Grandi, Milena Bonvissuto, Anna M. Lammardo, Enrica Riva, Marcello Giovannini

Aim of the study, subjects and methods

To investigate the relationships between maternal smoking habits and milk **fatty acid (FA)** composition, 95 mothers who gave birth to healthy, full-term infants were recruited to estimate their milk FA content and composition through 12 months of lactation. A mother was defined as a regular smoker (S) if consuming ≥ 5 cigarettes per day before being aware to be pregnant.

Pooled hind milk was collected over 24 hours at basal time (colostrum), 1 mo, 3 mos, 6 mos, 9 mos and 12 mos. Total lipids (TL) were measured **by the** microgravimetric method while milk FA were analysed with high-resolution capillary gas-chromatography.

Maternal demographic and anthropometric data were drawn from obstetric clinical charts. Infants' anthropometric data at birth were assessed with standardized techniques. Maternal dietary habits through the previous 3 months were assessed through a validated food-frequency questionnaire at delivery, 3, 6, 9 and 12 months. Statistics: chi square and non-parametric tests (Man-Whitney U test for between-group comparisons) for the non-Gaussian distribution of the data.

Results

Three mothers could not be properly classified. Among the remaining 92 mothers, 61 were non-smokers (non-S) and 31 smokers (S). Mothers still breastfeeding progressively decreased to 63 (43 non-S, 20 S) at 1 mo, 50 (35 non-S, 15 S) at 3 mos, 30 (22 non-S, 8 S) at 6 months, 16 (11 non-S, 5 S) at 9 months, 10 (7 non-S, 3 S) at 12 months. S-mothers were heavier at **prepregnancy** body weight and showed a higher

*Carlo Agostoni, Francesca Grandi, Milena Bonvissuto, Anna M. Lammardo, Enrica Riva, Marcello Giovannini, S. Paolo Hospital, Clinica Pediatrica U, Via A. Di Rudini 8, Milano, Italy 20112

weight gain during pregnancy. Nevertheless food dietary habits through the last months of pregnancy were similar in S and non-S mothers. Also infants' anthropometrics at birth were similar among the two groups. Nutrient intakes were similar for the last trimester of pregnancy. At the following assessments, S mothers showed lower values of energy, with a significant difference at 3 months (P = 0.02). No significant differences of lipid and polyunsaturated FA % intakes between the two groups were found at any assessment. TL content of colostrum was similar (16 g/L) in the two groups. A great within-group variability in milk fat content and composition was found afterwards. Milk **TLs** were higher in non-S mothers' milk at 1 mo (P = 0.01) and S mothers' milk at 9 mos (P = 0.10) and 12 months (P = 0.03).

As far as PUFA percentage (%) levels (proportions), linoleic acid was higher in S at 12 months (P = 0.05), arachidonic acid and alpha-linolenic acid patterns were almost parallel in the two groups, while docosahexaenoic acid was characterized by a trend towards lower (though non-significant) levels in NS at 3 and 9-12 months.

As a consequence of the parallel changes in TL content and % FA levels, the S-mothers' milk content (mg/L) of linoleic acid and alpha-linolenic acid was lower at 1 month (P = 0.03 and P = 0.02, respectively) and higher at 9 mos (P = 0.02 and P = 0.004, respectively), while docosahexaenoic acid levels were lower at 1 (P = 0.04) and 3 months (P = 0.02) compared to non-S mothers' milk.

Conclusions

Milk fat content and composition may be affected by maternal smoking habits, even if just partially forwarded during pregnancy. This cause-effect relationship seems to be partly independent from differences in dietary habits, particularly during the last months of pregnancy. During lactation, lower levels of energy intakes could have resulted in higher oxidation rates of polyunsaturated fatty acids, thus less available for transfert into human milk.

Future studies should better detail the independent associations of milk composition from smoking mothers with infants' growth and developmental outcomes.

Smoking habits should be **discouraged** during pregnacy for the **untoward** effects on fetal and neonatal outcome. When smoking is forwarded, interventions aimed at increasing the milk fat content and DHA levels should be considered. Smoking mothers should be supported to maintain adequate energy and nutrient intakes.

MAJOR BREAST MILK CAROTENOIDS OF HEALTHY MOTHERS FROM NINE COUNTRIES

Louise M. Canfield, Manping Liu, William J Goldman, Kathryn Pramuk*

INTRODUCTION

Carotenoids are pigments widely distributed in fruits and vegetables which have known biological activities. Three major carotenoids in serum and milk (α-carotene, β-carotene and β-cryptoxanthin) have provitamin A activity and thus offer a potential source of vitamin A for breastfed infants.[a] Dietary carotenoids are potent antioxidants and shown to enhance immune function.[b]

Objectives

To measure five major carotenoids (lutein/ zeaxanthin, cryptoxanthin, lycopene, α-carotene and β-carotene) and retinol in mature breast milk of healthy mothers in Australia, Canada, Chile, China, Japan, Mexico, Philippines, United Kingdom and United States.

Methods

Women were recruited during well baby care visits to pediatrician offices. Mothers were between 18-40 years of age, 1 - 12 months post- partum, did not take vitamins containing beta carotene or more than 8000 IU vitamin A, and breast fed at least 5 times/day. Participants returned on the following day to provide a complete breast expression during mid-afternoon (1 to 5 pm) from a single breast. An electric breast pump was used with disposable tubing. Samples were frozen at -70°C and shipped to research laboratory on dry ice. Carotenoids were determined by HPLC analysis and lipids by creamatocrit as previously described.[a]

*L. Canfield and M. Lui, University of Arizona, Department of Biochemistry, W.J. Goldman and K. Pramuk, Wyeth Nutritionals, Philadelphia, PA

Results

TABLE 1. MAJOR BREAST MILK CAROTENOIDS* FROM MOTHERS IN NINE COUNTRIES : μmol/ L, means	
Japan	0.29
Mexico	0.22
China	0.20
Australia	0.18
Chile	0.17
Canada	0.16
United Kingdom	0.16
Philippines	0.13
United States	0.11

*(Combined lutein/zeaxanthin, β-cryptoxanthin, lycopene, α- carotene and β-carotene)

Concentrations of the five major carotenoids varied widely between countries: the 95% confidence interval was 0.1 to 0.13 for US samples and 0.26 to 0.32 μmol/L for Japanese samples. Japanese mothers had the highest concentrations of total carotenoids and all individual carotenoids with the exception of lycopene, which was highest in the samples from the United Kingdom. In our study breastmilk of the US (Arizona sample) mothers had the lowest mean values of total carotenoids among the nine countries: a difference in dietary lycopene content may be responsible.

Discussion

Our data illustrates the diversity of carotenoid concentrations and patterns in breast milk of world populations. Because breast milk concentrations may reflect the carotenoid composition of the local diet, the unique patterns, which we observed may result in diverse long term effects on chronic disease. These associations as well as the relationship of carotenoid composition of breast milk to health benefit should be further investigated.

References

a. Canfield, L.M., et al. AJCN 66:52-61 (1997).
b. Bendich, A., J. Nutr. 119:112-115 (1989).

MATERNAL MILK COMPOSITION IS NOT ASSOCIATED WITH CURRENT DIETARY INTAKE, BUT WITH BODY COMPOSITION AFTER THREE MONTHS POST-PARTUM IN TWO MEXICAN REGIONS

Caire G, Bolaños AV, De Regil LM, Casanueva E and Calderón de la Barca AM*

INTRODUCTION

Although no relationships have been found between maternal fatness or dietary intake and milk composition in developed countries,[1] data from poorer populations show a positive relationship between maternal fat stores and milk lipid levels.[2] Mexico, a country in transition, presents different socio-cultural data between northwest (NW) and southeast (SE) regions. There are no studies about lactation performance and maternal nutritional status in NW region, while in the central region it is well known. This study examines the relationship of maternal body composition or dietary intake and milk composition in women from NW and central Mexico during the first 3 months of lactation.

Methods

The study was performed in Central (C region) and Northwest Mexico (Mexico City and Hermosillo, Sonora). Through a screening, 56 mothers (15.5 to 35 years old) met the criteria for inclusion, but only 15 in the NW region and 18 in the C region breastfed exclusively for the first 3 months postpartum (PP). All of the women signed in the informed consent. At the first and third month, body weight (BW), height, physical activity, dietary intake and milk composition were measured. Both milk production and total body water (TBW) were measured by the deuterium dilution technique dosing the mother.[3] Fat free body mass (FFBM = TBW/0.73) and body fat (BF = BW-FFBM) were calculated. Data of the two periods were analyzed by location. Spearman's correlation was used to test the relationships between maternal characteristics and milk composition.

*Caire G, Bolaños AV, De Regil LM and Calderón de la Barca AM, Centro de Investigación en Alimentación y Desarrollo, AC, Hermosillo, Sonora, México. Esther Casanueva, Instituto Nal. de Perinatología, México, DF

Student's *t* tests were used to identify differences between region and period.

Results and Discussion

Milk density and lipid concentration were strongly correlated at 15 days ($r=0.92$, $p < 0.01$) and 90 days ($r=0.98$, $p < 0.01$) post-partum. At 90 days PP, maternal fatness was related to milk energy ($r=0.44$, $p=0.01$) and lipid concentration ($r=0.442$, $p=0.01$). This is consistent with the results of Brown et al.[4] in Bangladeshi women at 3 months PP. During the same period, age was related to percentage body fat and lipid concentration in milk. None of the selected maternal variables was able to explain any of the variability in milk volume, lactose and protein concentrations. Milk composition was not associated with current dietary intake during any of the two periods.

Table 1. Body fat and milk composition at time of lactation*

	1 mo		3 mo	
	Northwest (n=15)	Central (n=18)	Northwest (n=15)	Central (n=18)
Body fat (%)	36.1 ± 6.9^{a}	30.3 ± 6.9^{b}	31.6 ± 5.4^{b}	27.4 ± 7.9^{ab}
Energy (kcal/g)	0.65 ± 0.2^{a}	0.54 ± 0.1^{b}	0.71 ± 0.2^{a}	0.57 ± 0.1^{b}
Protein (mg/mL)	13.7 ± 2.9^{a}	12.7 ± 2.2^{a}	11.8 ± 2.8^{b}	9.3 ± 1.1^{c}
Lactose (mg/mL)	59.2 ± 3.6^{a}	58.4 ± 3.4^{a}	60.2 ± 4.6^{ab}	61.6 ± 5.8^{b}
Lipid (mg/dL)	43.4 ± 23^{a}	29.8 ± 12^{b}	51.0 ± 19^{c}	33.4 ± 11^{b}

*Mean ± S.D., different letter superscript across columns is significantly different from each other, P<0.05.

Acknowledgement

The study was supported by the International Atomic Energy Agency (9381).

References

1. B. Lönnerdal, Effects of maternal dietary intake on human milk composition, J. Nutr. 116:499-513 (1985).
2. S.F. Villalpando, N.F. Butte, W.W. Wong, S. Flores-Huerta, M.J. Hernandez-Beltran, E.O. Smith, and C. Garza, Lactation performance of rural Mesoamerindians, Eur. J. Clin. Nut. 46:337-348 (1992).
3. W.A. Coward, R.G. Whitehead, M.B. Sawyer, A.M. Prentice, and J. Evans, New method for measuring milk intakes in breast-fed babies, Lancet 2:13 (1979).
4. K.H. Brown, N.A. Akhtar, A.D. Robertson, and M.G. Ahmed, Lactational capacity of marginally nourished mothers: relationships between maternal nutritional status and quantity and proximate composition of milk, Pediatrics 78:909-919 (1986).

VITAMIN A IN BREASTMILK AND HIV INFECTION IN WEST AFRICA

Katia Castetbon, Marie-Josée Thomas, Liliane Dubourg, Crépin Montcho, Olivier Manigart, Philippe Msellati, Philippe Van de Perre, and François Dabis* for the DITRAME Study Group

INTRODUCTION

An association between impaired maternal vitamin A status during pregnancy and increased risk of mother-to-child transmission (MTCT) of HIV has been documented in observational studies.[1] However vitamin A supplementation during pregnancy has not shown any efficacy in reducing the overall MTCT risk.[2, 3] WHO recommends vitamin A supplementation in all women within the first six weeks after delivery, complementary to the child supplementation[4]. Limited information is available on vitamin A in breastmilk of West-African women and its association with the risk of MTCT. We evaluated such a relationship in Côte d'Ivoire and Burkina Faso. A secondary objective was to assess the association between milk and maternal serum retinol levels at day 45 post-partum.

Methods

We measured milk retinol levels in HIV-infected women included in two randomised placebo-controlled trials conducted in Abidjan (Côte d'Ivoire) and Bobo Dioulasso (Burkina Faso). The DITRAME ANRS 049a trial evaluated the efficacy of a short-regimen of zidovudine during the last month of pregnancy in reducing the risk of MTCT. In the persent study women receiving zidovudine after the end of the trial with the same drug regimen were also considered. The DITRAME ANRS 049b trial studied the tolerance of vaginal cleansing with benzalkonium chloride. Milk samples were collected in one breast at day 45 post-partum and analyzed for retinol levels by an adapted HPLC technique. Creamatocrit was performed immediatly after collection at the local laboratories. Milk retinol levels were expressed as µg per % of fat in the milk. We

*Katia Castetbon, Francois Dabis, INSERM U.330, Universite Victor Segalen Bordeaux 2, France. Marie-Josee Thomas, Liliane Dubourg, Laboratoire de Biochimie, Centre Hospitalier Universitaire de Bordeaux, France. Crepin Montcho, CeDResS, CHU de Treichville, Abidjan, Cote D'Ivoire. Olivier Manigart, Philippe Van de Perre, Centre Muraz, Bobo Dioulasso, Burkina Faso. Philippe Msellati, Centre IRD de Petit-Bassam, Abidjan

present here preliminary results given the current estimation of the timing of pediatric HIV infection. Linear regression was used to estimate the association between milk and serum retinol both measured at day 45 post-partum. The association between milk retinol and overall risk of MTCT was estimated by a case-control study design. Cases were mothers of HIV-infected children diagnosed by early ($\leq$ 6 months) positive PCR or positive serology after 15 months of life. Controls were mothers of infants identified as uninfected two months after weaning or at the end of follow-up. Comparisons of milk retinol between transmitting and non transmitting mothers were realised by ANOVA.

Results

A significant and positive linear relationship between milk and serum retinol was found in 79 women, with a slope of 2.06 (95% CI : 0.91 – 3.22) and a correlation coefficient R of 0.37 (p=0.007). Women for whom milk retinol and infant HIV status were available had been included in the 049a trial for 248 of them and in the 049b trial for 80 of them. Overall, 171 women received zidovudine, 118 placebo and 39 benzalkonium chloride. There was no difference in the mean milk retinol between transmitting mothers (3.07 µg/% of fat ± 2.4, N= 82) and non transmitting mothers (3.25 µg/% of fat ± 2.4, N= 246) (p= 0.55). Women receiving zidovudine during pregnancy had higher milk retinol (3.59 µg/% of fat ± 2.3) than women who did not (2.80 µg/% of fat ± 2.3) (p=0.002). No difference was observed between mothers whom child deceased after day 45 (2.96 µg/% of fat ± 2.4, N=73) and mothers whom child was alive at the end of follow-up (3.28 µg/% of fat ± 2.4, N= 255) (p=0.31).

Discussion

These preliminary results need confirmation when all samples will have been processed and timing of HIV acquisition confirmed. The distinction between in utero, intrapartum and post-natal transmission will be particularly informative. The significant linear relationship between milk and serum retinol was in agreement with a previous study. The lack of association between milk retinol and overall risk of MTCT of HIV is compatible with the lack of association previously observed in the same population between maternal serum retinol during pregnancy and risk of MTCT. We will examine the significance of the higher milk retinol in women who received zidovudine during pregnancy by adjusting with potential confounding factors.

References

1. R. Semba, P. Miotti, J. Chiphangwi, et al., Maternal vitamin A deficiency and mother-to-child transmission of HIV-1, Lancet 343:1593-1597 (1994).
2. A. Coutsoudis, K. Pillay, E. Spooner, L. Kuhn, and H. Coovadia, The South African Vitamin A Study Group. Randomized trial testing the effect of vitamin A supplementation on pregnancy outcomes and early mother-to-child HIV-1 transmission in Durban, South Africa. AIDS 13:1517-1524 (1999).
3. W. Fawzi, G. Msamanga, D. Hunter, et al., Randomized trial of vitamin supplements in relation to vertical transmission of HIV-1 in Tanzania, J. Acquir. Immune Defic. Syndr. 23:246-254 (2000).
4. World Health Organization, Integration of vitamin A supplementation with immunization, Wkly Epidemiol. Rec. 74:1-6 (1999).

MILK TRANSFER VOLUME AT THE TIME OF MATERNAL PERCEPTION OF THE ONSET OF LACTATION

Donna J. Chapman and Rafael Pérez-Escamilla*

INTRODUCTION

Our previous research indicates that maternal perception of the onset of lactation is a valid, public health marker of lactogenesis stage II.[1] Little is known, however, about milk transfer volume at the time of maternal perception. The objective of this research is to characterize milk transfer volume per feeding at the time of maternal perception.

Methods

Data from a previous study of lactogenesis stage II following Cesarean delivery in a USA hospital were used in these analyses.[2] Three times daily, subjects were interviewed about their perception of the onset of lactation. When subjects reported that their milk had "come in", they were asked to state the time that this happened to the nearest hour. Infants were test weighed three times daily, after breastfeeding, using an electronic integrating balance (Sartorious, BP34). Milk transfer values were adjusted for insensible water loss and specific gravity.

Results

Forty-four subjects had test weight data completed within 6 hours of the time of maternal perception. Among these subjects, the mean milk transfer volume at the feeding closest to maternal perception was 19.4 ± 12.3 ml. There was no significant difference in milk transfer volume per feeding at the time of maternal perception between those who perceived the onset of lactation to be early (< 72 hours pp, 18.7 ± 10.1 ml/feeding,) or late (≥ 72 hours pp; 20.4 ± 15.2 ml/feeding).

* Department of Nutritional Sciences, University of Connecticut, Storrs, CT 06269-4017, djc@discovernet.net; rperez@canr.uconn.edu

Conclusions

The consistency in the amount of milk transferred at the time of maternal perception suggests that this event is strongly linked to the physiology of lactogenesis stage II, the ability of the infant to remove milk, or to a combination of these factors.

References

1. D.J. Chapman and R. Pérez-Escamilla, Maternal perception of the onset of lactation is a valid, public health indicator of lactogenesis stage II, J. Nutr. In press.
2. D.J. Chapman, The impact of breast pumping on the onset of lactogenesis stage II following Cesarean delivery: A randomized clinical trial. Doctoral dissertation. University of Connecticut, Storrs, CT, 1999.

INTRA-INDIVIDUAL CHANGES OF FATTY ACIDS IN BREAST MILK DUE TO THE SUCKLING STIMULUS

Teresa H.M. Da Costa & Marina K. Ito*

INTRODUCTION

Fore and hind milk vary in total fat content.[1] The suckling stimulus is known to increase the lipid content of milk within a feeding. We investigated the influence of the suckling stimulus in determining the composition of fatty acids in human milk, comparing the right and left breasts from lactation day 15 to 90, according to the suckling pattern established by the infant.

Methodology

Milk samples, from an exclusively breastfeeding mother, were manually collected before (fore) and after (hind) a feeding from both breasts in the morning (7:00 to 9:00) and afternoon (17:00 to 19:00) on days 15, 20, 25, 30, 45, 60, 75, and 90 of lactation. During the period of milk collection, the mother maintained a typical Brazilian diet and changed to a high intake of *n-3* fatty acid diet before and during lactation day 75. Total milk fat content (g/dl) was determined by the crematocricand the fatty acid composition by gas chromatography. The mother was 38 years old, with a BMI of 19.4 Kg/m^2 on lactation day 15. The female baby was born by normal delivery, weighing 2770g at birth. During the lactation follow-up period we asked the mother to use a pre-established form to maintin daily records of the time of beginning and end of each suckling session from each breast. A new feeding session was considered if separated by more than 30 minutes from the previous feed. The mother was asked to follow the pattern of suckling dictated by the infant, offering one or both breasts. Fore and hind milk were collected from both breasts independent of the presence of the sucking stimulus.

*Teresa H.M. DaCosta, University of Brasilia, C.P. 04511, CEP: 70919-970, Brasilia – DF, Brazil. Marina K. Ito, University of Brasilia, Department of Nutrition, CEP: 70910-900, Brasilia – DF, Brazil.

Results and discussion

Suckling time progressively decreased from about 11 minutes until lactation day 30 to 6 minutes thereafter. Mean (SD) fat content (g/dl) in the suckled breasts were 3.17 (1.74) and 1.29 (0.06) in the non suckled breasts. There was an inverse change in percent composition of medium chain (MCFA) and polyunsaturated fatty acids (PUFA). MCFA decreased from 23.14% on lactation day 15 to 12.97% on lactation day 90 while PUFA increased from 13.05% to 22.86%, respectively.

On days 15 and 20 only the right breast, and on day 30 and 60 only the left breast were suckled by the infant. When suckled and unsuckled breasts were compared there was a tendency toward a decrease of MCFA and an increase in monounsaturated and PUFA.

Figure 1 shows the content of EPA and DHA during the studied period. There was a decrease or no transfer at all of these fatty acids when the suckling stimulus was not present. These modifications indicate a local control of the suckling stimulus for the transfer of long chain unsaturated fatty acids. The addition of foods rich in *n-3* fatty acids in the mother's diet markedly increased the EPA and DHA content in milk (Figure 1).

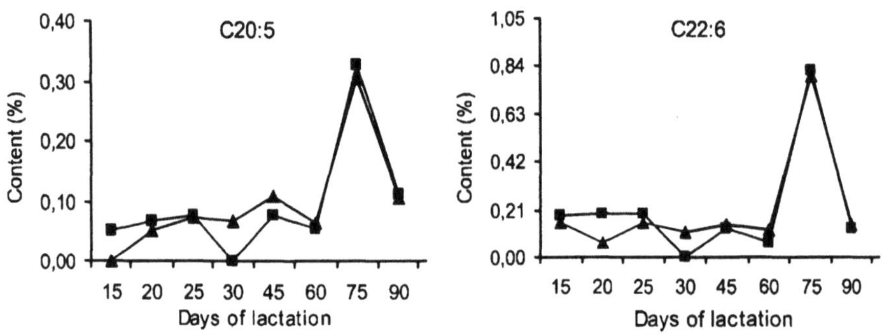

Figure 1 - Content (%) of eicosapentaenoic (EPA, 20:5) and docosahexaenoic (DHA, 22:6) acids during lactation on the right (_*_) and left (_*_) breasts. On day 15 and 20 only the left breast was suckled and on days 30 and 60 only the right breast was suckled. On day 75 the mother changed to a *n-3* fatty acid rich diet.

Conclusion

The increase in the total fat content was proportional to the time the breast was suckled. The results indicate that the suckling stimulus is not only a powerful local control in determining the quantity of milk fat transfer to the infant but may also be important in determining the quality of milk fat.

Funded in part by: CNPq grant n. 552985/96-3 and FAP-DF grant n. 193.000.119/96

Reference

1. R.G. Jensen, The lipids in human milk, Prog. Lipid Res. 35(1):53-92 (1996).

TRAINING ON BREASTFEEDING IN ITALY

A study on acceptance from participants

R. Davanzo, C. Pavan on behalf of The Breastfeeding Working Group

INTRODUCTION

The published information gives an inaccurate picture of the prevalence and duration of breastfeeding in Italy, leading to unjustified optimism and inaction.[1] Italy lacks of a national plan of the Ministry of Health or by UNICEF with objectives and targets to promote breast-feeding. In this context, the Breastfeeding Working Group (BFWG) of the Istituto per l'Infanzia of Trieste, which is a WHO Collaborating Centre for Mother and Child Health, acts on a national basis to promote and protect breastfeeding, trying to cover the existing educational gap on breastfeeding in Italy.

There is some evidence that the implementation of the 10 Steps to Successfull Breastfeeding of the Baby Friendly Hospital Initiative (BFHI) could increase the breastfeeding rates.[2] In a controlled, unpublished study we evaluated the effectiveness of training on hospital practices, degree of application of the The Ten Steps and breastfeeding rates at discharge, three and six months of age.[3] The impact of such a training program could be influenced by the extent to which participants evaluate and accept the contents of a course. In the present study we explore the acceptance of a 3 days' course on breastfeeding among Italian health workers.

Methodology

The BFWG includes 28 members from different italian institutions like Research Institutes, Universities, National Health System Hospitals, Local Health Authorities and from Private Practice. The Group is multidisciplinary, gathering nurses, midwives, lactation consultants, physicians, Masters in Public Health and Masters in Education. So far, the main activities of the BFWG have been research, training, scientific dissemination and advocacy.

*R. Davanzo, Department of Neonatology and Unit for Health Services Research and International Cooperation, Istituto per l'Infanzia, Via dell'Istria 65/1, 34100 Trieste, Italy (E-mail: davanzor @ burlo. trieste.it)

According to our actual program of work, training on breastfeeding for health workers of the Italian National Health System is a priority. From January 1996 to December 1999 we conducted thirty 18 hours courses to which 750 health workers involved in mother and child care took part. We modified the original 18 hours course on breastfeeding from UNICEF: we felt more functional to change the order of the Sessions suggested by UNICEF, putting physiology of lactation and practical problems before theoretical topics such as benefits of breastfeeding, the presentation of the WHO/UNICEF 10 steps for successfull breastfeeding and general recommendations from WHO. Moreover, we integrated a 2-hour session from the WHO's 40-hour course on counselling and another introductory session. The latter gives the opportunity to discuss with participants on the conflict between bottle and breast culture (Table 1).

We had perceived that such an approach could improve the acceptance from participants of the content and the values of the course itself.

The final product included 17 classroom sessions with up-to-date references to scientific literature and 2-hours clinical practices (Table 2).

At the end of each of the last sixteen courses, we (three teachers for each course) explored the opinions of participants on the site of the course with a questionnaire.

Table 1. Confrontation between the bottle culture and the breastfeeding culture

Bottle culture	Breastfeeding culture
Normalization	Flexibility Respect of the physiology Respect of the baby
Separation	Rooming-in; Bedding-in
Autonomy of the baby	Intimate closed relationship
Low value of house work	The care of the baby as a value
Consumerism	Breast milk as a nutritional liquid biologically And ecologically correct
The medical culture of intervening	Support to mother; Being able to refrain
Breast as a sexual object	Acceptance of the own body

Results

The percentage of returned questionnaires was 83% (331 out of 400).

The issue of BF was judged essential from 36% of responders and important from another 62.8%; 81.8% of the participants recognized a great practical usefullness, even if only 2 hours were devoted to clinical practice. Three days were considered an adequate duration for a course on BF by 65% of health workers. Teaching methods were appropriate for the great majority of participants (97.3%) and teaching materials (especially audiovisuals) were appreciated by 76.9% of responders.

Table 2. The programme of the breastfeeding course run by the BFWG of the Istituto per l'Infanzia of Trieste, Italy

1. Introductory session
2. Anatomy and physiology
3. Evaluation of a breastfeed
4. Physical problems of the breast
5. Epidemiology
6. WHO's Recommendation
7. The 10 Steps of the BFHI
8. Low milk intake
9. Breastfeeding History
10. Communication Skills
11. Breast Milk: composition and benefits
12. Babies with special needs
13. The breastfed baby crying
14. Refuse to breastfeed
15. Breast milk expression and storage
16. Advise in pregnancy
17. Support in the Community and the International Code of Marketing for Breast-milk subtitutes
18. Clinical Practice

Table 3. Opinions of participants on the characteristics of the course

Characteristic of the course	Judgement	Percentage
Theme	Essential	36 %
	Important	62.8 %
	Collateral	1.2 %
	Total	100 % (N: 261)
Practical usefullness	Great	81.8 %
	Modest	15.9 %
	Small	2.3 %
	Total	100 % (N: 265)
Duration	Too long	19.1 %
	Adequate	65 %
	Too short	15.9 %
	Total	100 % (N: 257)
Teaching methods	Appropriate	97.3 %
	Don't know	2.7 %
	Inappropriate	-
	Total	100 % (N: 263)
Teaching materials	Good	76.9 %
	Sufficient	23.1 %
	Bad	-
	Total	100 % (N: 264)

Conclusions

This analysis simply shows the good acceptance of a 3 days course on breastfeeding among Italian health workers. This is a substantial pre-requisite for the success of any training programme.

Of course, the effectiveness of training on breastfeeding should be witnessed by change of hospital practices, degree of application of the 10 Steps and increase of breastfeeding rates; this seems the case for Italy according to our recent unpublished data.[3]

References

1. A. Cattaneo, R. Davanzo, and L. Ronfani, Are data on the prevalence and duration of breastfeeding in Italy reliable? The case of Italy. Acta Paediatr. 89:88-93 (2000).
2. WHO. Evidence for the ten steps to successuful breastfeeding. Geneva: WHO (1996).
3. A. Cattaneo, and R. Buzzetti, On behalf of the Breastfeeding Research and Training Working Group. The effect of training on breastfeeding: a controlled study (unpublished).

BODY COMPOSITION OF LACTATING MOTHERS AT THE FIRST MONTH POST-PARTUM DEPENDS ON POST-GESTATIONAL BODY WEIGHT

[a]De Regil LM, [a]Bolaños AV, [a]Caire G, [b]Casanueva E and [a]Calderón de la Barca AM

INTRODUCTION

Lactation is one of the most energy demanding phases of the human reproductive cycle; therefore, maternal nutritional status can be affected by duration of breastfeeding. Studies on lactation have increased since the Baby-Friendly Hospital Initiative (BFHI) program was introduced. However, attention has been focused on the effect of lactation performance on the growth of the breast-fed infants with little investigation about the effect of breastfeeding on maternal nutritional status. This study analyzes the effect of lactation on maternal body composition in relation with anthropometrical, physiological and life style factors, in adolescent and adult women in two Mexican settings.

Methods

The study was performed in Mexico City and Hermosillo, Sonora. Subjects (27 adolescents and 29 adults breastfeeding exclusively for one month postpartum) were dosed with deuterium oxide (D_2O). Saliva was collected from mothers and babies. D_2O concentration was determined by infrared spectroscopy and maternal body water and milk production were estimated.[1] Milk composition (protein, fat and lactose) was determined. Height, pre- and post- gestational weights were taken. Dietary intake (24-h recall) and physical activity (recorded over 7 days) were estimated. Total energy expenditure was calculated using theoretical basal metabolic rate,[2] physical activity level (PAL) and milk energy output. Spearman coefficients were used to look for relationships between maternal body composition and the study variables. Two-sample student's *t* tests were used to identify differences ($p=0.05$) between *maternal maturity, and regions.[3]

*De Regil LM, Bolaños AV, Caire G and Calderón de la Barca AM, Centro de Investigación en Alimentación y Desarrollo, AC, Hermosillo, Son., México. Esther Casanueva, Instituto Nal. de Perinatología, México, DF

Results and Discussion

Mean body fat (33.2%) was related to pre- and post-gestational weight ($r=0.346$, $r=0.436$; $p < 0.01$, respectively There was no significant influence of current dietary intake, physical activity, milk quality and quantity on body composition. Maternal characteristics and lactation performance did not differ significantly between adolescents and adults. Region was a significant predictor of maternal characteristics and milk composition. In Hermosillo, the women were significantly taller, heavier (pre- and post-gestation), and their energy intake and milk lipids were higher than those from Mexico City. These results suggest that the setting in which a woman lives has an indirect influence on what lactation problems she faces, and consequently, how they are solved.

Table 1. Maternal characteristics at first month post-partum*

	Total sample n = 56	Hermosillo n = 34	Mexico City n=22
Antropometrical factors			
Height (m)	1.56 ± 0.05	1.58 ± 0.05 [a]	1.54 ± 0.05 [b]
Weight(kg)	59.2 ± 9.9	62.2 ± 9.6 [a]	54.7 ± 8.9 [b]
Pregestational weight (kg)	56.5 ± 6.46	56.5 ± 6.6 [a]	51 ± 7.4 [b]
Body fat (%)	33.2 ± 7.9	35.6 ± 7.6 [a]	29.1 ± 6.6 [b]
Life style factors			
Energy intake (kcal/d)	2078 ± 575	2261 ± 518 [a]	1916 ± 662 [b]
PAL	1.48 ± 0.14	1.47 ± 0.14 [a]	1.5 ± 0.15 [a]
Physiological factors			
Total energy expenditure (kcal/d)	2258 ± 751	2403 ± 614 [a]	2067 ± 898 [a]
Milk output (g/d)	577 ± 265	570 ± 243 [a]	587 ± 300 [a]
Energy (kcal/dL)	66 ± 16.4	71.4 ± 18.7 [a]	59 ± 9.45 [b]
Proteins (g/dL)	1.29 ± 0.22	1.31 ± 0.25 [a]	1.27 ± 0.19 [a]
Lactose (g/dL)	5.59 ± 1.13	5.8 ± 0.4 [a]	5.8 ± 0.32 [a]
Lipids (g/dL)	3.88 ± 1.72	4.45 ± 1.86 [a]	3 ± 1.04 [b]

*Mean ± SD, different letter superscript between the last two columns means significant differences, $P<0.05$.

Acknowledgement

The study was partially supported by the International Atomic Energy Agency (9381).

References

1. Calderón de la Barca AM, Bolaños AV, Caire Juvera G, Román Pérez R, Valencia ME, Casanueva E y, and Coward WA, Evaluación del consumo de leche humana por dilución con deuterio y detección por espectroscopia de infrarrojo, Perinatología y Reproducción Humana. 12 (13):142-150 (1998).
2. FAO/WHO/UNU EXPERT Consultation, Energy and Protein Requirements. World Health Organization, Geneva (1985).
3. SPSS. SPSSX user's guide. New York: McGraw-Hill (1983).

MILK-BORNE EGF AND NECROTIZING ENTEROCOLITIS IN NEONATAL RAT MODEL

Bohuslav Dvorak*, Debra L. McWilliam, Catherine S. Williams, Jessica A. Dominguez, Hana Holubec, Claire M. Payne

INTRODUCTION

Neonatal necrotizing enterocolitis (NEC) is the most common gastrointestinal (GI) disease of premature infants with excessive morbidity and mortality that afflicts 3,000 to 4,000 babies in the United States each year.[1] Many factors contribute to the development of NEC, mainly prematurity, enteral feeding, infectious agents and/or intestinal hypoxia-ischemia. Enteral feeding is nearly always a prerequisite for the development of NEC, but the exact mechanism of NEC pathogenesis is poorly understood. The protective role of maternal milk in NEC pathogenesis has been reported.[2] Various components of milk have been tested to protect the gut against NEC.[3] Epidermal growth factor (EGF) is a promising candidate for the treatment of NEC. Mammalian milk of many species contains high concentrations of EGF. Moreover, maternal milk is the major source of EGF for neonates during the suckling period.[4] In contrast, EGF is absent in all commercial infant formulas. Strong effects of exogenous EGF on healing of damaged gastrointestinal mucosa or on intestinal adaptation after injury are reported in a number of studies[5]. The aim of this study was to examine the effects of milk-borne EGF on the development of NEC in a neonatal rat model.

Methods

Newborn rats (Sprague-Dawley) were collected from their mothers immediately after birth to prevent suckling of maternal milk. Animals were fed with one of the following two diets: growth factor-free rat milk substitute (RMS)[4] or RMS supplemented with 500 ng/ml of rat EGF (RMS+EGF). For the first 48 hours, artificial rearing (AR) of neonatal rats was performed using a hand-feeding technique.[5] Rat pups were hand-fed 0.1 ml of diet every 3-4 hours using a silicone rubber tube. After 48 hours, the hand-feeding

*B. Dvorak, University of Arizona, 1501 N. Campbell Ave., Tucson, AZ, USA, 85724

method was replaced with mechanized artificial feeding[6]. In order to develop clinical and pathological signs similar to neonatal NEC, rats were stressed twice daily with asphyxia (100% nitrogen gas, 60 sec) and then with cold stress (4°C, 10 min). Animals that developed abdominal distention, respiratory distress and lethargy before the designated end of the experiment (96 hrs) were decapitated. At 96 hours, all surviving animals were decapitated and tissues collected. The small intestine and colon were first visually evaluated for typical signs of NEC such as discoloration of the small intestine and colon, intestinal hemorrhage, ileal distention, and stenosis. 1-2 cm section of the distal ileum and the entire colon were fixed in 70% ethanol, embedded in paraffin, sectioned, stained with H&E, and evaluated for histopathologic changes.

Results

Body weight of asphyxia/cold stressed neonatal rats in the RMS group gradually decreased during the four day experiment. In contrast, asphyxia/cold stressed rats fed with RMS+EGF diet were able to maintain the same body weight during the entire duration of experiment. Test for blood in strool was performed 72 hours after the beginning of the experiment. The RMS group exhibited a significantly stronger signal ($p<0.0001$) for the presence of blood in stools. Gross visual and histological evaluations of ileum have shown significantly higher incidence of NEC in RMS group compared to RMS+EGF group.

Conclusions

Milk-borne EGF improved the survival rate and markedly reduced the incidence of NEC in the rat model. Supplementation of RMS with EGF significantly improved body weight gain and reduced blood content in stool compared to rats fed only with RMS. These results clearly indicate a protective role of milk-borne EGF on the development of neonatal NEC.

Acknowledgement

This work was supported in part by The University of Arizona VP for Research and Graduate Studies Award (B.D.), The University of Arizona Undergraduate Biology Research Program (J.A.D.), and Program Project NIH HD 26013 (B.D.).

References

1. A. M. Kosloske, Epidemiology of necrotizing enterocolitis, Acta Paediatr. 396:2-7 (1994).
2. A. Lucas and T.J. Cole, Breast milk and neonatal necrotising enterocolitis, Lancet 336:1519-1523 (1990).
3. M. S. Caplan, and W. MacKendrick, Necrotizing enterocolitis: a review of pathogenetic mechanisms and implications for prevention, Pediatr. Pathol.13:357-369 (1993).
4. B. Dvorak, C.S. Williams, D.L. McWilliam, H. Shinohara, J.A. Dominguez, R.S. McCuskey, A.F. Philipps, and O. Koldovsky, Milk-borne epidermal growth factor modulates intestinal transforming growth factor-levels in neonatal rats, Pediatr. Res. 47:1-7 (2000).
5. M.K. Jones, M. Tomikawa, B. Mohajer, and A.S. Tarnawski, Gastrointestinal mucosal regeneration: role of growth factors, Front. Biosci. 4:D303-309 (1999).
6. B. Dvorak, and R. Stepankova, Effects of dietary essential fatty acid deficiency on the development of rat thymus and immune system, Prostaglandins Leukotrienes and Essential Fatty Acids 46:183-190 (1992).
7. B. Dvorak, D.L. McWilliam, C.S. Williams, J.A. Dominguez, N.W. Machen, R.S. McCuskey, and A.F. Philipps, Artificial Formula Induces Precocious Maturation of the Small Intestine of Artificially Reared Suckling Rats, J. Pediatr. Gastroenterol. Nutr. 31:162-169, (2000).

CONCENTRATIONS OF INTERLEUKIN-10 IN PRETERM MILK

Fituch CF[1], Palkowetz KH[2], Hurst N[1], Goldman AS[2], Schanler RJ[1]

Despite the protective effects of human milk against necrotizing enterocolitis (NEC), the incidence of NEC is highest in the extremely low birthweight (ELBW) infant, and only minimally decreased with the feeding of human milk. These observations raise a concern that concentrations of bioactive factors may differ in the milk obtained from mothers delivering ELBW infants compared with more mature infants. One such factor that may be affected is interleukin-10 (IL-10), a cytokine that downregulates inflammation and inhibits the production of proinflammatory cytokines. A deficiency of IL-10 in mice causes growth retardation, anemia, and chronic enterocolitis. IL-10 has been found in human milk samples from mothers of full-term infants. We hypothesized that low and/or varying concentrations of IL-10 in preterm human milk might contribute to the development of NEC in the ELBW infant. We compared the concentrations of IL-10 in milk samples obtained from mothers of infants in three groups: 1) 23-27 weeks gestational age (GA), 2) 32-36 weeks GA, and 3) full-term infants 38-42 weeks GA. Milk samples were collected fresh during the first and second weeks postpartum and every two weeks thereafter. IL-10 was quantified using an ELISA sandwich method.

Results: IL-10 was present in milk from 5 of 7 mothers of infants 23-27 weeks GA, all 5 mothers of infants 32-36 weeks, and all 5 mothers of full-term infants. There was a trend toward decreasing IL-10 concentrations with advancing postnatal age. Concentrations of IL-10 tracked through lactation; most mothers with detectable IL-10 initially continued to have detectable milk levels through 2 months postpartum. These preliminary results suggest that IL-10 is present in preterm milk and that milk IL-10 concentrations from mothers of ELBW infants has more variability than milk obtained from mothers of more mature infants. The relationship between IL-10 and other bioactive factors in human milk and the development of NEC warrant further exploration.

[1]Baylor College of Medicine, Houston, TX, [2]University of Texas Medical Branch, Galveston, TX

LONGITUDINAL CHANGES IN MILK SODIUM/POTASSIUM RATIO IN WOMEN WITH SERIOUS INFECTION IN THE POSTPARTUM PERIOD

Jennifer C. Georgeson, Yusuf Ahmed, Suzanne M. Filteau, Andrew M. Tomkins*

INTRODUCTION

Subclinical mastitis, as defined by raised milk sodium/potassium (Na/K) ratio (>0.6), is common among several populations and is associated with poor infant growth (Filteau *et al.* in press) and, in HIV-infected women with high milk HIV viral load, a risk factor for mother to child transmission.[1, 2]

Previous research suggests that subclinical mastitis is likely to be due to a number of causes including poor lactation practice, micronutrient malnutrition, local infections and systemic infection.[3] In this longitudinal pilot study we investigated whether high milk Na/K ratios were prevalent among women admitted to hospital with severe postpartum infections and whether ratios decreased on antibiotic and other appropriate treatment.

Methods

Spot breastmilk samples were collected from each breast of 22 lactating Zambian women suffering from systemic infection in the postpartum period and 4 lactating women with breast abscesses. Samples were taken daily whilst admitted to the University Teaching Hospital, Lusaka, Zambia and analyzed for Na/K ratio by flame photometry.

Results

Systemic infection caused a bilateral rise in breastmilk Na/K, whereas localized infection caused a unilateral rise. The correlation coefficient between left and right

*J.C.Georgeson, S.M.Filteau, A.M.Tomkins, Centre for International Child Health, Institute of Child Health, London, UK. Y.Ahmed, University Teaching Hospital, Lusaka, Zambia

breasts in patients with systemic infection was 0.87 ($P < 0.001$), whereas the correlation coefficient between left and right breasts of patients with breast abscesses was 0.17 ($P= 0.45$). On treatment of the systemic and localized infection the levels of breastmilk Na/K decreased rapidly, as can be seen in the table below.

	Admission: Geometric mean Na/K (95%CI) n	Discharge: geometric mean Na/K (95%CI) n
Puerperal sepsis	0.8 (0.4-1.5) n=12	0.4 (0.2-0.7) n=12
Malaria	2.1 (0.5-9.6) n=5	0.6 (0.4-2.1) n=5
Miscellaneous	0.5 (0.3-0.8) n=5	0.4 (0.2-0.8) n=5
Breast abscesses*	12.4 (8.16-18.92) n=3	0.17, 16.25 n=2

*Abscessed breast only. The Na/K values for breast abscesses showed large variability therefore individual values and not 95% confidence intervals have been listed.
NB/ Na/K ratio, normal range: 0.1- 0.6.

Discussion

Postpartum systemic infections in lactating Zambian women causes a large bilateral rise in breastmilk Na/K ratio, which rapidly decreases back to normal on treatment of the infection. However, further work is needed to determine whether HIV viral load also increases with a rise in Na/K ratio during postpartum systemic infection and whether it will decrease on treatment.

References

1. L.A. Guay, D.L. Hom, F. Mmiro, E.M. Piwowar, S. Kabengera, J. Parsons, C.N. Ndugwa, L. Marum, K. Olness, P. Kataaha, and J.B. Jackson , Detection of human immunodeficiency virus type 1 (HIV-1) DNA and p24 antigen in breastmilk of HIV infected Ugandan women and vertical transmission, Pediatrics 98:438-444 (1996).
2. P. Van de Perre, A. Simonon, D-G. Hitimana, F. Dabis, P. Msellati, B. Mukamabano, J-B. Butera, C. Van Goethem, E. Karita, and P. Lepage, Infective and anti-infective properties of breastmilk from HIV-1-infected women, Lancet 341:914-918 (1993).
3. J.F. Willumsen, F.M. Filteau, A. Coutsoudis, K.E. Uebel, M-L. Newell, and A.M. Tomkins, Subclinical mastitis as a risk factor for mother-infant HIV transmission, in: *Short and long term effects of breastfeeding*, B.Koletzko, ed., Kluwer Academic/Plenum Publishers 478:211-224 (2000).

INCREASED BREASTFEEDING RATES IN ITALY

Marcello Giovannini, Giuseppe Banderali, Carlo Agostoni, Silvia Scaglioni, Marco Silano, and Enrica Riva

Aim of the study, subjects and methods

The aim of this ongoing longitudinal study was to evaluate whether any change in the breastfeeding rate has occurred in Italy in the last five-year period and to assess the implememntation in Italy of the WHO "ten steps" to sucessful breastfeeding. Two cohorts of 2192 and 3249 mother-infant pairs were recruited at random from all the healthy singleton term newborns recorded in the National Birth Register during November 1995 and November 1999, respectively. Survey planned interviews **with** the mothers were **conducted** with a telephone questionnaire at 1, 3, 6, 9 and 12 months after delivery. Well-trained personnel carried out the interviews. The type of breastfeeding was classified according to the WHO definitions (WHO, Geneva, 1996).

Results

The participation rate was 73.0% in 1995 and 75.4% in 1999. The rate of initiation of breastfeeding was higher in 1999 (89.0%, 95% confidence interval, CI, 87.8% to 90.2%) than in 1995 (85.3%, 95% CI, 83.6% to 87.0%) ($p<0.0001$). The rate of full (exclusive or predominant) breastfeeding was also higher in 1999 (78.0%, 95% CI, 76.4% to 79.6%) than in 1995 (71.8%, 95% CI, 69.6% to 74.0%) ($p<0.0001$). Three months after delivery the rate of breastfeeding was 41.8% (95% CI, 39.4% to 42.2%) in the cohort 1995 and 62.3% (95% CI, 60.4% to 64.2%) in the cohort 1999 ($p<0.0001$). The corresponding rates of full breastfeeding were 37.3% (95% CI, 34.9% to 39.7%) and 44.6% (95% CI, 42.6% to 46.6%), respectively ($p<0.0001$). As far as the changes in the

*Marcello Giovannini, Giuseppe Banderali, Carlo Agostoni, Silvia Scaglioni, Marco Silano, and Enrica Riva, S. Paolo Hospital, Pediatrics Dept. 5, Via A Di Rudini 8, Milano, Italy 20122

Table. Percentage of implementation of the WHO "ten steps" to successful breastfeeding in Italy

STEP	1995%	1999%	Change 1999-1995	P*
1	Not evaluated	Not evaluated	Not evaluable	-----
2	Not evaluated	Not evaluated	Not evaluable	-----
3	63.9	67.9	+ 4.0	< 0.01
4	12.4	21.3	+ 8.9	< 0.0001
5	51.4	55.7	+ 4.3	< 0.01
6	33.0	42.9	+ 9.9	< 0.0001
7	23.7	33.6	+ 9.9	< 0.0001
8	44.3	52.4	+ 8.1	< 0.0001
9	47.7	66.4	+18.7	< 0.0001
10	Not evaluated	Not evaluated	Not evaluable	-----

* chi-square test

WHO "ten steps" to successful breastfeeding (WHO, Geneva, 1998) at all the evaluable steps increased rates were found (Table).

Conclusion

The present findings show a significant increase of the breastfeeding rate in Italy during the last five years. In particular a major increase has been found as far as the breastfeeding rates at 3 months, together with an improvement in implementation of the WHO "ten steps" to successful breastfeeding. The next data, available at December 2000, might **confirm** the increasing trends also of the total duration of breastfeeding. Campaigns aimed at further improving nursing should be promoted especially regarding the WHO recommendations.

A BREASTFEEDING SUPPORT STRATEGY IN AN HIV ENDEMIC AREA

Goga AE[1,2], Bland R[1], Van Rooyen H[3], Rollins NC[1,2] and Coovadia HM[1,2]

INTRODUCTION

By the end of 1998 UNAIDS estimated that of the 33 million persons worldwide infected with HIV, 95% were living in developing countries. Reducing mother to child transmission of HIV (MTCT) in such countries is a public health priority. The estimated risk of postnatal MTCT (through breastfeeding) in infants breastfed until 15 months, is 16%. However, most MTCT studies have not distinguished between exclusive breastfeeding and mixed feeding. Coutsoudis et.al[1] observed that exclusive breastfeeding (EBF) may carry a lower risk of MTCT than mixed feeding (MF) (involving 2500 mother-baby pairs) will be conducted in a rural area of KwaZulu Natal, South Africa, to determine the impact of a breastfeeding counseling and support strategy that promotes exclusive breastfeeding, on mother to child transmission of HIV. This abstract reports our experiences during the early stages of the development of this breastfeeding counseling and support strategy.

Objectives

The primary objective was to develop a community-based breastfeeding counseling and support strategy that increases breastfeeding rates in the population. The secondary objectives were to determine the short and long-term impact of training on the knowledge, attitudes and skills of community-based breastfeeeding counselors, and to develop guidelines that facilitate consistent high quality breastfeeding counseling.

Methods

Fifteen potential field workers (with no previous health training) attended a selection course which tested ability to assimilate and recall information, work as part of a team, problem solve and counsel. Eight women were selected and trained as

[1]Africa Centre for Population Studies and Reproductive Health, Mtubatuba, South Africa
[2]Department of Paediatrics and Child Health, Nelson R Mandela Medical School, University of Natal, South Africa
[3]School of Psychology, University of Natal, Pietermaritzburg, South Africa

breastfeeding counsellors (BC). Training comprised a 2-day workshop on research and community and household entry; a modified version of the WHO/UNICEF Breastfeeding Counselling: A Training Course; a 2-day workshop on counselling skills; and home visits. A Field Guide based on the WHO/UNICEF Breastfeeding Counselling: A Training Course was developed as a training tool and a reference document in the field. In an area with high HIV prevalence, the training package and Field Guide had to include up to date information on breast conditions and feeding during episodes of mastitis or bleeding nipples. Furthermore mechanisms for rapid testing of the infants HIV status so that appropriate feeding advice could be given at four to six months had to be developed, and information on complementary feeding included in the guide.

The performance - knowledge, attitudes (KA) and counselling skills (CS) of BC were assessed pre-course (Pr), immediately post-course (P_1) and 2 months post-course (P_2).

Results

Of 8 BC, 4 had university degrees, and 4 were school leavers. Mean pre- and post-course scores were: 52% (Pr - KA); 80% (P_1 - KA); 80% (P_2 - KA); 70% (P_2 -CS). Course scores did not correlate with academic qualification. After initial training, BC "intervened" rather than "counselled". However, ongoing supervision showed that community-based BC could acquire powerful counselling skills. The main difficulties encountered during home visits related to beliefs that: breastmilk alone is insufficient; baby is constipated or thirsty and needs additional fluids; mother has isilwane (a spirit that affects her breastmilk) and therefore cannot exclusively breastfeed and that baby has inyoni (leading to incessant crying), and needs traditional medicines.

Conclusions

The main lessons learned from developing the BCCS were that: non-health trained field workers can change knowledge and attitudes but acquiring counselling skills requires practice and takes a longer time. Resources need to be invested in selection, training and ongoing supervision of breastfeeding counsellors for the strategy to be successful. Several future challenges exist, such as defining the ethical responsibility of BC regarding health-related issues, and addressing community-beliefs that hinder exclusive breastfeeding. The study is ongoing and results will be forth-coming.

Acknowledgements

Nomantshali Mtshali – the Field Manager of the Breastfeeding Counsellors
Jane Lucas - WHO Consultant assisting in the development of the Field Guide.

References

1. A. Coutsoudis, K. Pilllay, E. Spooner, L. Kuhn, and H.M. Coovadia, Influence of infant feeding patterns on early mother to child transmission of HIV-1 in Durban, South Africa: a prospective cohort study, Lancet 354 (Aug 7):471-476 (1999).

BREAST FEEDING IN PRETERM INFANTS AFTER HOSPITAL DISCHARGE

Patricia Green, Lisette Jehn, Dare Desnoyers, Nan Peterson, Frank Greer*

Historically there is a high rate of failure to breastfeed preterm infants after hospital discharge. The purpose of this study was to determine breastfeeding rates 6 weeks after hospital discharge in 2 groups of infants --Very Low Birth Weight (VLBW, Bt Wt < 1500 gm) and Low Birth Weight (LBW, Bt Wt 1500-2000 gm). A previous study in 1987 showed a 20% breastfeeding rate at 6 weeks after hospital discharge.

In 1995 we instituted a Standard of Care for Breastfeeding program designed to increase breastfeeding success among mothers of preterm infants. Interventions included instructing mothers how to pump and store breast milk within 4 hours of delivery, reinforcing the importance of pumping 8 times a day to maintain milk supply, allowing breastfeeding moms to initiate breast feeding by 32 weeks gestation if baby is medically stable and showing readiness cues, and breastfeeding exclusively for 1 to 2 weeks prior to introduction of bottle feedings. Interventions within the past 5 years include introducing lactation consults for moms with premature infants and the use of nipple shields.

We wanted to determine if the 1995 Standard of Care had improved breastfeeding rates after hospital discharge. We hypothesized that at least 50% of preterm infants who were discharged on human milk feedings would still be successfully breastfeeding (defined as > 50% of feedings directly from the breast) 6 weeks after hospital discharge.

Infants were included in the study if they were admitted to the Special Care Nursery with birth weight < 2000 gm and were breastfeeding at time of discharge. Consent was obtained from eligible mothers (VLBW 29/30, LBW 27/30) either in the hospital prior to discharge or by follow-up letter 4 to 6 weeks after discharge. Telephone interviews were conducted with consenting mothers regarding their breastfeeding experience while in the hospital and at home. Breastfeeding was defined as successful if an infant was receiving 50% or more of his/her calories from breastfeeding (not including bottles of breast milk) weeks after discharge. We hypothesized that > 50% of preterm infants included in the study would be successfully breastfeeding 6 weeks after hospital discharge. Data was collected from medical records during hospitalization and retrospectively after discharge.

Of surviving infants who were discharged, 40/49 (82%) VLBW and 37/47 (79%)

*Patricia W. Green, General Clinical Research Center, University of Wisconsin, Madison, Wisconsin. Lisette Jehn, Student University of Wisconsin, Madison, WI. Dare Desnoyers, Nurse Manager Meriter Hospital SCN, 202 S. Park St. Madison, WI, Nanette Peterson, General Clinical Research Center, University of Wisconsin, Madison, WI. Frank Greer, Neonatologist, University of Wisconsin, Madison, WI

LBW received some breast milk during hospitalization. 56/60 eligible mothers were surveyed. Breastfeeding rates were at or close to our hypothesized 50% rate with 15/29 (52%) VLBW and 13/27 (49%) LBW infants "successfully" breastfeeding 6 weeks after discharge. Average VLBW was 1194.7 gm (range 832 - 1496gm); average gestational age was 29.4 weeks (range 26 - 35wks). 9 mothers stopped pumping during hospitalization, leaving 30/49 (61%) infants who were still breastfeeding at discharge. Average length of hospital stay was 48.1 days (range 19-88 days); average discharge weight was 2012 gm (range 1508 - 3410gm). Of those 30 infants, 15/29 were still receiving at least 50% of their oral intake by breast 6 weeks after hospital discharge; one mother was unreachable. More importantly 27/29 (93%) were still receiving breast milk by either breast and/or bottle. An impressive 79% (23/29) of mothers were still breastfeeding or attempting breastfeeding (defined as putting infant to breast at least once per day) 6 weeks after taking their infants home. 7/29 (24%) VLBW infants were exclusively breastfeeding. Average birth weight for LBW infants was 1790.1 gm (range 1508 - 2000gm); average gestational age was 33.1 weeks (range 30 - 36 wks). In this group 7 mothers stopped pumping during hospitalization so 30/47 (64%) infants were discharged home still breastfeeding. Average length of stay was 17.9 days (range 7 - 33 days); average discharge weight was 1937 gm (range 1630 - 2266gm). Of those 30 mothers, one was unreachable and 2 refused consent. 13 of the remaining 27 infants were still receiving at least 50% of their oral intake by breast 6 weeks after hospital discharge. Again, in this group a high percentage of babies' 24/27 (89%) were still receiving breast milk by either breast and/or bottle. In this group 70% (19/27) were still breastfeeding or attempting breastfeeding (defined as putting infant to breast at least once per day) 6 weeks after taking their infants home. 6/27 (22%) LBW infants were exclusively breast-feeding.

We have improved breastfeeding rates of premature infants (52% VLBW and 49% LBW) in the first 6 weeks after hospital discharge as compared with rates in 1987. Despite differences in length of hospital stay and degree of prematurity, there were no differences in breastfeeding at 6 wk between the two groups. Even though the LBW infants were born with greater maturity and weight, they did not prove to be more successful breastfeeders at home. We postulate that the extra support provided to the VLBW infants during their longer NICU stays resulted in their ability to breastfeed as successfully as the larger and older babies.

Although we nearly reached our 50% "successful breastfeeding" goal, there is much room for improvement. Mothers expressed deep regret and guilt at their "failure" to breastfeed and yet continued to have the commitment to pump and provide breast milk for their babies. Use of the nipple shield has been shown to improve milk transfer in the premature. Mothers also need support and follow-up after discharge. Making sure that mothers have access to breast pumps and baby scales for home use have proven useful. A clearly written feeding plan at discharge for mothers to follow along with follow-up phone calls and visits to the Lactation Clinic should be encouraged. There continues to be a need to educate, support, and encourage mothers to provide breast milk for their premature infants.

PHYSIOLOGY OF BREASTMILK EXPRESSION USING AN ELECTRIC BREAST PUMP

Peter E. Hartmann,[1] Leon R. Mitoulas,[1] and Lyle C. Gurrin[2]

INTRODUCTION

There has been a recent increase in both the use and variety of breast pumps. Although current breast pump designs are based on attempts to simulate infant sucking actions, formal evaluations of breast pumps are lacking with assessments of efficiency being limited to user responses and testimonials. We have developed a procedure for the objective determination of breast pump efficiency and have investigated milk removal from one breast over a 5-minute period in 30 women using an electric breast pump (vacuum pattern of the Medela Classic; Medela AG, Baar, Switzerland). These data were then compared to breastfeeding characteristics.

Methods

Breastfeeding characteristics for each mother were determined by collecting milk samples ($\leq$ 1 mL) before and after each feed from each breast by either manual breast pump (Medela AG) or by hand expression. Milk yield was determined for each breast at each feed by test weighing the infant over a 24 – 28 h period using a 'Baby Weigh' scale (Medela AG) and standardized to 24 h. The amount of milk available prior to milk expression was determined using the 24 h data with the degree of fullness method of Daly et al. (1993) and by direct measurement of breast volume using the Computerized Breast Measurement System (Daly et al., 1992). Milk expression was conducted using a computer controlled breast pump, which was able to control and record the vacuum applied to the breast. The amount of milk removed, the rate of milk removal and milk fat content were recorded for each 30s. An infrared thermographic camera provided data on

[1]Department of Biochemistry, The University of Western Australia, Nedlands, WA 6907, Australia and [2]Women and Infants Research Foundation, King Edward Memorial Hospital, Subiaco,WA 6008, Australia

changes in breast temperature and a video camera recorded the session and any comments from the mother.

Results

The mean (± SD) volume of milk removed (60.6 ± 39.0 mL) differed greatly between mothers (range 15.2 – 170.8 mL) but was not significantly different from the mean (± SD) volume of milk consumed per feed by the infant (70.3 ± 39.2 mL, range 4 – 194 mL). However, mean (± SD) breastfeeding time (16.6 ± 10.5 min) was greater than the expression period (5-min). The change in the fat content of milk (± SD) during the pumping session (25.9 ± 17.6 g/L) was not significantly different to that for a breastfeed (26.5 ± 12.5 g/L). The rate of milk removal differed greatly between mothers ($P = 0.0001$ and with time ($P = 0.0001$). The mean (± SD) rate of milk removal for he first 30 s was 14.8 ± 14.3 mL and remained constant for the first 2.5 min before slowly decreasing. Whole breast temperature did not change during the expression period however nipple temperature (± SD) increased from 33.6 ± 0.3 °C to 34.2 ± 0.6 °C ($P = 0.0008$) during the expression period.

Discussion

Breast pump efficiency can be assessed provided maternal breastfeeding characteristics and the amount of milk in the breast available to be expressed are known. In this regard we found the proportion of available milk expressed carried greatly between mothers indicating different levels of expressing success. However, mean expressing and breastfeeding characteristics were similar when compared on a 'per feed/expression' basis indicating breast emptying to a similar degree even though mean breastfeeding time was greater than the expressing time.

Acknowledgements

This study was supported by Medela AG, Women and Infants Research Foundation and the Lotteries Commission of Western Australia. All studies were approved by the Human Research Ethics Committee of The University of Western Australia.

References

S.E.J. Daly, A. Di Rosso, R.A. Owens, and P.E. Hartmann, Degree of breast emptying explains changes in the fat content, but not fatty acid composition, if human milk, Exp. Physiol. 78:741-755 (1993).

S.E.J. Daly, J.C. Kent, D.2q. Huynh, R.A. Owens, B.F. Alexander, K.C. Ng, and P.E. Hartmann, Determination of short-term breast volume changes and the rate of synthesis of human milk using computerized breast measurement, Exp. Physiol. 77:79-87 (1992).

THE EFFECT OF DIETARY DHA SUPPLEMENTATION ON HUMAN MILK CYTOKINES

Joanna S. Hawkes, Dani-Louise Bryan, Maria Makrides, Mark A. Neumann and Robert A. Gibson*

BACKGROUND

Some expert nutrition committees have recommended LCPUFA supplementation for lactating women. As the n-3 PUFA have been recognized as regulatory agents for cytokine and eicosanoid production it is important to define the effect such dietary changes have on the immunomodulating potential of human milk. The aim of this study was to examine the effects of dietary docosahexaenoic acid (DHA) supplementation on the human milk cytokines IL-6, TNFα, TGFβ1 and TGFβ2.

Methods

Subjects and Dietary Intervention

Healthy mothers of full term babies were assigned to one of 3 groups from day 3 post-partum: HiDHA, 600mg DHA+140mg eicosapentaenoic acid (EPA)/day (n=40); LoDHA, 300mg DHA+70mg EPA/day (n=40); placebo, no DHA or EPA (n=40).

Fatty Acid Analysis

Breast milk samples (2mL) were collected each day for 5 days in the 4th week of the study and analyzed for fatty acids as described previously.[1]

*JH, MM and RG, Child Nutrition Research Centre, Child Health Research Institute, D-LB, Department of Paediatrics and Child Health, Flinders University of South Australia, South Australia, Australia.

Cytokine Assays

An additional sample (50mL) expressed at 4 weeks was centrifuged and the aqueous fraction stored at -80°C for analysis of cytokines by ELISA as described previously.[2]

Results

3.1. Breast Milk Fatty Acids and Human Milk Aqueous Phase Cytokines (IL-6, TNFα, TGFβ1 and TGFβ2) at 4 Weeks

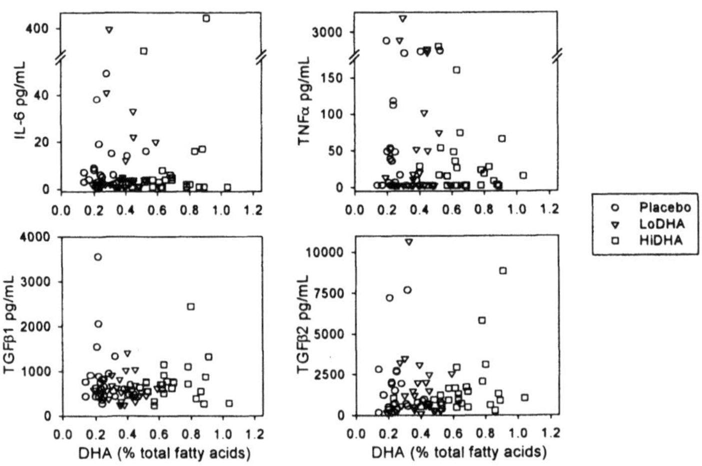

Figure 1. Cytokine versus DHA levels in human milk samples collected after 4 weeks of the dietary supplementation. Placebo n=26, LoDHA n=25, HiDHA n=28. A one-way ANOVA model was used to determine fatty acid differences between diet groups. Cytokine values were log transformed and significant differences between dietary groups calculated by Kruskal-Wallis ANOVA. Correlations between breast milk DHA level and cytokine concentration were analysed using Spearman's rank correlation coefficient.

There was no difference in mean rank concentration for any of the breast milk cytokines measured between any of the groups. In addition, there were no correlations between breast milk DHA and cytokine concentration.

Conclusions

Within the limits of our study we could detect no difference in mean rank concentration of cytokines between dietary groups. Our results indicate that increased maternal consumption of n-3 LCPUFA does not cause perturbations in breast milk cytokines beyond the variability reported in unsupplemented women.[2]

References

1. M. Makrides, M.A. Neumann, and R.A. Gibson, Effect of maternal docosahexanoic acid (DHA) supplementation on breast milk composition, Eur. J. Clin. Nutr. 50:352-357 (1996).
2. J.S. Hawkes, D-L. Bryan, M.J. James, and R.A. Gibson, Cytokines (IL-1β, IL-6, TNFα, TGFβ1 and TGFβ2) and prostaglandin E_2 in human milk during the first three months postpartum, Pediatr. Res. 46:194-199 (1999).

INITIATION OF PUMPING AND MILK WEIGHT IN MOTHERS OF NONNURSING PRETERM INFANTS

Hill, PD[1], Aldag, JC[2], Chatterton RT, Jr.[3]

Studies of lactating mothers who must use mechanical expression to initiate and maintain their milk supply are limited. No studies were found that have examined timing of first breast pumping after birth and subsequent milk volume in mothers of preterm infants using mechanical expression. Thus, the purpose of this secondary data analysis examined the amount of milk in grams mechanically expressed weeks 2 through 5 after birth by mothers of preterm infants in relation to the first initiation of breast pumping and frequency of pumping.

Demographic differences were tested using t-tests for interval variables and Fisher Exact tests for nominal variables. Raw data are reported for means and standard deviations for milk weight, weeks 2 through 5. Milk weight data during week 1 were not used as mothers entered the study at various points in time the first week after birth. Two-way analysis of variance was used to test group differences for time to pump (early ≤ 48 hours vs. late > 48 hours) and frequency of pumping (low, range = 2.60 to 6.14 vs. high, range = 6.25 to 8.10 times per day). One-way analysis of variance with post-hoc test LSD was used to test differences in milk weight among the four groups. The alpha level was set at .05.

The convenience sample consisted of 39 mothers who were primarily white (n = 29, 74.4%), married (n = 33, 84.6%), and who delivered vaginally (n = 21, 53.8%). Mean birth weight and gestational age of the 50 infants were 1058 ± 254 grams and 27.2 ± 1.5 weeks, respectively. For early (n = 20) vs. late initiators of pumping (n = 19) there were no statistically significant differences for maternal age, education, marital status (married vs. not married), income (< \$50,000 vs. ≥\$50,000), singleton vs. multiple births, previous breastfeeding experience, and infant gestational age. The mean for the low frequency pumping group was 4.90 ± .92, and for the high group, 7.00 ± .62.

Mothers who were in the high frequency pumping group had a significantly higher mean daily milk weight in grams, $F(1, 35) = 8.84$, $p < .001$, as compared to

[1]University of Illinois at Chicago, Chicago, IL, [2]University of Illinois at Peoria, Peoria, IL, [3]Northwestern University, Chicago, IL, USA

mothers in the low frequency pumping group (631.72 ± 323.67 vs. 319.17 ± 291.51, respectively). There was no significant milk weight difference for early vs. late initiators of pumping, $F (1, 25) = 2.61$, $p = .11$, but the interaction was significant, $F (1, 25) = 4.09$, $p = .05$, for time of initiation of pumping by frequency of pumping.

Group differences are reported below. Mothers in the low frequency/late initiating group had significantly lower milk weight in grams compared to the low frequency/early initiating group, high frequency/early initiating and high frequency/late initiating groups. In contrast, mothers who were in the low frequency/early initiation group were similar with respect to milk weight compared to the high frequency/early initiating and high frequency/late initiating groups. For mothers in the high frequency/early initiating and high frequency/late initiating groups, the milk weights were similar. While this was a small sample of mothers who were mechanically expressing milk for their preterm infants, milk weight for mothers who were in the low frequency/late initiating group expressed a dramatically smaller amount of milk over time compared to mothers in the other three groups. Mothers had a distinct advantage in expressing more milk over time if they mechanically expressed their milk frequently, regardless of initiation time of pumping, or initiated early and pumped a low frequency.

In our study, mothers in the low frequency group pumped daily an average of 4.90 (range = 2.60 – 6.14) times, whereas the high frequency group pumped an average of 7.0 (range = 6.25 – 8.10) times daily. In this study, time of initiation was not a factor for mothers who mechanically expressed their milk an average of more than 6.25 times daily.

The group of mothers who initiated mechanical expression prior to 48 hours after birth and pumped a 'low' frequency had similar milk weight means over time as did mothers who pumped more frequently regardless of time of initiation. This finding is surprising in light of what is known from research and clinical practice in that frequency of breast stimulation is positively related to milk production and removal. In this study, breast stimulation within the first 48 hours following birth seemed to play an important role in milk production from weeks 2 through 5 weeks postpartum. Unfortunately, there is little in the literature concerning why early initiation of breast stimulation in the first two days could be critical to milk production. Might there be a delicate interplay between the hormonal milieu and removal of milk?

Unfortunately the sample size and small numbers in the cells did not allow us to examine the data to determine if pumping prior to 24 hours after birth made a difference in milk weight over time. Clearly, larger studies of both pumping and naturally feeding infants are needed to confirm the findings in this study as well as to examine the effect of earlier initiation (< 24 hrs) on milk production.

Whatever the physiologic mechanism between time of initiation and frequency of breast stimulation, milk weight from week to week is highly correlated. This suggests that the level of milk production in the early weeks after birth is likely to be maintained at that level in subsequent weeks. Thus, it behooves clinicians to get mothers of preterm infants off to the best start possible if they intend to provide exclusively human milk to their preterm infant, as this group is at risk for diminished milk production over time. This may mean not only encouraging frequent breast pumping but also 'early' initiation of mechanical expression.

Milk Weight in Grams Weeks 2 through Week 5 by Low and High Frequency Pumping Groups

	Initiation of Pumping	
	≤ 48 hrs.	≥ 48 hrs.
Low frequency	14,753 ± 9,982.29	5,059 ± 3,266.72*
	(n = 8)	(n = 12)
High frequency	17,288 ± 8,110.91	18,374 ± 11,177.07
	(n = 12)	(n = 7)

- p > 05, statistically different from each of the other three groups
- Funded by NIH, NINR (R55 NR04118-01A1), UIC, College of Nursing, & Medela, Inc.

POSTPARTUM HOME VISITS INCREASE DURATION OF EXCLUSIVE AMONG INEXPERIENCED BREASTFEEDING WOMEN

Judy M. Hopkinson, E. O'Brian Smith, Rosa Acosta, Daisy Handal, Sandra Lopez, Nancy F. Butte, and Janet Rourke*

Exclusive breastfeeding is recommended but uncommon in the United States. We undertook a randomized study of the efficacy of telephone consultation (T) and home visits (H) to increase the duration of exclusive breastfeeding among low-income Hispanic women in Houston, Texas. At project initiation, WIC breastfeeding rates were 43% with < 1% exclusive breastfeeding. Data collected retrospectively from 105 accompanied by children under 12 months of age attending well child visits at the local health clinic indicated that at 3 months postpartum 42% were breastfeeding and 6% were breastfeeding exclusively. The pre-intervention control group (PC) consisted of a subset of twenty of those interviewed who matched the intervention group for sites of delivery and prenatal care, zip code, parity, maternal age, and singleton, term, healthy infant at birth. Of these, 8 (40%) were breastfeeding at 3 months and 1 (5) was breastfeeding exclusively at 3 months. A concurrent control group (CC) was selected from women interviewed six months after initiating the intervention study. Nineteen women met the demographic criteria listed above and delivered the index child within the previous 6 months. Rates of breastfeeding (52% and exclusive breastfeeding (10%) at 3 months in the CC groups did not differ significantly from those in the PC group. Therefore, data were combined for analyses. The intervention groups of 121 mothers of healthy, singleton, and term infants were enrolled in hospital. Mothers were assisted with breastfeeding in hospital and assigned to home-visit (H) (n=63) or telephone consultation (T) (n-58) groups by random number table after discharge. There were no differences between intervention and control groups for site of delivery, prenatal care, zip code, parity, birth weight or age. Introduction of routine supplements was examined by Cox regression analysis. Both H and T groups withheld supplements for a longer time than controls ($p<0.003$). There was an interaction between group and prior breastfeeding

*Judy M. Hopkinson, E. O'Brian Smith, Rosa Acosta, Daisy Handal, Nancy F. Butte, USDA/ARS Children's Nutrition Research Center at Baylor College of Medicine, Houston, TX 77030. Janet Rourke, Texas Department of Health, Austin, TX 78756

(p<0.01) such that home visits were more effective than phone calls for inexperienced women (p<0.001) (fig), but not for experienced breast feeders. At 3 months, 80% of study participants were breastfeeding, and 38% were breastfeeding exclusively.

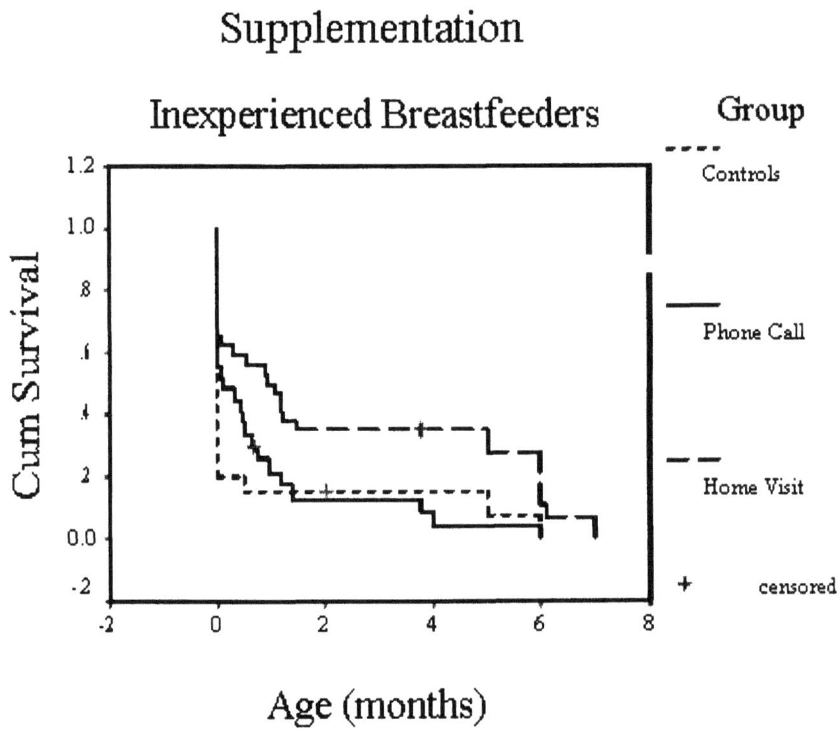

Conclusions: Home visits by experienced breastfeeding counselors aid in the acquisition of critical breastfeeding skills. Once breastfeeding skills are acquired, telephone consultation and home visits are equally effective methods for increasing the duration of exclusive breastfeeding with subsequent children.

INTRODUCTION OF SOLIDS AND FORMULA TO BREASTFED INFANTS AND ITS ASSOCIATION WITH PATTERN AND DURATION OF BREASTFEEDING

A. Hörnell,[1] Y. Hofvander[*], E. Kylberg[*]

This summary is based on two papers: Acta Paediatrica, 2001;90(5):477-82 and Pediatrics, 2001;107 (3), www.pediatrics.org/cgi/content/full/107/3/38

AIM

The aim of the work was to describe the introduction of solids and formula to breastfed infants, and to study associated changes in breastfeeding.

Subjects and methods

This study, based on daily recordings of infant feeding, comprised 506 infants from Uppsala, Sweden. All mothers had had previous breastfeeding experience of ≥ 4 months, and were planning to breastfeed the index child for ≥ 6 months.

Results

Accustoming the infants to solids was a lengthy process, the longer the younger the infant at introduction. Life-table analyses showed a median duration of 28 days from the first introduction until the infant ate > 10 mL of solids/occasion daily. It took a median of 46 days before the infants ate ≥ 100 mL of solids in a single day. In contrast, 32% of the infants given formula consumed ≥ 100 ml the first time it was given, and 66% did so within a week, regardless of infant age.

The change in breastfeeding frequency after the start of solids or formula was studied in two groups starting with solids/formula from ≥ 2 months of age. Group *Solid* (n=66) never/sporadically received formula and started with solids at a median

[1]Corresponding author: A. Hörnell, present address: Department of Food and Nutrition, University of Umeå, SE-901 87 UMEÅ, Sweden.
[*]Unit for International Maternal and Child Health – IMCH. University Hospital, Entrance 11, SE-751 85 Uppsala, Sweden.

age of 4.5 months. Group *Formula* (n=63) started with regular formula feeds (> 1/day) at a median age of 3.9 months (and solids at a median age of 4.3 months). Introduction of solids had little impact on breastfeeding frequency, while the start of regular formula feeds was associated with a rapidly declining frequency (Fig. 1).

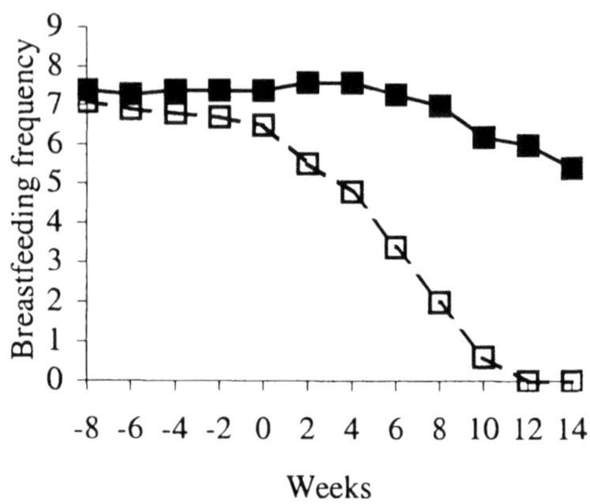

Figure 1. The median breastfeeding frequency in the *Solids* and *Formula* groups, from 8 weeks before the start of solids or regular formula feeds until 14 weeks after the start. The 14-day period when solids and regular formula feeds were started is set to zero. ■ = group *Solids*, □ = group *Formula*. Reprinted with permission from the American Academy of Pediatrics.

Among the infants never given formula, age at introduction of solids showed no association with total breastfeeding duration. Age at start of regular (but not occasional) formula feeds was significantly associated with breastfeeding duration, even after adjustment for infants birth year, pacifier use and maternal education.

Conclusion

Health care personnel and parents need to be aware that accustoming breastfed infants to solid food is a lengthy process, and that physical maturation of the infant is important. It is also essential to consider what consequences the (usually) more abrupt introduction of formula might have on breastfeeding.

Acknowledgements

The project was part of "The WHO Multinational Study of Breastfeeding and Lactational Amenorrhoea", and received financial support from the UNDP/UNFPA/WHO/World Bank Special Programme of Research, Development and Research Training in Human Reproduction, World Health Organization.

GENDER DIFFERENCES IN ENERGY INTAKE OF NEWBORNS

Marina K. Ito, Juliana Da Cunha, Teresa H. M. Da Costa*

INTRODUCTION

Lipid is the major energy source for the breastfeeding newborn. It is well known that a number of factors influence the lipid composition of human milk.[1] These include lactational period, maternal diet, duration of a feed, time of collection, feeding interval, gestational age, parity, maternal age and adiposity. In general, although boys are born bigger than girls and their growth curve maintain the difference, no report seems to indicate a relationship between breast milk lipid composition and the gender of the baby. In a single feed, the lipid content of the milk increases from fore to hindmilk, and, for our knowledge, sexual differences have not been documented.

Method

In a cross-sectional, observational study, primiparous women of middle to low social economic class who gave birth to term, healthy, adequate for the gestational age babies participated in the study. On the day 15 of exclusive breast-feeding, fore and hind milk from a single feed was collected in the morning period (betweem 7:00 and 9:00 a.m.). Lipids were analysed by a modified crematocrit method, using Bligh & Dyer[2] for the extraction of lipids. Analysis of variance of the results were performed.

Results and Discussion

Mean lipid content of our samples was high (4.8 ± 2.1 g/ dl), compared to reported values, due probably to the efficiency of the extraction method adopted. Distribution of lipids in fore and hindmilk and infant's gender according to maternal body

*Marina K. Ito & Teresa H. M. Da Costa, Departamento de Nutrição, Universidade de Brasília, Brasília, DF, Brazil. Juliana Cunha, Universidade Católica, Brasília, DF, Brazil.

mass index (BMI), are shown in Table 1. Mean lipid contents (g/ dl) were 4.0 ± 1.7 in boy's and 4.3 ± 1.3 in girl's foremilk, respectively, and they increased to 5.8 ± 1.6 in boy's and 4.8 ± 1.4 in girl's hindmilk (Figure 1). This means that fat concentration increased by 44% in boy's and by 13% in girl's hindmilk. General linear model analysis of variance, considering weight of the infant and maternal BMI as covariates, revealed a statistically significant association of hindmilk fat with child sex ($p = 0.009$) and maternal BMI ($p = 0.03$).

Table 1. Distribution of fore and hind milk fat and infant's gender, according to mother's body mass index (BMI)

Maternal BMI	Infants (*n*)		Milk fat (g/ dl)	
(kg/ m^2)	boys	girls	Foremilk	hindmilk
Below 20	6	2	3.8 ± 1.3	5.5 ± 2.4
20 to 24.9	23	18	4.0 ± 1.7	5.9 ± 2.2
25 and over	16	8	4.9 ± 1.8	5.9 ± 1.8

The mechanism involved in the increase of lipid content during a feed is not yet clear. The gender difference observed in our hind milk lipid concentration may be related to the stronger suckling power of boys compared to girls, allowing them to take in more energy in a single feed. Considering that infant weight did not account for the difference observed, the higher energy taken up by boys may be part of the explanation.. As proposed by Dewey,[3] in well-nourished population, child seems to dictate the amount of milk consumed and thus the intake of energy. Our findings suggest that infant's sex may be another factor affecting the lipid (i.e. energy) content in human milk. These findings should be further investigated at different times of the day and lactational period in a carefully designed protocol.

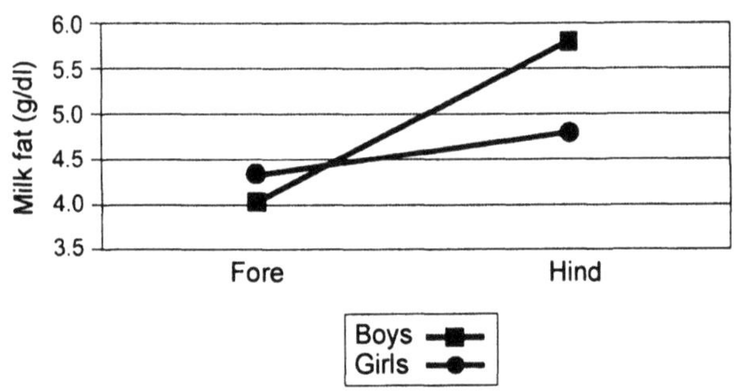

Figure 1. Lipid content in fore and hind milk, according to infant's sex.

References

1. R. G. Jensen. The lipids in human milk, Prog. Lipid Res. 35:53-92 (1996).
2. E. G. Bligh, and W. J. Dyer, A rapid method of total lipid extraction and purification, Can. J. Bio. Physiol. 37:911-917 (1959).
3. K. G. Dewey,, M. J. Heining, L.A. Nommsen, and B. Lonnerdal, Maternal versus infant factors related to breast milk intake and residual milk volume: the DARLING study, Pediatrics 87:829-837 (1991).

IgA IN HUMAN MILK AND DEVELOPMENT OF COW'S MILK ALLERGY IN THE BREAST-FED INFANT

Kirsi-Marjut Järvinen, Seppo Laine, Anna-Liisa Järvenpää, and Hanna Suomalainen*

INTRODUCTION

The effect of breast-feeding in reducing the incidence of atopic disease and cow's milk allergy (CMA) in the first years of life still remains to be settled. This could partly be explained by varying concentrations of protective factors in mothers' milk. Much of the passive protection in human milk is due to the IgA present in highest concentrations in the colostrum, which subsequently falls progressively to a basal level of 0.2 to 0.3 g/L.[1] After ingestion, maternal milk IgA antibodies have been suggested to passively protect the infant by binding to IgA receptors on the luminal surface of enterocytes, thereby decreasing or preventing antigen entry across the immature gastrointestinal epithelium.[2]

The purpose of the study was to determine the content of total and cow's milk-specific IgA in human milk in relation to the development of CMA in the breast-fed infant.

Subjects and Methods

87 *mothers and their infants were followed prospectively. **How often were milk specimens collected? t 1 year, 48 mothers (69% atopic) had an infant with CMA, verified by clinical cow's milk challenge, 8 (38% atopic) had a baby who had had protracted infantile colic but no CMA, and 31 (23% atopic) had a healthy infant. Breast milk samples were collected during the breast-feeding period (2 days to 7 months). Radial immunodiffusion and enzyme immunoassay were used to measure total and cow's milk-specific IgA, respectively.

The data were analyzed using analysis of variance for repeated measurements

*Kirsi-Marjut Järvinen and Hanna Suomalainen, Helsinki University Central Hospital, Dept. of Dermatology, Helsinki, Finland 00029. Seppo Laine, Tampere University Hospital, Dept. of Clinical Microbiology, Tampere, Finland 33521. Anna-Liisa Järvenpää, Helsinki City Maternity Hospital, Helsinki, Finland 00610.

with mothers as *the unit of analysis. Logarithmic transformation was applied to values of IgA prior to model fitting. The logarithm of a baby's age was used as such as an explanatory variable. The starting model was: ln(age) as the random part and the systemic one: ln(IgA)~ ln(age)* (1allergy+ 1mother's atopy+ 1symptoms), where 1character is an indicator variable for character. The unimportant terms (significance >5%) were excluded.

Results

An analysis of variance for repeated measurements gave an estimated (geometric mean) value for IgA on the first day of 0.38 g/L (0.24-0.82) for the mothers whose baby developed CMA, and 0.82 g/L (95% CI 0.99-1.51) for the mothers whose infant did not develop CMA ($p<0.05$).[3] The levels of IgA antibodies dropped in both groups during the course of breast-feeding, but owing to the lower starting value, the drop was smaller and statistically insignificant in the mothers whose babies developed allergy. *Infants who developed CMA were significantly more likely to have concentration of total IgA antibodies in milk below 0.25 g/L, when measured between 6 days and 4 weeks postpartum (sensitivity 0.55, specificity 0.92, odds ratio 14.7 [95% CI 3.1-70.2], $p<0.001$). The levels of IgA antibodies to cow's milk in the milk from the mothers of the infants with CMA and from the healthy controls were comparable.

The levels of IgA antibodies in milk were comparable in the mothers of a baby with infant colic but no CMA and in those with a healthy infant. The contents of total and cow's milk-specific IgA antibodies in human milk did not correlate with maternal atopy.

Conclusion

IgA antibodies in human milk may play a role in development of oral tolerance to food antigens in the breast-fed. Low total IgA content of mother's milk may thus predispose an offspring to develop CMA.

References

1. E. Savilahti, V. M. Tainio, L. Salmenperä, P. Arjomaa, M. Kallio, and J Perheentupa, Low colostral IgA associated with cow's milk allergy, Acta Paediatr. Scand. 80:207-1213 (1991).
2. W. A. Walker, in: Immunology of Breast Milk, edited by P.L. Ogra PL and D. Dayton D (Raven Press, New York), pp. 227-234 (1979).
3. K.-M. Järvinen, S. Laine, A.-L. Järvenpää, and H. Suomalainen, Does low IgA in human milk predispose the infant to development of cow's milk allergy?, Clin. Exp. Allergy (2000).

ABNORMAL CELLULAR COMPOSITION OF HUMAN MILK AND DEVELOPMENT OF COW'S MILK ALLERGY IN THE BREAST-FED

Kirsi-Marjut Järvinen, Pamela Österlund, and Hanna Suomalainen*

INTRODUCTION

The purpose of this study was to determine the leukocyte composition of human milk, as compared to the health status and development of food allergies in a suckling infant.

Subjects and Methods

Altogether 61 breastfeeding mothers and their infants were followed up prospectively from birth for one year. Of the mothers, 36 (59%) had atopic constitution. Human milk samples were collected and their leukocyte subsets were evaluated by flow cytometry and light microscopy.

Results

After the follow-up, 39 mothers had an infant with cow's milk allergy (CMA), 10 had an infant with atopic dermatitis without CMA, and 12 had a healthy infant. In the breast milk of mothers with an infant with CMA, the number of macrophages was significantly smaller than in those with an infant without CMA ($p=0.02$). Mothers with a high number of neutrophils in their milk (>20%) had significantly more often an infant with CMA ($p=0.02$). Eosinophils were detected only in breast milk of mothers with an infant with CMA. Eosinophil cationic protein could be measured in breast mik in correlation with the number of eosinophils in milk.

*Kirsi-Marjut Järvinen, Pamela Österlund, and Hanna Suomalainen, Helsinki University Central Hospital, Department of Dermatology, Helsinki, Finland 00029

Conclusion

Our results suggest that disturbed cellular composition of breast milk might have a role in the development of CMA in the breastfed.

FATE OF MILK BORNE ERYTHROPOIETIN IN SUCKLING RATS

Pamela J. Kling,* Amy L. Gilbert, Suzanne H. Dubuque, Bohuslav Dvorak, Catherine S. Williams, James G. Grille, Suann S. Woodward, Otakar Koldovsky

INTRODUCTION

Mammalian milk is a rich source of hormones and growth factors which survive proteolytic degradation and are absorbed intact in the neonate.[1] Human milk contains erythropoietin (Epo), the primary endocrine hormone responsible for erythropoiesis.[2] Recombinant human erythropoietin (rhEpo) may be an effective treatment for the anemia of prematurity but traditional routes of administration, such as intravenous or subcutaneous dosing have thus far provided disappointing results. In animals, milk borne Epo stimulates erythropoiesis.[3] Additionally, Epo receptors have been found in gastrointestinal tract, where *in vitro* studies show mitogenic and stimulatory effects of Epo on gastrointestinal cells.[4,5] We hypothesized that milk borne Epo is distributed both to local GI tissues and to the systemic circulation in the Sprague Dawley suckling rat.

Methods, Results

We evaluated the protective effects of rat milk or rat milk substitute (RMS) formula[6] compared to saline by incubating ^{125}I rhEpo with rat gastric and small intestinal juices in simulated *in vitro* digestion at physiologic pH levels. From 10-day-old rats, gastric juices were collected and incubated with rhEpo at pre-prandial and postprandial stomach pH in the 3 vehicles listed above. Similarly, intestinal juices were collected and incubated at intestinal pH 7.4. When measured by acid precipitation and confirmed by RIA, rat milk and RMS formula protects rhEpo from digestive juices better than saline ($p<0.0002$).

*P.J. Kling, University of Arizona, 1501 N. Campbell Ave, Tucson, AZ, USA, 85724

Two hours after ^{125}I rhEpo and milk feeding in 10-day old suckling rats, label obtained from gastric tissue and lumen co-migrated on SDS/PAGE with intact ^{125}I rhEpo at 36.5 kD, further supporting that the acid precipitation technique estimates intact rhEpo after feeding. Acid precipitation studies were used to determine tissue distribution and fate of physiologic doses of enterally fed ^{125}I rhEpo.[7] Suckling rats were tube fed ^{125}I rhEpo, followed by harvesting of GI luminal contents, GI tissues, blood and peripheral tissues. Tissue and luminal contents from stomach, duodenum, proximal jejunum, mid-jejunum and ileum were all harvested separately, and intact rhEpo was located throughout the gastrointestinal tract between 1-4 hours after feed. After harvesting gastrointestinal tissues and their luminal contents 2 hours after feed, rhEpo in milk was better protected from *in vivo* degradation than in saline, $p<0.05$.

Throughout the gastrointestinal tract, the ratio of intact counts compared to total counts was higher in the luminal contents compared to its corresponding tissue wall. With rhEpo distribution throughout the gastrointestinal tract and with the pattern of highly intact rhEpo in the lumen supports local cellular rhEpo metabolism, as opposed to luminal proteolytic degradation. Intact rhEpo also reaches peripheral organs, including liver, spleen, kidney, and brain, all tissues with rhEpo receptors. Between 1-4 hours after feed, 5% of total administered dose was found intact in the plasma, while another 8-10% of total administered dose was localized to bone marrow, percentages comparable to that seen following parenteral administration.[8] These patterns of localization and degradation of rhEpo after acute administration support both systemic absorption and gastrointestinal cellular processing. As Epo is distributed to tissues containing Epo receptors, both locally and distally, further investigation into the potential roles of milk borne Epo are necessary.

Acknowledgement

This work was supported in part by Arizona Disease Control Research Commission 1-272 (P.J.K.), Arizona and Southwest Affiliates of American Heart Association AZGS-37-96 and SW-GS-16-98 (P.J.K.), University of Arizona Undergraduate Honors Research Grant (A.L.G.), The University of Arizona Undergraduate Biology Research Program (A.L.G.), and Program Project NIH HD 26013 (O.K., B.D., P.J.K.).

References

1. O. Koldovsky, The potential physiological significance of milk-borne hormonally active substances for the neonate, J. Mamm. Gland Biol. Neoplasia. 1:317-322 (1996).
2. P.J. Kling, T.M. Sullivan, R.A. Roberts, A.F. Philipps, and O. Koldovsky, Human milk as a potential enteral source of erythropoietin, Pediatr. Res. 43:216-221 (1998).
3. R.D. Carmichael, J. LoBue, A.S. Gordon, Neonatal erythropoiesis. I. Peripheral blood erythropoietic parameters: Data suggest erythropoietin transfer via maternal milk, Endocrine Regulations 26:83-88 (1992).
4. A. Okada, Y. Kinoshita, T. Maekawa, M. Hannan, C. Kawanami, M. Asahara, M. Matsushima, K. Kishi, H. Nakata, Y. Naribayashi, T. Chiba, Erythropoietin stimulates proliferation of rat-cultured gastric mucosal cells, Digestion 57:328-332 (1996).
5. S.E. Juul, A.E. Joyce, Y. Zhao, D.J. Ledbetter, Why is erythropoietin present in human milk? Studies of eythropoietin receptors on enterocytes of human and rat neonates, Pediatr. Res. 46:263-268 (1999).

6. B. Dvorak, R. Stepankova, Effects of dietary essential fatty acid deficiency on the development of rat thymus and immune system, Prostaglandins Leukotrienes and Essential Fatty Acids 46:183-190 (1992).
7. A.F. Philipps, R.K. Rao, G.G. Anderson, D.M. McCracken, M. Lake, O. Koldovsky, Fate of insulin-like growth factors I and II administered orogastrically to suckling rats, Pediatr. Res. 37:586-592 (1995).
8. J.L. Spivak, B.B. Hogans BB, The *in vivo* metabolism of recombinant human erythropoietin in the rat, Blood 73:90-99 (1989).

BREASTFEEDING PATTERNS AND MENSES RETURN: FINDINGS FROM RESEARCH ON LAM

Miriam Labbok, Veronica Valdes, Ricardo Aravena[1]

INTRODUCTION

Much has been learned in the past 30 years concerning the physiological relationship of lactation and lactational infertility. Studies of maternal seemed to indicate that maternal nutritional status, parity and age were key.[1] More recent work has indicated that characteristics of the infant feeding that may explain more of the difference.[2,3,4] A recent study of the Lactational Amenorrhea Method (LAM) allowed study of the return of menses in two similar populations in urban Santiago Chile. Data from this prospective research were re-analyzed to assess patterns of breastfeeding and menses return.

Methods

Women were recruited postpartum at the Universidad Catolica de Chile hospital and asked to participate in a prospective study of breastfeeding and LAM. The control cohort of 313 women were followed monthly by a OB/GYN/Pediatrician pair, and received the normal support for breastfeeding. Following this, 412 women were similarly recruited and received guidance on LAM and special support for breastfeeding in the monthly visit. Details of the study design are published elsewhere.[5, 6,]

Findings

Analysis of the study population revealed no significant differences between the Control and Intervention groups in age, % with higher education, maternal weight, percent of infants male, or birth weight. Since breastfeeding patterns varied between the two groups, menses return was assessed in the fully breastfeeding women separately. A large difference in the timing of menses return was noted. (Table 1). There was a small

[1]Georgetown University/USAID, Washington, DC, USA, Universidad Catolica de Chile, Santiago, Chile

but significant difference in breastfeeding frequency and average interval between feeds ($p<0.05$) in months, 2, 5, and 6. (Fig. 1).

Table 1. Return of Menses among Exclusively Breastfeeding Women: Cumulative Percent by Six-Month Life Table

	1	2	3	4	5	6	month
Control Group	0	4.7	18.6	37.3	47.5	55.1	
LAM Group	0	2.4	8.7	12.7	16.5	18.8	

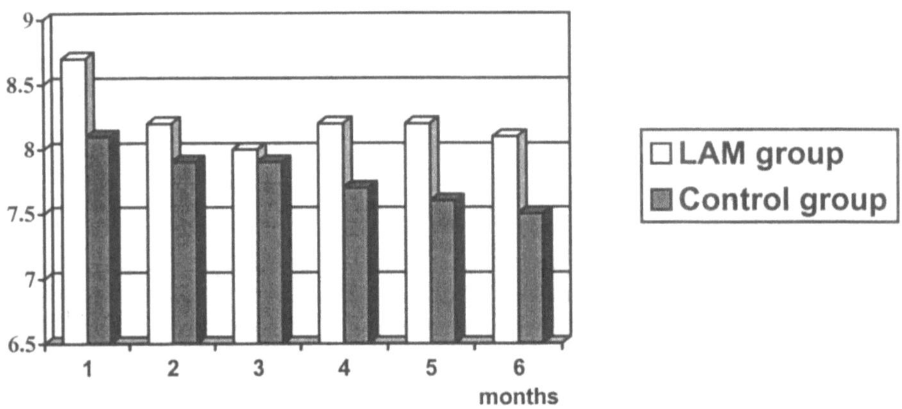

Fig. 1.Average Frequency of Breastfeeds by Month and Group.

Discussion and Conclusions

Exclusively breastfeeding women in the control and LAM groups manifested differing rates of menses return. Increased support for optimal breastfeeding as required for LAM apparently resulted in slight, but significant, differences in breastfeeding patterns. Small differences in numbers of feeds, and in the intervals between feeds, seem to be associated with significant differences in rates of menses return. It is reasonable to conclude that when intervals between feeds approach the same duration as the physiological impact of each feeding episode on hormonal responses (about 2-4 hour), even slight differences can make a profound impact on the return of hormonal cycling, and hence, the return of menses. At about 8 feeds per day, we might be at the edge of the physiological response, and small shifts could have large impacts at the population level.

References

1. H. Delgado, E. Brineman, A. Lechtig, J. Bongaarts, R. Martorell, and R.E. Klein, Effect of maternal nutritional status and infant supplementation during lactation on postpartum amenorrhea, Am. J. Obstet. Gynecol.. Oct. 1;135(3):303-307 (1979).
2. R. Gray, O. Campbell, R. Apelo, S. Eslami, H. Zacur, R. Ramos, J. Gehret, and M. Labbok, The Risk of Ovulation during Lactation and Use of Breastfeeding for Fertility Regulation, Lancet 335:25-29 (1990).
3. C.C. Tay, A.F. Glasier, and A.S. McNeilly, Twenty-four hour patterns of prolactin secretion during lactation and the relationship to suckling and the resumption of fertility in breast-feeding women, Hum. Reprod. May; 11(5):950-955 (1996).
4. A.S. McNeilly, Lactational amenorrhea, Endocrinol. ,Metab. Clin. North Am. Mar; 422(1):59-73 (1993).
5. A. Perez, M. Labbok, and J. Queenan, A Clinical Study of the Lactational Amenorrhea Method for Family Planning, Lancet 339:968-970 (1992).
6. V. Valdés, A. Pérez, M.H. Labbok, E. Pugin, I. Zambrano, and S Catalan, The Impact of a Hospital and Clinic-based Breastfeeding Promotion Program in a Middle Class Urban Environment, J. Trop. Peds. 139:142-151 (1993).

POST-MARKETING STUDY OF THE LACTATIONAL AMENORRHEA METHOD (LAM): Impact of Putting LAM in Women's Hands

Miriam Labbok and Anne Peterson[1]

INTRODUCTION

The Lactational Amenorrhea Method for family planning, based on the physiology of lactational infertility, has undergone extensive clinical study. Concern remains, among clinicians and demographers alike, that this is a behavior-based method and is therefore unreliable. This study was undertaken to observe method use under conditions that would more approximate use outside of a research setting.

The objective of this effort was to assess the use and efficacy of the Lactational Amenorrhea Method (LAM) with reduced numbers of client-provider contacts. A co-sponsored Multicenter Study of LAM was carried out to test the efficacy and acceptability of the method under these "post-marketing" conditions, with investigator-initiated contact only twice: at the time of intake and then again at month 7 postpartum. This approach was selected to provide an assessment of LAM's use, efficacy, and performance that more closely reflects the prevailing conditions of these populations during actual use.

Methods

Three hundred and sixty-two (362) subjects were recruited through centers that had participated in the previous, more contact-intensive studies (refs.) Using a cooperatively developed protocol, data were gathered prospectively on at least 10 and up to 50 LAM acceptors at 9 sites, and entered and cleaned on site. Data were further cleaned and analyzed at the Georgetown University Institute for Reproductive Health (IRH) and the Department of Nutrition at the University of Connecticut. Using country-

[1]Prepared at Georgetown University with support from USAID Cooperative Agreements with the Institute for Reproductive Health, Georgetown University, Linkages, Academy for Educational Development (AED), the World Health Organization, and south to South Foundation. The views expressed are those of the authors and do not necessarily reflect that of Georgetown, Linkages, USAID, WHO or South to South. Peer reviewed paper accepted for publication, *Contraception*.

level and pooled data, descriptive statistics and life tables were produced. The subjects were counseled on LAM and asked to return only if one of the 3 criteria no longer applied.

Findings

LAM efficacy in this sample is 100% as there were no pregnancies at any of the participating sites. Other relevant findings include a high level of satisfaction with the method and more than two-thirds had successfully switched to another method by month 7 postpartum.

There was apparent increase in family planning use stimulated by LAM acceptance. Importantly, of the women who had never used family planning prior to LAM, 63.0% went on to use another method of family planning in a timely manner.

Discussion and Conclusions

With significantly reduced client contact, the results remain similar to previous studies: The efficacy, levels of acceptance and the distribution of follow-on family planning methods were similar. More than half of the women were able to use LAM for the full 6 months and 65.2% of women amenorrheic at 6 months. These numbers are also consistent with those in the previous protocol of this study: 68.4% of those women were amenorrheic through 6 months. Women are able to use LAM effectively without extensive counseling or follow-up, with a high level of user satisfaction and efficacy.

This study reconfirmed the findings of previous work on LAM, including that the method can be used effectively in a wide range of ethnic and cultural situations and service-delivery settings, with limited counseling, and that duration of use and satisfaction were high both in industrialized and developing countries.

Clearly, when LAM is put in women's hands, they are able to use the method appropriately, without extra strain on clinical services.

References

1. Labbok MH, Hight-Laukaran V, Peterson AE, Fletcher V, von Hertzen H, and Van Look PFA. Multicenter Study of the Lactational Amenorrhea Method (LAM): I. Efficacy, Duration, and Implications for Clinical Application. Contraception 55:327-336 (1997).
2. Hight-Laukaran V, Labbok M, Peterson AE, Fletcher V, von Hertzen H, and Van Look PFA. Multicenter Study of the Lactational Amenorrhea Method (LAM): II. Acceptability, Utility, and Policy Implications. Contraception 55:337-346 (1997).
3. Peterson AE; Pérez-Escamilla R; Labbok MH; Hight V; von Hertzen H; and Van Look P. Multicenter Study of the Lactational Amenorrhea Method (LAM) III: Effectiveness, Duration and Satisfaction with Reduced Client-Provider Contact. Accepted, Contraception, (1999).

EFFECT OF EXERCISE ON SECRETORY IgA AND LACTOFERRIN CONCENTRATIONS IN HUMAN MILK

Cheryl A. Lovelady, Christie L. Phipps, Ciara J. Bradley, Kate F. Synnott, and Cissy M. Geigerman*

INTRODUCTION

Exercise may enhance immune function and reduce risk of infection; however, exhaustive exercise may lead to immunosuppression.[1] Researchers have reported a correlation between strenuous exercise and a reduction in salivary IgA.[2] Gregory et al[3] investigated the effect of maximal, exhaustive exercise on immunoglobulins in breast milk. They reported sIgA concentrations to be lower 10 and 30 minutes after intense exercise but, by minute 60, they returned to resting values. However, women in this study exercised to exhaustion. Therefore, these results may be applicable only to those lactating women who engage in very strenuous activities at maximal intensities. The purpose of this study was to determine if 30 minutes of moderate exercise affects immunological properties of breast milk.

Methods

Healthy, nonsmoking, exclusively breastfeeding women (n=11) were recruited at three months postpartum. They reported exercising aerobically at least 30 minutes per day, three days per week, for the past six weeks. In our Human Performance Laboratory, they participated in an exercise session and a rest session, with the order of the sessions randomized. The exercise session was 30 minutes of exercise at 75% of predicted maximum heart rate and the rest session was 30 minutes of sitting rest. Milk samples were collected before, 10, and 60 minutes following exercise and rest by completely emptying both breasts with an electric breast pump. Samples were frozen at -70° until analysis. Milk was thawed and defatted by ultracentrifugation. Lactoferrin and sIgA were measured using ELISA.

*University of North Carolina at Greensboro, Greensboro, NC 27402-6170

Results

Table 1 shows that levels of sIgA and lactoferrin were similar between exercise and rest sessions and over time.

Table 1. Concentrations (g/L) of sIgA and lactoferrin before and after rest and exercise [a]

	Minute 0	Minute 10	Minute 60
Rest sIgA	1.46 ± 0.25	1.45 ± 0.22	1.47 ± 0.27
Exercise sIgA	1.61 ± 0.24	1.45 ± 0.23	1.63 ± 0.29
Rest lactoferrin	0.55 ± 0.05	0.51 ± 0.04	0.59 ± 0.06
Exercise lactoferrin	0.59 ± 0.03	0.60 ± 0.06	0.62 ± 0.06

[a]Means and standard error of the means.

Conclusion

There were no significant differences in sIgA or lactoferrin concentrations between rest and exercise sessions or within each session over time. These results suggest that 30 minutes of moderate aerobic exercise can increase cardiovascular fitness without affecting immunological properties of breast milk.

References

1. J.A. Woods, J.M. Davis, J.A. Smith, and D.C. Nieman, Exercise and cellular innate immune function, Med. Sci. Sports Exer. 31:57-66 (1999).
2. A.K. Blannin, P.J. Robson, N.P. Walsh, and A.M. Clark, The effect of exercising to exhaustion at different intensities on salivary immunoglobulin A, protein and electrolyte secretion, Int. J. Sports Med. 19:547-552 (1998).
3. R.L. Gregory, J.P. Wallace, L.E. Gfell, J. Marks, and B.A. King, Effects of exercise on milk immunoglobulin A, Med. Sci. Sports Exer. 29:1596-1601 (1997).

BIRTH WEIGHT AS A FACTOR OF LACTATION INITIATION IN POLISH MATERNITY HOSPITALS

Krystyna Mikiel-Kostyra and Joanna Mazur*

During the last decade the WHO/UNICEF Baby Friendly Hospital Initiative (BFHI) has had an impact on mother-newborn care after birth in thousands of hospitals worldwide. The 10 steps to successful breast-feeding became standard for good practice. However, the implementation of recommended procedures to the care of ill and prematurely born infants is very limited.

In Poland the newborn birth weight has appeared the most significant determinant of lactation initiation while in hospital.

The objective of the study was to examine the association between birth weight and breastfeeding initiation in Polish maternity hospitals with references to type of delivery, neonatal problems, skin-to-skin contact after birth, rooming-in.

Subject and Methods

The national survey on hospital practices was conducted by hospital staff in 427 maternity wards (95,6%) in 1995. The information on 11748 mothers-newborns pairs included: duration and complication of pregnancy, type of delivery, birth weight, neonatal problems (birth asphyxia, disorders with adaptation, jaundice, infections, congenital malformations or others), feeding method and breastfeeding promoting practices. The sample comprised 650 newborns under 2500g (5,5%).

Statistical Analyses

Univariate logistic regression model was estimated to calculate the probability of breastfeeding initiation for different birth weight exact points.

Bivariate logistic regression model was used to compare different groups of newborns and explain the additional influence of other dichotomous variables: type of

*National Research Institute of Mother and Child, Warsaw, Poland

delivery, neonatal problems and rooming-in and skin-to-skin contact.

Multivariate logistic regression model with removing variables in backwards stepwise procedure was applied to identify independent factors for breastfeeding initiation and estimate the odds ratio (OR) with 95% confidence interval (CI).

Analyses were performed using SPSS 8.0 package.

Results

In total 97,2% newborns were ever breast-fed while in hospital. The distribution of analysed factors in weight groups is presented in table I.

Table 1. Main sample characteristics by weight groups

Birth weight (g)	N	Breastfed (%)	Caesarean delivery (%)	Neonatal problems (%)	Skin-to-skin contact (%)	Rooming-in (%)
<= 1000	9	0,0	22,2	55,6	0,0	0,0
1000 – 1499	44	13,6	50,0	50,0	9,1	4,5
1500 – 1999	128	43,8	35,9	35,2	34,4	11,7
2000 – 2499	469	87,2	27,1	33,0	56,5	40,3
2500 – 2999	1951	97,9	15,8	20,5	76,3	58,8
>= 3000	9147	98,8	13,7	15,7	79,5	61,8
Total	11748	97,2	15,0	17,6	77,2	59,7

Bivariate logistic regression models showed that caesarean delivery and neonatal problems reduced probability of breastfeeding initiation in the group of newborns under 2000g with the same proportion for different birth weight. Above 2000g the influence of these two practices on breastfeeding initiation decreased. The practice of rooming-in and skin-to skin contact after birth increased the probability of breastfeeding and were more significant with decrease of birth weight.

Multivariate logistic regression model showed that the independent prognostic factors for breastfeeding initiation were: birth weight (increase in continuous scale), rooming-in [OR=5,5; CI(OR)=3,9-7,9], skin-to-skin contact [OR=3,4; CI(OR)=2,6-4,5] and lack of neonatal problems [OR=2,1; CI(OR)=1,6-2,7].

Conclusion

Implementation of procedures supportive for breastfeeding to hospital care of low birth weight infants can increase breastfeeding initiation.

MILK OLIGOSACCHARIDES VARY WITHIN INDIVIDUALS AND DURING LACTATION

D.S. Newburg, P. Chaturvedi, C.D. Warren, M. Altaye, A.L. Morrow, G.M. Ruiz-Palacios, and L.K. Pickering*

INTRODUCTION

Specific human milk oligosaccharides inhibit specific microbial pathogens.[1] As soluble homologs or analogs of host receptors for pathogens, milk oligosaccharides, especially fucosylated neutral oligosaccharides, may act as decoys to protect infants against disease.[2–4] To study a relationship between specific human milk oligosaccharides and disease in breast-fed infants, it is necessary to know the levels of specific oligosaccharides present in milks of individuals during lactation. Variation in total oligosaccharides, specific α1,2-linked oligosaccharides, and activities of fucosidases and fucosyltransferases has been reported.[5, 6] This study measures variation in individual fucosylated oligosaccharides in milks of individual mothers over the course of lactation.

Methods

Ninety milk samples were obtained from 12 women from 1 to 52 weeks of lactation. Oligosaccharides were quantified by HPLC using authentic standards. The concentration of 2-linked oligosaccharides was calculated as the total of 2'-FucLac, LNF-I, LDFH-I, and LDFT; 3/4-linked oligosaccharides as the total of 3-FucLac, LNF-II/III and LDFH-II. Covariation among oligosaccharides was determined by correlation analysis.

Results

In 11 of 12 donors, fucosyloligosaccharides containing α1,2-linked fucose were prevalent. The mean total oligosaccharide concentration in their milk declined from ~ 9 g/L for the 1st 14 weeks to ~ 4 g/L at 1 year of lactation. α1,2-linked and α1,3/4-linked

*L.K. Pickering, Eastern Virginia Medical School, Norfolk VA 23510

(without 2-linkages) oligosaccharide concentrations displayed distinct patterns. Ratios of α1,2-linked to α1,3/4-linked fucosyloligosaccharide concentrations undergo exponential decay over the course of lactation, defined by the equation $Y=4.85x10^{-0.013x}$ (r=0.99).

The donor whose milk was initially devoid of α 1,2-linked fucosyloligosaccharides, but contained α1,3- and α1,4-linked fucose, displayed inverse oligosaccharide ratios. The ratio of 2-linked fucosyloligosaccharides to those devoid of 2-linked fucose was close to zero until the 38th week of lactation, whereupon it rose toward unity by the 49th week.

Discussion

The types of individual oligosaccharides in milk did not change over the first 6 months of lactation, but the concentrations displayed greater variation than had been previously reported, both between mothers and over lactation.[7] Variation of human milk oligosaccharide expression has been attributed to differences in 2-fucosyltransferase activity. The difference between the dominant oligosaccharide pattern of the 11 donors and that of the one mother who initially lacked 2-linked fucose moieties is reminiscent of differences between the dominant secretor genotypes, who express 2-fucosyltransferase in milk (and other secretions), and recessive non-secretors, who are genetically unable to express this enzyme. Consistent with this interpretation is the tendency of the individual α1,2-fucosyloligosaccharide concentrations to covary among themselves over the course of lactation, implying that they are controlled by a single Fucα1,2-transferase. Individual α1,3/4-fucosyloligosaccharide concentrations covaried among themselves, implying control by a single Fucα1,3/4-transferase or coordinated control by separate Fucα1,3- and Fucα1,4-transferases. After 26 weeks, the concentrations of these two families covaried, coinciding with a shift toward the simultaneous synthesis of equal amounts of both families. This change did not coincide with ablactation, and may be characteristic of human lactation.

Our findings imply that changes in α1,2- and α1,3/4-fucosyltransferase activities underlie the variations in concentrations of individual milk oligosaccharides. Because some of these oligosaccharides protect against pathogens, such variation may result in differential abilities of milk from different mothers or different stages of lactation to protect breast-feeding infants. Fucosyloligosaccharide synthesis and the relationship between milk oligosaccharide levels and infant health warrant further study. Support by HD13021.

References

1. Newburg, D.S. (1996) Oligosaccharides and glycoconjugates in human milk: Their role in host defense. *J. Mammary Gland Biol. Neoplasia* **1**, 271-283.
2. Cervantes, L.E., Newburg, D.S. and Ruiz-Palacios, G.M. (1995) α1-2 Fucosylated chains (H-2 and Lewis[b]) are the main human milk receptor analogs for *Campylobacter*. *Pediatr. Res.* **37**, 171A.

3. Ruiz-Palacios, G.M., Cervantes, L.E., Newburg, D.S., Lopez-Vidal, Y. and Calva, J.J. (1992). *In vitro* models for studying *Campylobacter jejuni* infections. In Nachamkin, I., M.J. Blaser and L.S. Tomkins (eds.), *Campylobacter jejuni. Current Status and Future Trends.* American Society for Microbiology, Washington, D.C. pp. 176-183.7
4. Crane, J.K., Azar, S.S., Stam, A. and Newburg, D.S. (1994) Oligosaccharides from human milk block binding and activity of the *Escherichia coli* heat-stable enterotoxin (STa) in T84 intestinal cells. *J. Nutr.* **124,** 2358-2364.
5. Viverge, D., Grimmonprez, L., Cassanas, G., Bardet, L. and Solere, M. (1990) Discriminant carbohydrate components of human milk according to donor secretor types. *J. Pediatr. Gastroenterol. Nutr.* **11,** 365-370.
6. Wiederschain, G.Y. and Newburg, D.S. (1995) Human milk fucosyltransferase and α-L-fucosidase activities change during the course of lactation. *J. Nutr. Biochem.* **6,** 582-587.
7. Coppa, G.V., Gabrielli, O., Pierani, P., Catassi, C., Carlucci, A. and Giorgi, P.L. (1993) Changes in carbohydrate composition in human milk over 4 months of lactation. *Pediatrics* **91,** 637-641.

PHYSICIAN ATTITUDES ABOUT BREASTFEEDING

MaryAnn O'Hara, David Grossman, and Lorna Rhodes*

BACKGROUND

Despite mounting evidence of the health importance of breastfeeding, physicians demonstrate persistent, suboptimal knowledge about and support for breastfeeding.[1] Educational interventions can increase physician breastfeeding knowledge, and yet not necessarily improve behavior.[2] This suggests the need for a richer understanding of physician attitudes towards breastfeeding, including how those attitudes relate to utilization of data and clinical practice.

Purpose

To identify modifiable barriers to physician support for breastfeeding.

Methods

Pediatricians were observed during well-infant visits (n=20) and interviewed in-depth (n=30) about their perspectives on infant health supervision. Participants (after pilot interviews) were blind to research focus on breastfeeding to avoid bias toward normative responses and to gain a unique view of the place of breastfeeding within the broader context of infant health supervision. Key informants were selected to include a diversity of participants, practice types, populations served, and theoretical perspectives. Enrollment continued until we achieved thematic saturation. Interviews and office visits were audiotaped and transcribed, yielding approximately 600 pages of narrative texts that were qualitatively analyzed by a multidisciplinary research team.

Results

All participants described themselves as supportive of breastfeeding. However, analysis of the narratives revealed pervasive physician ambivalence about both their role

*MaryAnn O'Hara (Robert Wood Johnson Clinical Scholars Program and Department of Family Medicine), Davis Grossman (Department of Pediatrics, and Lorna Rhodes (Department of Anthropology). University of Washington, Seattle, Washington, USA 98195

Table 1. Sample contrasting pediatrician (MD) approaches to infant feeding vs. sleep

Area of contrast	Breastfeeding (BF)	Sleep
1) MD behavior	Hesitancy to encourage BF	Directive advice: - "Put babies to sleep alone & awake" - "Don't nurse in bed, to sleep, at night"
2) Fear of causing guilt	Paramount in tempering BF support	Rarely mentioned
3) Assumed type of issue	Lifestyle choice	Health imperative (infant development)
4) Comments on data about the issue	Data unfamiliar yet mistrusted ("Strong evidence" per AAP[a])	Data unfamiliar yet rarely questioned (Cultural norms. Scan data available)

[a]American Academy of Pediatrics

and the health importance of breastfeeding. Most doubted their ability to influence breastfeeding, citing inadequate counseling skills and lack of contact with mothers when breastfeeding decisions occur. Whereas participants gave few examples of how they help women to breastfeed, most talked in detail about how they try to avert maternal guilt by reassuring mothers that breastfeeding does not impact infant health. In contrast to hesitancy to encourage breastfeeding, physicians commonly gave directive advice about sleep without expressing concern about the impact on parental feelings (See Table 1). Most advised sleep practices that can undermine breastfeeding, such as admonishing against night feedings, nursing children to sleep, and co-sleeping. A minority of participants ("outliers") demonstrated consistent support for breastfeeding. This minority attributed their confidence and motivation to several factors: role models (who breastfed and/or demonstrated how to help women breastfeed), feedback from women who expressed sadness rather than guilt about not breastfeeding, and positive personal experience breastfeeding.

Conclusions

Despite physicians' self-reported support for breastfeeding, this study revealed pervasive ambivalence and potentially counterproductive practices. Physicians displayed contrasting approaches to breastfeeding vs. other central issues like sleep, including differential attention to perceived parental feelings. Based on direct recommendations from participants as well as qualitative analysis of their dialogue and practices, we propose five strategies for medical training/systems to improve physician motivation and ability to facilitate breastfeeding: 1) mentor physicians in both motivational interviewing and management of breastfeeding problems, 2) schedule clinic visits during the predictable periods when breastfeeding decisions and difficulties occur, 3) critique cultural vs. scientific bases for components of infant health supervision, 4) explore and tailor care to individual families' needs and perspective, and 5) enable physicians to fully breastfeed their own children.

References

1. R. Schandler, K. O'Connor, and R. Lawrence, Pediatricians' practices and attitudes regarding breastfeeding promotion, Pediatrics 103(3): p. e35 (1999).
2. D. Psiaki and C. Olson, Design and evaluation of breastfeeding for medical students, J. Trop. Med. 28:85-88 (1982).

MOTHERS WITH ASTHMA:

Does breastfeeding increase the asthma risk in their children?

Doris Oberle, Erika von Mutius, Rüdiger von Kries*

BACKGROUND

The association between breastfeeding and the development of atopic diseases in childhood remains controversial. While several studies, including prospective ones, found a reduction in the risk of atopic disease in children who were exclusively breastfed for at least 4 months, one recent study suggested an increased risk for asthma among breastfed children of mothers with asthma.[1] The aim of our study was to test the hypothesis that breastfed children of mothers with asthma have a higher risk for childhood asthma than non-breastfed children of asthmatic mothers.

Methods

Cross-sectional survey data for 9644 children entering school (aged 5 - 6 years) in two Bavarian rural districts (Upper Palatinate and Lower Bavaria) were analyzed. The data were obtained from the parents using a written questionnaire including ISAAC core questions on atopic disease and a number of additional questions on a wide range of determinants of atopic disease. The main outcome measures were lifetime prevalence of physician-diagnosed asthma and wheeze (ISAAC questions). Children were considered to have physician-diagnosed asthma if their parents reported that the diagnosis 'asthma' had been made at least once or that their child had been diagnosed with asthmatic, spastic or obstructive bronchitis more than once by a practitioner. Wheeze was assessed by parental report. Differences between prevalence of atopic disorders for non-breastfed and breastfed children within subgroups were tested using chi-square statistics. In addition, tests for homogeneity of odd ratios were computed. Crude and adjusted odd ratios were calculated using chi-square tests and unconditional logistic regression analysis.

*Doris Oberle, Rüdiger von Kries, Institute for Social Pediatrics and Adolescent Medicine, Ludwig-Maximilians-University, Heiglhofstr. 63, 81377 München, Germany, Erika von Mutius, University Children's Hospital, Lindwurmstr. 4, 80337 München, Germany

Results

Analyses revealed higher odds ratios for the relation between breastfeeding and childhood asthma when the mothers were asthmatic than when they were not. The adjusted odds ratio (aOR) for children with asthmatic mothers was 2.37 (95%-CI:1.29-4.33) compared with an aOR of 1.11 (95%CI: 0.86-1.44) for children without asthmatic mothers. Adjustment was made for the number of older siblings, parental education, family history of atopic disease except for maternal asthma and farming. Maternal asthma status was identified as an effect modifier for the association between breastfeeding and childhood asthma (Breslow-Day test: Chi-Square=6.209 with p=0.013). This result supports the notion that an adverse influence of breastfeeding on childhood asthma will only apply to children with asthmatic mothers. For wheezing, a similar, but insignificant effect was observed. The aOR for children with asthmatic mothers was 1.32 (95%CI: 0.86-2,01); the aOR for children whose mothers did not have asthma 1.03 (95%CI: 0.91-1.17). In children of mothers with hay fever or eczema, breastfeeding did not account for an increased asthma risk. Among children whose mothers had asthma, the prevalence of physician-diagnosed asthma was even lower among breastfed farmers' children than among non-breastfed farmers' children, but the numbers were too small for further statistical analysis.

Conclusions

Our data confirm an increased risk for physician-diagnosed asthma related to breastfeeding in children of asthmatic mothers, whereas the results were less clear regarding the symptoms of asthma (wheeze). It has been suggested that the reduced asthma risk in children of farmers may result from increased bacterial antigen exposure associated with the living conditions on a farm. During early childhood, farm-specific antigens may skew the immune response towards the T helper 1 (Th1)-pathway leading to a correct cytokine balance. **Breastfeeding is known to protect against infections in early childhood, and infection influences the maturation of the immune system by inducing a systemic and non-specific switch to Th1 activity. We therefore hypothesized that the increased asthma risk related to breastfeeding in children of asthmatic mothers would be absent in farmers' children as they benefited from early exposure to immunological stimuli in the farming environment and experienced adequate priming of Th1 activity.** As to physician-diagnosed asthma, our results appear compatible with this concept, whereas the data regarding wheeze do not. However, the numbers are small precluding any conclusive statistics.

References

1. Wright, A.L., Holberg, C.J., et al, Maternal asthma alters relation to infant feeding to asthma childhood. In Short and long term effects of breastfeeding on child health. Edited by Berthold Koletzko et al., Kluwer Academic/Plenum Publishers, (2000).

PREGNANCY INTENTIONS AND BREAST-FEEDING SUCCESS: A CROSS-CULTURAL RELATIONSHIP?

Rafael Pérez-Escamilla and Bridget Chinebuah

INTRODUCTION

Half of pregnancies in the USA and over 40% in developing countries are unplanned. Studies conducted in New York[1] and Peru[2] suggest that unplanned pregnancy is a risk factor for poor breast-feeding outcomes. The purpose of this project was to find out if this association is also present in Ghana, West Africa.

Methods

Women participating in the 1993 Demographic and Health Survey (DHS) from Ghana who had children during the 3 years preceding the survey were included in the analyses if they had retrospective pregnancy intentions data [planned vs. unplanned pregnancy (mistimed or unwanted)] and current breast-feeding status information. Child age-adjusted multivariate logistic regression analyses were conducted to examine the association between pregnancy intentions and the likelihood of breastfeeding at the time of the survey (n = 1882).

Results

Women with planned pregnancies were significantly more likely to breastfeed (Odds Ratio = 1.41, 95% Confidence Interval = 1.05-1.88) than their counterparts with unplanned pregnancies. Other risk factors for not breastfeeding at the time of the survey were: older child, higher formal education, hospital delivery, and urban residence.

*Department of Nutritional Sciences, University of Connecticut, Storrs, CT 06269-4017, rperez@canr.uconn.edu.

Conclusion

Results from Ghana, New York State, and Peru suggest that pregnancy intentions should be taken into account by breast-feeding promotion programs.

References

1. Dye TD, Wojtowycz MA, Aubry RH, Quade J, and Kilburn H. Unintended pregnancy and breast-feeding behavior, Amer. J. Public Health. 87:1709-1711 (1997).
2. Perez-Escamilla R, Cobas JA, Balcazar H, and Holland Benin M. Specifying the antecedents of breast-feeding duration in Peru through a structural equation model, Public Health Nutrition 2:461-467 (1999).

EFFECTS OF MILK-BORNE PHYSIOLOGICAL CONCENTRATIONS OF INSULIN-LIKE GROWTH FACTORS-I OR –II (IGF-I, -II) UPON GROWTH IN THE ARTIFICIALLY FED (AR) SUCKLING RAT

Anthony F. Philipps#, Bohuslav Dvorak, Pamela J. Kling, James G. Grille, Cathy S. Williams, Abdul M. Fellah, Robert S. McCuskey*, and Otakar Koldovský+

Insulin-Like Growth Factors-I and –II are potent mitogenic peptides that are synthesized by many cell types in mammals (Sara and Hall, 1990, Styne, 1998). These hormones appear to be important in growth regulation of the organism, including during perinatal life (Cohick, et al., 1993, Philipps, et al., 1988). This presumption is strengthened by recent observations of mice bearing nonfunctional genes for IGF-I, -II (Baker, et al., 1993, Lau, et al., 1994), since these have significant growth restriction as well as a high perinatal mortality rate. Attention has also recently focused upon the presence of IGF's in milk of many species in biologically relevant concentrations. Studies from our laboratories (Philipps, et al., 1991) as well as others (Donovan, et al., 1990) have shown that IGF-I and –II are present in the milk of at least several species. IGF's are present in most biological fluids in close association with binding proteins and it has been suggested that these binding proteins play a significant role in modulation of IGF-receptor interaction and as a potential reservoir for IGF in the circulation. The serum concentration of IGFBP, particularly BP2 in the suckling, may be responsive to levels of nutritional sufficiency.

Over the past several years, we have utilized a suckling rat model of neonatal nutrition, with particular reference to premature human babies, to explore whether or not milk-borne IGF's have any significant biological effects (Philipps, et al., 1997). In this preparation, suckling rats were fed via gastrostomy tube over 3 to 4 day periods of time, hence the term "artificially reared" (AR). The milk used was a rat milk substitute (RMS) with or without the addition of rh IGF-I in concentrations approximately 10X above physiological. These studies showed that animals fed the IGF supplement RMS had improved somatic growth as well as that of liver and brain, more normal serum IGF-I

#Department of Pediatrics, University of California Davis, Sacramento, CA 95817 and Departments of Pediatrics and *Anatony and Cell Biology, University of Arizona, Tucson, AZ 85718
+Deceased

levels and evidence of enhanced intestinal development. Studies by other investigators have also pointed toward a role for milk-borne IGF in stimulating intestinal development (Staley, et al., 1998, Lemmey, et al., 1991).

More recently, we have studied the effects of either rh IGF-I or –II in the same suckling rat design, except for the use of IGF in amounts similar to those found in native milk. In separate experiments we also investigated the potential for these peptides to be absorbed into the circulation. Weights and skeletal growth were similar between the groups at the end of the study period. Liver weight was increased over controls in the IGF-II but not the IGF-I fed rats. Both IGF-I and –II were associated with increased jejunal weight and protein content in comparison to the controls. Measurements of intestinal cell growth (crypt depth and enterocyte migration) were also increased in both groups of IGF-supplemented animals. Serum glucose concentrations, IGF-I, -II and IGF BP-2 levels were highest in dam fed rats and lowest in the RMS group and intermediate in both IGF fed groups. Milk-borne IGF-I but not –II could be detected intact in portal venous blood from these sucklings (Philipps, et al., 2000). We conclude that milk-borne IGF-I and –II, when given in physiological concentrations, induce stimulation of intestinal growth in suckling rats and that IGF-II has a specific effect on hepatic growth. The AR suckling rat may be useful as a model in studying the potential benefit of supplementing artificial human formulas with IGF for premature babies.

References

W.S. Cohick, and D.R. Clemmons, The insulin-like growth factors, Ann. Rev. Physiol. 55:131-153, 1993.

V. Sara, and K. Hall, Insulin-like growth factors and their binding proteins, Physiol. Rev. 70:591-614, 1990.

D.M. Styne, Fetal growth: emerging concepts in perinatal endocrinology, Clin. Perinatol. 2:917-938, 1998.

A.F. Philipps, B. Persson, K. Hall, M. Luke, A. Skottner, T. Sanongen, and V. Sara, The effects of biosynthetic insulin-like growth factor-I supplementation on somatic growth, maturation, and erythropoiesis on the neonatal rat, Pediatr. Res. 23:298-305, 1988.

M.M.H. Lau, C.E.H. Stewart, Z. Lui, I.I. Bhatt, O, Rotwein, and C.L. Stewart, Loss of the imprinted IGFZ/cation-independent mannose 6-phosphate receptor results in fetal overgrowth and perinatal lethality, Genes Dev. 8:2953-2963, 1994.

J. Baker, J.P. Liu, E.J. Robertson, and A. Efstratiodis, Role of insulin-like growth factors in embryonic and postnatal growth, Cell 75:73-82, 1993.

A.F. Philipps, J.M. Wilson, R.K. Rao, D.M. McCraken, and O. Koldovský, Presence of insulin-like growth factors and their binding proteins in rat milk; in Raizoda MK, LeRoith D (eds) Molecular Biology and Physiology of Insulin and Insulin-like Growth Factors, Plenum Press, NY, pp. 179-186, 1991.

S.M. Donovan, and J. Odle, Growth factors in milk as mediators of infant development, Annu. Rev. Nutr. 14:147, 1994.

A.F. Philipps, G.G. Anderson, B. Dvorak, C.S. Williams, M. Lake, A.V. LeBouton, and O. Koldovský, Growth of artificially fed infant rats: effect of supplementation with insulin-like growth factor-I, Am. J. Physiol. 272 (Regulatory Integrative Comp. Physiol. 41): R1532-R1539, 1997.

M.D. Staley, A. Gibson, J.F. Herbein, C.E. Grosvenor, and C.R. Baumrucker, Rat milk and dietary long arginine 3 insulin-like growth factor-I promote intestinal growth of newborn rat pups, Pediatr. Res. 44:512-518, 1998.

A.B. Lemmey, A.A. Martin, L.C. Read, F.M. Tomas, P.C. Owens, and F.J. Ballard, IGF-I and the truncated analogue des-(1-3) IGF-I enhance growth in rats after gut resection, Am. J. Physiol. 260 (Endocrol. Metab. 23) E213--219, 1991.

A.F. Philipps, J. Grille, B. Dvorak, P. Kling, and O. Koldovský, Gastrointestinal absorption of receptor active IGF-I and –II into the portal circulation of suckling rats, J. Pediatr. Gastro. Nutr. 31:128-135, 2000.

SPEECH IN HONOR OF THE ISRHML ERLICH-KOLDOVSKÝ - YOUNG INVESTIGATOR AWARD SEPTEMBER 18, 2000, TUCSON, AZ, USA

Anthony F. Philipps*

I am honored to give this brief address in honor of one of the scientists whose names grace this award, Professor Otakar Koldovský. Dr. Koldovský was a friend and colleague of mine for the last 10 years of our joint careers in the Department of Pediatrics within the University of Arizona here in Tucson. I first met him in 1977 at the annual meeting of the Perinatal Research Society in Mont Tremblant, Quebec, Canada. I remember this meeting very well since, as a very junior assistant Professor attending this meeting as a guest, I was impressed by three things about the man; Otakar's shiny pate, his booming voice and his encyclopedic knowledge of his subject. Little did I know that 10 years later, he would be interviewing me for the position as Section Chief in Neonatology and Nutritional Sciences and as a potential colleague in research at the University of Arizona. As you can see from the photograph (that his friends picked out as representative of his "professional "photographs) he and I shared a keen interest in bowties, among other things. His untimely death in April of 1998 left all whom he had worked with and known with a tremendous sense of loss. I am pleased to tell you that Professor Koldovký's wife, Eva Koldovský, is here this evening to share in the presentation of the award.

Otakar Koldovský was born March 31, 1930 in Olomouc, Moravia, and the son of an ophthalmologist and his wife. The family moved to Prague where Otakar and his brother, Pavel, were raised and educated. Otakar attended Charles University and studied at the Institute of Physiology within the Czechoslovak Academy of Sciences under the direction of Professor Krěcěk. He then went into medical practice in Karlovy Vary, and, for a time, worked with coal miners in the mines in that region. He came back to Prague to do graduate work with Professor Krěcěk, studying the ontogenetic development of the intestine and intestinal function. During this period he became acquainted with another graduate student interested in this area, Peter Hahn. These two graduate students

*Anthony F. Philipps, Department of Pediatrics, University of California Davis, Sacramento, CA 95817

maintained a life-long friendship and collegial relationship. During this period Professor Koldovský won the CIBA Prize in Physiology and accepted it on a trip to London.

Further work in the area of intestinal development led to a fateful 1965 trip to the United States, where he met and worked with Professor Norman Kretchmer at Stanford University. During the unrest that followed in his native country, Dr. Koldovský emigrated to the U.S. in 1968, initially working at Stanford and then achieving a faculty position at the University of Pennsylvania. He married in 1971 and continued his work related to intestinal enzyme systems maturation. In 1980, he moved to Tucson to take a position as Professor of Pediatrics and Physiology at the University of Arizona College of Medicine. During this period he also developed many close ties with investigators in the Departments of Pharmacology and Anatomy and Cell Biology.

Otakar Koldovský by this time had also begun to delve into the mechanism underlying changes in intestinal maturation during the suckling period. He began to realize the importance of mammalian milk for its growth promoting properties, exclusive of the known nutrients present. This led him to develop the concept of "Biologically Active Substances" ("BAS") in milk and to explore what these substances might be and how they might act using artificially reared animals fed in the presence or absence of a variety of growth factors (the 'pup in the cup" model). Ultimately, these studies and his prior work account for over 300 scientific papers, reviews and book chapters, all related to intestinal development, highly important subjects for the understanding of nutrition of the human infant. Professor Koldovský was awarded the American Academy of Pediatrics' Nutrition Award and the Harry Shwachman Award in Pediatric Gastroenterology and Nutrition and was also invited to lecture at scientific meetings and academic institutions all over the world.

At the memorial service held in his honor shortly after his death, many came to share their thoughts about having known and studied under this remarkable man. What clearly stood out was his dedication to learning and to the students interested in learning more about nutrition. He imbued all of us with a sense of enthusiasm, and impressed upon all of us the necessity for scientific honesty and rigor in all of our studies. I believe these lessons have been well learned under his direction and that they truly form his legacy to the scientific community as a whole.

Otakar Koldovský, M.D., Ph.D. (1930-1998)

ANTIBODY RESPONSE TO HAEMOPHILUS INFLUENZAE TYPE B (HIB) IN CHILDREN WITH INVASIVE HIB INFECTION IN RELATION TO DURATION OF EXCLUSIVE BREASTFEEDING

Sven Arne Silfverdal[1]; Lennart Bodin[1]; Marina Ulanova[3]; Mirjana Hahn-Zoric[3]; Lars Å.Hanson[3]; Per Olcén[4]

Key words: Breastfeeding, *Haemophilus influenzae* infection, Hib, antibody response, IgG2

INTRODUCTION

In an earlier case-control study we found a protective effect of breastfeeding on invasive *Haemophilus influenzae* type b (Hib) infection beyond the period of breastfeeding itself.[1] In our ecologic study we also found an indication of a protective effect of breastfeeding at a population level.[2] The present investigation was performed in order to analyze the serum antibody response to Hib in children with invasive Hib infection in relation to age, diagnosis, smoking, maternal anti-Hib antibodies and duration of breast-feeding.

Subjects and Methods

From a prospective study, conducted between 1987 and 1992 in Örebro, Sweden, sets of sera were obtained from 30 children with invasive Hib infection below 6 years of age and their mothers. Sixteen children were 18 months or older and 14 below that age. The duration of breastfeeding was registered in weeks based on the mother's recall and checked with records at the Baby Well Clinic. Long duration of exclusive breastfeeding was defined as 13 weeks or more while short duration was equal to 12 weeks or less.

Assays for anti-Hib antibodies of the IgG1, IgG2, IgA and IgM isotypes were performed with sera taken as a mean at 5, 26 and 82 days after onset of the Hib disease employing the Elisa technique.

Non-parametric statistics were used in the basic analysis. In the older age group multiple regression analysis was done on logarithmic values of IgG1 and IgG2 anti-Hib

Departments of [1]Pediatrics, [2]Biostatistics and Epidemiology, [4]Microbiology and Immunology, Örebro Medical Centre Hospital, Örebro, [3]Department of Clinical Immunology, Göteborg University, Göteborg, Sweden

antibody levels with age, duration of breastfeeding, diagnosis, passive smoking and maternal antibody levels as explanatory factors.

Findings

Children 18 months or older with 13 weeks or more of exclusive breastfeeding had significantly higher anti-Hib antibody levels of the IgG1, IgG2, IgA and IgM isotypes than those with less than 13 weeks of exclusive breastfeeding (Figure 1). The difference between the breastfeeding groups was significant for the IgG2 isotype in the acute disease as well as in early convalescence (Table 1).

In regression analyses the association between the duration of exclusive breastfeeding and the anti-Hib IgG2 level was significant when breastfeeding, type of Hib infection, maternal anti-Hib antibody level and age were used as explanatory factors ($p=0.038$).

Discussion

In the present study, the antibody level of the IgG2 subclass was strongly associated with the duration of breastfeeding. Other studies have documented that IgG2 production is dependent on IFN-γ.[3-6] Human colostrum and mature milk are rich in IFN-γ and IFN-γ producing cells.[7,8] A possible explanation for a selective stimulatory effect of breastfeeding on the IgG2 anti-Hib antibody response may be through the IFN-γ in the milk.

In conclusion, this study indicates that breastfeeding enhances the specific IgG2 anti-Hib response to invasive Hib infection years after the termination of breastfeeding.

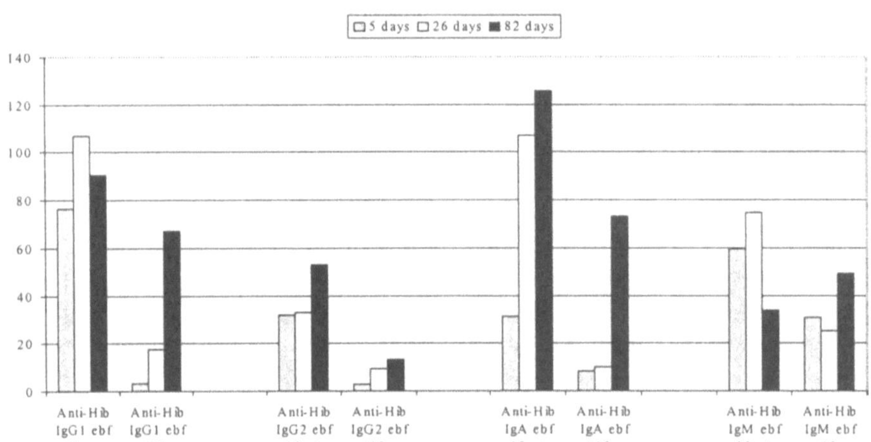

Median serum levels in % of standard are shown. Three serum samples were taken; during acute illness, and in early and late convalescence at an average of 5, 26 and 82 days after onset of invasive Hib disease.

Figure 1. Serum anti-Hib antibody levels in children with invasive Hib disease aged 18 months or older in relation to duration of exclusive breastfeeding.

Table 1. Serum anti-Hib antibody levels in children with invasive Hib disease aged 18 months or older in relation to duration of exclusive breastfeeding.

	0 – 12 days				15 – 35 days				36 – 155 days			
	N	Median	Range	P[1]	n	Median	Range	P[1]	n	Median	Range	P[1]
IgG1												
Ebf[2]<13 w	10	3.3	1 – 81		9	18	1 – 106		6	67	16 – 91	
13 + w	6	77	0 – 117	0.22	3	107	96 – 142	***0.02***	4	91	23 – 115	0.48
IgG2												
Ebf< 13 w	10	2.8	0 – 7		9	9.5	0 – 51		6	13	3.0 – 74	
13 + w	6	32	0 – 151	***0.02***	3	33	31 – 101	***0.04***	4	53	6 - 145	0.26
IgA												
Ebf< 13 w	10	8.3	2.6-155		9	10	2 – 87		6	73	7 - 192	
13 + w	6	31	1 – 261	0.21	3	107	16 – 272	0.34	4	126	36 - 230	0.48
IgM												
Ebf< 13 w	10	31	1 – 121		9	25	11 – 83		6	49	14 - 105	
13 + w	6	60	7 – 114	***0.04***	3	75	69 – 115	0.21	4	34	17 - 62	0.61

[1]P-value for testing difference between breastfeeding groups, Mann-Whitney's U.
[2]Ebf = duration of exclusive breastfeeding in weeks dichotomized to less than 13 weeks and 13 weeks or more respectively.

References

1-2, 4 and 5. SA Silfverdal (1997, 1999), R Inoue (1995), Y Kawano (1996), see Hanson et al. Immune system modulation by human milk in this volume.

3. Y. Kawano, T. Noma, and J. Yata. Regulation of human IgG subclass production by cytokines. IFN-γ and IL-6 act antagonistically in the induction of human IgG1 but additively in the induction of IgG2. J. Immunol. 153:4948-4958 (1994).

6. C. Servet-Delprat, J.M. Bridon, O. Djossou, S.A. Yahia, J. Banchereau, and F.Briere. Delayed IgG2 humoral response in infants is not due to intrinsic T or B cell defects, Int. Immunol. 8:1495-1502 (1996).

7. BA Eglinton, DM Roberton, and AG Cummins. Phenotype of T cells, their soluble receptor levels, and cytokine profile of human breast milk, Immunol. Cell Biol. 72:306-313 (1994).

8. AS Goldman, S Chheda, R Garofalo,, and FC Schmalstieg. Cytokines in human milk: properties and potential effects upon the mammary gland and the neonate, J. Mammary Gland Biol. Neoplasia 1:251-258 (1996).

ESTIMATED BIOLOGICAL VARIATION OF THE MATURE HUMAN MILK FATTY ACID COMPOSITION

Ella N. Smit, Ingrid A. Martini, E.Rudy Boersma, Frits A.J.Muskiet[1]

INTRODUCTION

Human milk is considered to be the single best food for babies and young infants. One of the reasons relates to its specific fatty acid (FA) composition. Although some scientists tend to stress the uniformity of human milk,[1, 2] Jensen in his recently published review[3] pointed out that the range in milk FA contents is wide and that there are insufficient reliable data showing the ranges of the biologically important FA. We estimated the biological variation of the 27 major FA in mature human milk, using samples collected over the last 25 years in different countries.

Materials and Methods

We studied 465 mature milk samples derived from The Netherlands (n=222), the Caribbean Region (Belize, Curaçao, Dominica, St. Lucia, St Vincent, Surinam; n=159), Jerusalem (n=63), Tanzania (n=11) and Pakistan (n=10). Their FA compositions were analysed in the same laboratory by a single capillary gas chromatographic technique. Analytical coefficients of variation (CV_{anal}) were taken from the series-to-series precision.[4] Inter-individual biological variation (CV_{biol}) was calculated from the observed variation (CV_{obs}) using $CV_{biol} = (CV^2_{obs} - CV^2_{anal})^{1/2}$.

Results

The figure depicts the calculated CV_{biol} for the 27 FA that were quantified in the 465 milk samples. The largest CV_{biol} was observed for 20:5ω3 (120.8%), followed by 22:6ω3 (69.1%) and 24:1ω9 (63.9%). The smallest CV_{biol} were observed for 16:0 (13.4%), 18:1ω9 (18.8%) and 18:0 (22.8%). The CV_{biol} were somewhat different among the various countries, but the patterns were remarkably similar (results not shown).

[1]University Hospital Groningen, Dept. CMC-V, Room Y1147, P.O. Box 30.001, 9700 RB Groningen, The Netherlands

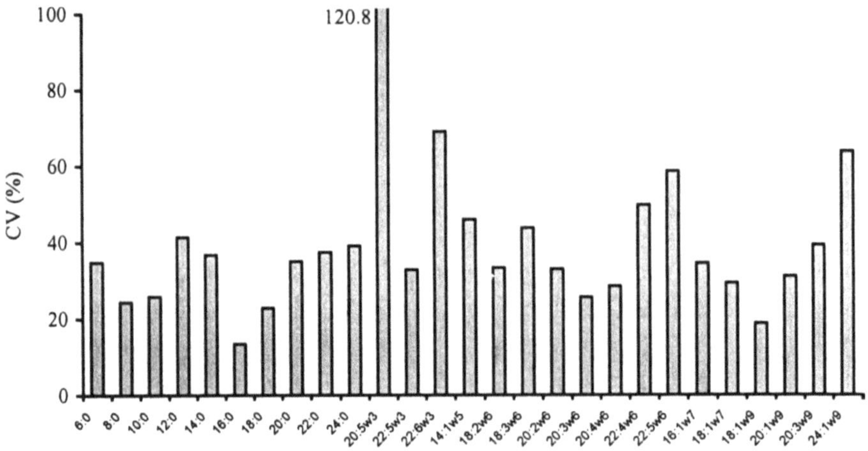

The biological variation of different milk fatty acids.

Discussion

It is generally accepted that milk FA content is to a certain extent dependent on the maternal diet. Factors such as gestational age, parity and postpartum time, and diseases also affect milk FA composition. The observed variation of breast milk FA composition does not reflect the real biological variation, because it also harbours analytical variation.

The high CV_{biol} of 20:5ω3 and 22:6ω3 were not unexpected, since many investigators have demonstrated the high variability of these FA in breast milk. The two quantitatively most important FA 16:0 (palmitic acid) and 18:1ω9 (oleic acid) have the lowest CV_{biol}. Linoleic acid (18:2ω6; LA), the most abundant PUFA, exhibited a CV_{biol} of 33.4%. Since LA derives exclusively from the diet, and milk LA is reported to be sensitive to LA intake, one might have expected a larger CV_{biol}. The rather low CV_{biol} of the other important FA of the ω6 series, 20:4ω6 (28.5%), is in accordance with observations by others that milk 20:4ω6 is not much affected by dietary changes.[3]

The observed large CV_{biol} of a number of important milk FA and uncertainty regarding appropriate maternal diet raise questions about the selection of a 'gold standard' for human milk FA composition.

References

1. B. Hall, Uniformity of human milk, Am. J. Clin. Nutr. 32:304-312 (1979).
2. B. Koletzko, I. Thiel, P.O. Abiodun, The fatty acid composition of human milk in Europe and Africa, J. Pediatr. 120:S62-70 (1992).
3. R.G. Jensen, Lipids in human milk, Lipids 34(12):1243-1271 (1999).
4. G. van der Steege, F.A.J. Muskiet, I.A. Martini, N.H. Hutter, E.R. Boersma, Simultaneous quantification of total medium- and long-chain fatty acids in human milk by capillary gas chromatography with split injection, J. Chromatogr. Biomed. Appl. 415:1-11 (1987).

CONTENT OF CONJUGATED LINOLEIC ACIDS, *cis*-9, *trans*-11-18:2 and *trans*-10,*cis*-12-18:2, IN BREAST MILK FROM BRAZILIAN WOMEN

Association with milk composition and diet

Alexandre G. Torres, Flávia Meneses, and Nádia M. F. Trugo*

INTRODUCTION

Conjugated linoleic acids (CLA) are found in food products from ruminants, and the *c*-9,*t*-11 and *t*-10,*c*-12 isomers show many biological effects (Pariza *et al.*, 2000). Data on CLA contents in human milk are scarce. The determination of human milk CLA content and its interaction with other fatty acids, in populations with different dietary habits could be of concern to the mother-infant dyad. The aims of the present work were to determine the fatty acid contents of milk from underprivileged Brazilian nursing women, with special reference to CLA, and to investigate the association of human milk CLA contents with other milk fatty acids, and also with the intake of dairy fat.

Subjects and Methods

Breast milk samples were obtained from 13 healthy women (24-120 d post-partum) attending a public day-care clinic in Rio de Janeiro – Brazil, after informed consent. The study protocol was approved by the Ethical Committees of the Institutions involved. For fatty acid analysis, lipid extracts were submitted to transesterification (Kramer *et al.*, 1997), and the methyl esters were analyzed by capillary gas chromatography. CLA were identified by means of standards and confirmed by mass spectrometry.

Results

Dairy fat contributed (mean ± SE) to 38.5 ± 5.7 % of the fat intake, which represented 34.1 ± 1.6 % of the energy intake of the volunteers. The dietary 18:2n-6

*Laboratório de Bioquímica Nutricional e de Alimentos, Departamento de Bioquímica, Instituto de Química, Universidade Federal do Rio de Janeiro, Rio de Janeiro, RJ, Brazil

and 18:3n-3 intakes were (mean ± SD), respectively, 3.4 ± 0.8 and 0.33 ± 0.05 (kcal%). Breast milk fat was 33.5 ± 4.8 g/l (mean ± SE). Sixty-seven fatty acids were identified, including odd and branched chain, and conjugated linoleic acids. The contents of CLA (mean ± SD, weight %) were: *c*-9,*t*-11-18:2, 0.49 ± 0.17; *t*-10,*c*-12-18:2, 0.02 ± 0.01; *c,c*-18:2, 0.02 ± 0.01; *t,t*-18:2, traces. Partial correlation analysis showed that milk *c*-9,*t*-11-18:2 was associated with dairy fat intake (r= 0.63), controlling for 18:1, 18:2n-6 and *t*-10,*c*-12-18:2 in milk.

Table 1. Associations between human milk CLA content and fatty acid-related variables (partial correlation analysis)

Variables		Covariates	r	n
*Trans*10,*cis*-12-18:2	18:1	d18:1[a]; 18:2n-6; *c*-9,*t*-11-18:2	-0.61	12
*Trans*10,*cis*-12-18:2	18:1/18:0	d18:1; 18:2n-6; *c*-9,*t*-11-18:2	-0.48	12
*Trans*10,*cis*-12-18:2	Milk fat	dfat[b]; *c*-9,*t*-11-18:2; MCFA[c]; 18:2n-6	-0.49	12
cis-9,*trans*-11-18:2	20:3/18:2	d18:2[d]; lac[e]; *c,c*-18:2	-0.49	13

[a] dietary 18:1; [b] dietary fat; [c] medium chain fatty acids; [d] dietary 18:2; [e] lactational period.

Discussion and Conclusions

Four CLA isomers (*c*-9,*t*-11; *t*-10,*c*-12; *c,c*; *t,t*) were identified in the milk of Brazilian donors including *c,c* isomers that, to our knowledge, have not been reported to be present in human milk. The *c*-9,*t*-11-18:2 content of the milk from Brazilian women was similar to that of donors from Australia and from USA (Idaho) (Jensen, 1999). A positive association was found between dairy fat intake and the *c*-9,*t*-11-18:2 content in milk. Milk *t*-10,*c*-12-18:2 was negatively associated with milk fat content, possibly as a consequence of its inhibitory effect on lipogenesis. Medium chain fatty acid contents seemed not to be influenced by *t*-10,*c*-12-18:2, as indicated by correlation analysis. Milk *t*-10,*c*-12-18:2 seemed to negatively affect the content of 18:1n-9 and the 18:1n-9/18:0 ratio, which can be used as an indirect marker of stearoyl-CoA desaturase activity. These associations are in agreement with *in vitro* experiments in which the level of stearoyl-CoA desaturase gene expression was reduced by *t*-10,*c*-12-18:2 (Pariza *et al.*, 2000).

Acknowledgements

The authors thank Dr. R. Jensen for the invaluable suggestions concerning fatty acid analyses, Dr. R. Adlof for the generous gift of CLA standards, and CNPq, FAPERJ, FINEP and CAPES (Brazil) for financial support.

References

Jensen RG, Lipids in human milk, Lipids 34:1243-1271 (1999).

Kramer JKG, Fellner V, Dugan MER, Sauer FD, Mossoba MM, and Yurawecz MP, Evaluating acid and base catalysts in the methylation of milk and rumen fatty acids with special emphasis on conjugated dienes and total *trans* fatty acids, Lipids 32(11):1219-1228 (1997).

Pariza MW, Park Y, and Cook ME, Mechanisms of action of conjugated linoleic acid: evidence and speculation, Proc. Soc. Exp. Biol. Med. 223:8-13 (2000).

ERYTHROCYTE MEMBRANE FATTY ACID COMPOSITION OF BRAZILIAN NURSING WOMEN

Effect of the lactational period

Alexandre G. Torres, Flávia Meneses, and Nádia M. F. Trugo*

INTRODUCTION

Fatty acid composition of phospholipids in mature erythrocyte membranes are directly affected by alterations in the intermediate metabolism, transit of substrates between organs and tissues, and by changes in the dietary fat intake. During lactation, maternal fatty acid metabolism is altered to fulfill the demand for milk synthesis and, as lactation progresses, further endocrine and metabolic alterations occur in the maternal organism. All these changes might affect cell membrane composition, which could be of concern to maternal health, depending on the quality and extent of such alterations. Besides, data on fatty acid composition of erythrocyte membrane in lactating women with different patterns of fat and energy intakes, nutritional status and lactational periods are scarce. The aims of the present work were to determine erythrocyte fatty acid composition of Brazilian low-income nursing women and to evaluate the relationship between the lactational period and erythrocyte fatty acid composition.

Subjects and Methods

Blood samples were taken, after an overnight fast, from 24 healthy nursing women (20 - 120 days post-partum), attending a public day-care clinic in Rio de Janeiro – Brazil. The study protocol was approved by the Ethical Committees of the Institutions involved and the volunteers gave their informed consent. Washed erythrocytes were submitted to direct transesterification (Lepage & Roy, 1986). The resulting methyl esters were analyzed by capillary gas chromatography (Omegawax-320, Supelco, Co.) and the relative quantities (weight %) of the fatty acids determined.

*Nadia M.F. Trugo, Laboratório de Bioquímica Nutricional e de Alimentos, Departamento de Bioquímica, Instituto de Química, Universidade Federal do Rio de Janeiro, Rio de Janeiro, R.J., Brazil.

Results

Nineteen fatty acids were identified and quantified (weight %) in the samples. Polyunsaturated fatty acid (PUFA) contents (mean ± SD) were: 18:2n-6, 10.5 ± 1.3; 20:2n-6, 0.4 ± 0.2; 20:3n-6, 1.4 ± 0.3; 20:4n-6, 12.3 ± 1.8; 22:4n-6, 2.8 ± 0.5; 22:5n-6, 0.7 ± 0.5; 20:5n-3, 0.4 ± 0.3; 22:5n-3, 1.7 ± 0.7; 22:6n-3, 4.7 ± 1.7; total n-6, 27.9 ± 2.5; total n-3, 6.9 ± 2.2; total PUFA, 34.8 ± 3.3; n-6/n-3 ratio, 4.4 ± 1.1. The influence of the lactational period on erythrocyte membrane fatty acid composition was investigated by correlation analysis, and the results are shown in Table 1.

Table 1. Significant correlation coefficients between lactational period and erythrocyte membrane fatty acids.

Erythrocyte membrane	Pearson[a] or Spearman[b] correlation			Partial correlation[c]	
Fatty acid	r	p	n	r	n
18:2n-6	0.52[a]	0.0116	23	0.56[d]	11
22:6n-3	-0.42[b]	0.0464	24	-0.71[e]	11

[c] Covariates: [d] age, 18:2n-6 and 18:3n-3 intake (kcal%), and erythrocyte 18:1n-9, 18:2n-6, 20:4n-6, 22:4n-6 and total saturated and n-3 fatty acids; [e] age, 18:2n-6 and 18:3n-3 intake (kcal%), and erythrocyte 18:1n-9, 18:2n-6, 20:5n-3 and total saturated fatty acids.

Discussion and Conclusions

Erythrocyte PUFA composition of the volunteers was similar to that of North American (Henderson *et al.*, 1992) and Australian (Makrides *et al.*, 1996) nursing women, suggesting that this composition may be less variable in nursing women than in mixed populations. The lactational period seemed to affect erythrocyte membrane fatty acid composition. 18:2n-6 increased whereas 22:6n-3 decreased with the progress of lactation. Similar results were found for human milk in a longitudinal study (Jackson *et al.*, 1994). Our results show that lactation and its progress may affect the erythrocyte membrane fatty acid composition. These effects could be the result of endocrine and metabolic alterations associated with lactation in adipose tissue mobilization.

Acknowledgements

The authors thank CNPq, FAPERJ, FINEP and CAPES (Brazil) for financial support.

References

Henderson RA, Jensen RG, Lammi-Keefe CJ, Ferris AM, and Dardick KR, Effect of fish oil on the fatty acid composition of human milk and maternal and infant erythrocytes, Lipids 27(11):863-869 (1992).

Jackson MB, Lammi-Keefe CJ, Jensen RG, Couch SC, and Ferris AM, Total lipid and fatty acid composition of milk in women with and without insulin-dependent diabetes mellitus, Am. J. Clin. Nutr. 60(3):353-361 (1994).

Lepage G e Roy CC, Direct transesterification of all classes of lipids in a one-step reaction, J. Lipid Res. 27:114-120 (1986).

Makrides M, Neumann MA, and Gibson RA, Effect of maternal docosahexaenoic acid (DHA) supplementation on breast milk composition, Eur. J. Clin. Nutr. 50:352-357 (1996).

TGFα IS ASSOCIATED WITH THE MILK FAT GLOBULE MEMBRANE

Carol L. Wagner, John E. Baatz, Donna W. Forsythe, Gabriel A. Virella, and Stuart Patton *

INTRODUCTION

Human milk plays a significant role in postnatal gut maturation, and may serve as a vehicle for transmitting developmental signals from mother to neonate. Various growth factors, cytokines, and gut peptides have been isolated from human milk, often in quantities that exceed maternal serum levels. There exists an interrelated system in which compartmentation of milk's components lead to controlled release of nutrients and metabolites to the breastfed infant. Various proteins are differentially sequestered within the compartments of human milk. These proteins become sequentially processed within the gastrointestinal tract. The majority of studies on human milk have focused on the aqueous (defatted) fraction of whole milk rather than the various compartments that include fat. Yet, the fat compartment composed of the milk fat globule with its associated plasma membrane appears to provide a unique delivery system of bioactive substances to the newborn gut. A growth factor that is present in all compartments of human milk is transforming growth factor-alpha (TGFα), a member of the epidermal growth factor (EGF) family. TGFα has biological activity in its transmembrane as well as its mature, cleaved form. As a gut peptide, TGFα contributes to the growth-promoting properties of human milk through activation of EGF receptors (EGFr) of human fetal small intestinal cells *in vitro*. Higher concentrations of TGFα are present in the fat compartment of human milk. It is unknown if TGFα in human milk fat is associated with the milk fat globule membrane (MFGM). Based on the transmembrane structure of TGFα, we hypothesized that TGFα would be present in the milk fat globule (MFG), and with further isolation, associated with the MFGM.

*Carol L. Wagner, M.D., John E. Baatz, Ph.D., Donna W. Forsythe, and Gabriel A. Virella, M.D., Ph.D., Medical University of South Carolina, Charleston, South Carolina, 29425. Stuart Patton, Ph.D., University of California, San Diego, CA, 92103.

Methods

Milk samples were processed immediately at room temperature to insure adequate separation of the milk fat globule (MFG) and the milk fat globule membrane (MFGM). A portion of each sample was centrifuged at 1700Xg for 20 minutes at 4°C to isolate the fat from the aqueous fraction. A portion of the MFG (labeled low-speed MFG) was retained before isolation of MFGM.

Isolation of the MFG

Following the methods of Patton and Huston (1987), 15-mL PBS at 37°C and 2.5-g sucrose were placed in a 50-mL centrifuge tube to which 35-mL freshly expressed breast milk were added. To allow separation of the MFG from whole milk, the milk mixture was then centrifuged at 20-25°C at 1500 X g for 20 minutes. The MFG was removed and washed in PBS (37°C). The MFG was resuspended in Na taurocholate (12-mM) PBS for emulsification to attain a lipid concentration of 4-5%.

Isolation of the MFGM

To achieve separation of the MFGM from the MFG, a portion of the MFG preparation was exposed to sodium taurocholate (1 mM) for 2 min at 37°C, then centrifuged at 80,000 X g at 4°C for 1 hr. After discarding the supernatant, the centrifuge tube was wiped clean with tissue to reduce contamination with residual supernatant, taking care to leave the pellet intact. The MFGM pellet was then reconstituted in 1 mL PBS.

Measurement of TGFα Concentration and Molecular Mass Profile

TGFα concentrations in the MFG and MFGM preparations were determined by RIA (Peninsula Laboratories, Belmont, CA). The samples also were analyzed for TGFα molecular mass profile using a standard method for Western blot analysis.

Results

The concentration of TGFα (mean ± S.D.) was 310 ± 214 pg/mL in whole milk samples, 599 ± 361 pg/mL in low-speed MFG, and 187 ± 78 pg/mL in resuspended MFGM. In the residual MFG devoid of membrane, TGFα was detected in trace amounts in 3 of 11 samples and not detected in the remaining samples. Western blotting for TGFα in MFG (low-speed) and resuspended MFGM revealed the characteristic band pattern (light band at 6-kD and prominent bands at 14, 21, 30 and 46-kD) found with whole milk and fat fractions. TGFα was detected in the isolated MFGM but not in the residual MFG samples.

Conclusions

Our findings support the premise that TGFα is associated with the MFGM within the milk fat globule. It is the MFGM itself rather than the lipid component of the MFG that appears to be the source of TGFα in human milk fat. What role TGFα-associated MFGM plays in neonatal gut cell signaling and maturation remains to be determined.

Funded in part by NIH grant #3M01RR01070-20S3A1.

BACTERIAL SCREENING OF HUMAN MILK AND BOWEL DISEASE IN PREMATURE INFANTS

Andrea Willeitner, Gert Lipowsky, and Helmut Küster

BACKGROUND

Bacteria in human milk are predominantly apathogenic, but potential pathogens have been found in about 1 out of 5 human milk samples.[1, 2] To protect prematures, many neonatologists advocate human milk to be regularly screened for bacteria and, in case of contamination, to be heat-treated or even discarded [3-6] However, heat treatment of human milk causes significant damage to immunoglobulins, enzymes, growth factors and live cells that are known to protect against a wide range of diseases.[7-9] On the other hand, the potential of heating human milk to prevent infection has never been shown.[3]

Methods

The purpose of this study was to evaluate the benefits or risks of screening human milk and subsequent heat-treatment in regard to mortality and gastrointestinal complications by comparing two different screening policies of human milk: In 1998, pumped milk from all breastfeeding mothers of admitted very low birth weight infants under 1500g was weekly sampled for bacteria. Milk was then treated according to our lab's recommendations. Starting from January 1999, we primarily fed untreated human milk and sampled milk only on indication, i.e. when gastrointestinal problems occurred.

Results

Treatment of Mothers' Milk

In 1998, results of routine screening required Holter's pasteurization (62.5°C, 30 min.) in 59% of cases. Untreated milk could be given in 27% of cases, whereas milk had to be discarded in 14%. In 1999, all pumped milk could be fed untreated, because

*Andrea Willeitner, MD, University of Munich, Dept. of Pediatrics, Dr. von Haunersches Kinderspital, Lindwurmstrasse 4, D-80337, Muenchen, Germany, *Email:* awilleitner@lrz.uni-muenchen.de

results of clinically indicated milk cultures and their comparison to the results of the simultaneously sampled stool cultures did never suggest milk-borne infection.

Microbiological Results

Table 1. In 1998, when screening was a routine procedure, the number of milk samples was 10 fold higher than in 1999, when milk was only examined on indication. In both years, *Staphylococcus epidermidis* was prevailing, but, in 1999, more gram-negative organisms were cultivated. However, the bacteria found in mothers' milk were never identical to those isolated from simultaneously cultured stool specimens.

Germ	1998 (N = 113)	1999 (N = 11)
No growth	6 (5 %)	-
Staph. Epidermidis	96 (85 %)	5 (46 %)
Staph. Aureus	6 (5 %)	3 (27 %)
Gram negative species	5 (4 %)	3 (27 %)

Outcomes before (1998) and after (1999) cessation of weekly routine bacterial screening of mothers' milk:

Table 2. In 1998, when pasteurization rate was 59%, mortality and gastrointestinal morbidity were higher than in 1999, when predominantly untreated human milk was fed.

Outcome (Infants)	1998 (N = 66)	1999 (N = 63)
Mortality	10 (15 %)	4 (6 %)
NEC	9 (14 %)	0 (0 %)
Enteritis	7 (11 %)	2 (3 %)

Conclusion

A policy of routine bacterial screening of human milk for premature infants leads to unnecessary pasteurization or withdrawal of breastmilk and seems to be useless, as it does not prevent gastrointestinal infectious disease and can deteriorate rather than improve overall outcome.

References

1. L. Carroll, M. Osman, D.P. Davies, and A.S. McNeish, Bacteriological criteria for feeding raw breast-milk to babies on neonatal units, Lancet. 12:732-733 (1979).
2. A.I. Eidelman, and G. Szilagyi, Patterns of bacterial colonization of human milk, Obstet. Gynecol. 53:550-552 (1979).
3. P.L. Ogra, and D.K. Rassin, Human Breast Milk. In: Remington JS, Klein JO, eds. Infectious diseases of the fetus & newborn infant. 4 ed. Philadelphia, Pennsylvania, W.B.Saunders; 1995, pp 108-139.
4. D.C. Davidson, and C. Roberts, Bacteriological monitoring of human milk, Arch. Dis. Child. 54:760-764 (1979).
5. L. Carroll, M. Osman, D.P. Davies, and A.S. McNeish, Bacteriological criteria for feeding raw breast-milk to babies on neonatal units, Lancet 2:732-733 (1979).
6. K.B. Botsford, R.A. Weinstein, K.M. Boyer, C. Nathan, M. Carman, and J.B. Paton, Gram-negative bacilli in human milk feedings: quantitation and clinical consequences for premature infants, J. Pediatr. 109:707-10. (1986).
7. J.E. Ford, B.A. Law, V.M. Marshall, and B. Reiter, Influence of the heat treatment of human milk on some of its protective constituents, J. Pediatr. 90:29-35 (1977).
8. C.L.J. Paxson, and C.C. Cress, Survival of human milk leukocytes, J. Pediatr. 94:61-64 (1979).
9. I. Narayanan, K. Prakash, N.S. Murthy, and V.V. Gujral, Randomised controlled trial of effect of raw and holder pasteurised human milk and of formula on incidence of neonatal infection, Lancet 2:1111-1113 (1984).

DETERMINANTS OF MILK SODIUM/POTASSIUM RATIO AND VIRAL LOAD AMONG HIV-INFECTED SOUTH AFRICAN WOMEN

Juana F. Willumsen, Anna Coutsoudis, Suzanne M. Filteau, Marie-Louise Newell, Andrew M. Tomkins*

Although HIV can be transmitted from mother to infant through breastmilk,[1] breastfeeding remains a key component of maternal and child health policy. Therefore, it is crucial to understand factors contributing to breastmilk HIV transmission, especially in Africa where HIV is most prevalent. In South Africa, exclusive breastfeeding was associated with a lower rate of transmission than was mixed breastfeeding with other foods.[2]

High breastmilk HIV viral load was found among women with subclinical mastitis[3, 4] defined biochemically as raised milk Na/K ratio. Breastmilk viral load is thought to be a major determinant of postnatal transmission of HIV-1.[5, 6] To date breastmilk viral load has been determined in only single samples from each woman without regard to differences between breasts or across time. We have investigated risk factors for subclinical mastitis and milk viral load in both breasts of HIV-infected women at three times postpartum.

Methods

Spot breastmilk samples were collected from each breast of 145 HIV-1-infected lactating South African women at 1, 6 and 14 weeks postpartum and analysed for Na/K ratio by flame photometry and for cell-free viral load by RNA PCR. Multiple regressions were used to determine factors associated with raised Na/K or viral load at each time. Explanatory variables tested were: socioeconomic status, maternal age and health, antenatal CD4 and CD8 count, mode of delivery, infant birthweight, sex, gestational age, and age, weight and feeding mode (exclusive breast feeding or mixed breastmilk plus other foods) at time of milk sampling. Since breasts appeared to behave largely independently, breast, not woman, was the unit of analysis in multiple regressions.

*J.F. Willumsen, S.M. Filteau, M.-L. Newell, A.M. Tomkins, Institute of Child Health, London, U.K. A. Coutsoudis, University of Natal, Durban, South Africa

Results and Conclusions

The prevalence of raised milk Na/K ratio was high as described previously.[4] Bilateral raised Na/K decreased with time whereas unilateral remained fairly constant and more common than bilateral. Breastmilk viral load was often below detection level in one breast, but not in the other.

The available explanatory variables did not predict milk Na/K ratio well, never achieving R^2 more than 7%. At 1 week postpartum only, exclusive breastfeeding was associated with significantly lower Na/K ratio than was mixed feeding ($P = 0.05$). At 6 and 14 weeks the main contributor to milk Na/K was infant percent weight gain between the two time points, suggesting either that healthy infants who suckled better emptied the breast more effectively and prevented milk stasis or that infants of women who breastfed most effectively grew most rapidly. The explanatory variables were better predictors of milk viral load with R^2 of 26% at 1 week, 11% at 6 weeks and 18% at 14 weeks. CD4 count during pregnancy was an important predictor of milk viral load at all times. Feeding mode and milk Na/K ratio had complex interactive effects on milk viral load which varied with time. At 1 week increased Na/K was associated with increased viral load in the mixed-feeding but not in the exclusively breastfeeding, group. At 6 weeks increased Na/K was associated with increased viral load in both feeding groups and at 14 weeks only in the exclusively breastfeeding group.

Subclinical mastitis and shedding of virus in breastmilk varied considerably between breasts and over time within an individual woman. Thus it is not surprising that milk sodium concentration or viral load measured in a single sample have not shown strong associations with HIV transmission.[3, 5, 6] Breastmilk transmission of HIV may be highest in early lactation.[1] Our results suggest that in early lactation exclusive breastfeeding may protect against HIV transmission by decreasing the prevalence of subclinical mastitis and by preventing the increased viral load associated with raised milk Na/K ratio. Later lactation requires further study and maternal health interventions which increase CD4 counts are likely to be important.

References

1. P. G. Miotti, T. E. T. Taha, N. I. Kumwenda, R. Broadhead, L. A. R. Mtimavalye, L. Van der Hoeven, J. D. Chiphangwi, G. Liomba, and R. J. Biggar. HIV transmission through breastfeeding: a study in Malawi, J. Amer. Med. Assoc. 282:744-749 (1999).
2. A. Coutsoudis, K. Pillay, E. Spooner, L. Kuhn, and H. M. Coovadia, Influence of infant-feeding patterns on early mother-to-child transmission of HIV-1 in Durban, South Africa: a prospective cohort study, Lancet 354:471-476 (1999).
3. R. D. Semba, N. Kumwenda, D. R. Hoover, T. E. Taha, T. C. Quinn, L. Mtimavalye, R. J. Biggar, R. Broadhead, P. G. Miotti, L. J. Sokoll, L. Van der Hoeven, and J. D. Chiphangwi, Human immunodeficiency virus load in breast milk, mastitis and mother-to-child transmission of human immunodeficiency virus type 1, J. Infect. Dis. 180:93-98 (1999).
4. J. F. Willumsen, S. M. Filteau, A. Coutsoudis, K. E. Uebel, M. L. Newell, and A. M. Tomkins, Subclinical mastitis as a risk factor for mother-infant HIV transmission, in: *Short and long term effects of breast feeding on child health*, edited by B. Koletzko, K. Fleischer Michaelsen, and O. Hernell (Kluwer Academic/Plenum, London, 2000), pp. 211-223.
5. P. Van de Perre, A. Simonon, D.-G. Hitimana, F. Dabis, P. Msellati, B. Mukamabano, J.-B. Butera, C. Van Goethem, E. Karita, and P. Lepage, Infective and anti-infective properties of breastmillk from HIV-1-infected women, Lancet 341:914-918 (1993).

6. L. A. Guay, D. L. Hom, F. Mmiro, E. M. Piwowar, S. Kabengera, J. Parsons, C. Ndugwa, L. Marum, K. Olness, P. Kataaha, and J. B. Jackson, Detection of human immunodeficiency virus type 1 (HIV-1) DNA and p24 antigen in breast milk of HIV-1-infected Ugandan women and vertical transmission, Pediatrics 98:438-444 (1996).

INDEX